BIOLOGICAL SHAPE ANALYSIS

Proceedings of the 3rd International Symposium

BIOLOGICAL SHAPE ANALYSIS

Proceedings of the 3rd International Symposium

University of Tokyo, Japan 14 – 17 June 2013

Pete E. Lestrel
University of California, Los Angeles, USA
editor

World Scientific

NEW JERSEY • LONDON • SINGAPORE • BEIJING • SHANGHAI • HONG KONG • TAIPEI • CHENNAI

Published by

World Scientific Publishing Co. Pte. Ltd.
5 Toh Tuck Link, Singapore 596224
USA office: 27 Warren Street, Suite 401-402, Hackensack, NJ 07601
UK office: 57 Shelton Street, Covent Garden, London WC2H 9HE

British Library Cataloguing-in-Publication Data
A catalogue record for this book is available from the British Library.

BIOLOGICAL SHAPE ANALYSIS
Proceedings of the 3rd International Symposium

ISBN 978-981-4704-18-2

Printed in Singapore

The Third International Symposium of Biological Shape Analysis

June 14-17, 2013
University of Tokyo, Japan

Organizers: H. Iwata, H. Tatsuta, Y. Yoshioka, S. Ninomiya and P. E. Lestrel

Symposium Participants

[1] Seishi Ninomiya	[12] Ryosuke Iritani	[23] Yoshihiro Morishita
[2] Terra Triwahyuni	[13] Wataru Anzai	[24] Kazuo Takahashi
[3] Yasushi Mitani	[14] Masaru Mashiko	[25] Masashi Komori
[4] Rongling Wu	[15] Dagmar Lestrel	[26] Peter Claes
[5] Katleen Daniels	[16] Norman MacLeod	[27] Won Moon
[6] Reena Khullar	[17] Jonathan Clark	[28] Jitendra Paliwal
[7] Pete Lestrel	[18] Nobuhiro Minaka	[29] Wei Guo
[8] Abdul Kadir	[19] Soichiro Kawabe	[30] Yosuke Yoshioka
[9] Haruki Tatsuta	[20] Ken-ichi Hironaka	[31] Kyosuke Yamamoto
[10] Norikuni Kumano	[21] Yoshinari Yonehara	[32] Yuta Shintaku
[11] Hiroyoshi Iwata	[22] Koji Noshita	

PREFACE

I am again pleased to be in the position of writing the preface for the proceedings of the Third International Symposium of Biological Shape Analysis (ISBSA). This 3rd proceeding represents the continuation of two former symposia; namely, the 1st ISBSA held at the Agricultural Center in Tsukuba, Japan on June 3-6, 2009 [1], and the 2nd ISBSA held Naha, Okinawa, Japan on September 4-7, 2011 [2]. The current proceedings, the 3rd ISBSA was held at the University of Tokyo, Yayoi campus on June 14-16, 2013 and continues the trend set by the first two symposia. Twenty-three presentations from the podium were offered of which fifteen are published here. Both previous symposiums were notable successes in that the papers represented a wide diversity of academic disciplines as well as wide-ranging in subject matter pertaining to the central theme of biological shape analysis. Moreover, it should be emphasized that all three symposia represent a departure from the usual emphasis in science on specialization by bringing very diverse disciplines together.

As argued in the earlier volumes, and worth re-iterating here, the biological form is composed of attributes such as size, shape, color, inner and outer structure, etc. Nevertheless, it is the shape among these attributes that plays a major role in all of the biological sciences. Shape is defined here as the outline or boundary of an organism or part of an organism. This has been generally viewed in 2-space ($\mathbb{R}^2$) but increasing studies are beginning to extend it into 3-space ($\mathbb{R}^3$). The measurement of shape (in contrast to size) is not a trivial endeavor. It has become increasingly apparent over the last few decades that the conventional measurement approach (CMA) has drawbacks that make it an inadequate approach for quantifying complex irregular shapes of the type encountered in biology [3]. This has led to a number of alternative approaches such as Fourier Descriptors, wavelets, Riemann curves, surface mapping, neural networks, etc., to name a few.

As indicated, the diversity of subject matter dealing with shape analysis was clearly evident in the first two proceedings (Fig. 1) and this has continued in the 3rd symposium. Fig. 1 displays the diversity of the disciplines involved and their subject matter, which have dealt with biological shape analysis in one form or another.

The papers presented at the 3rd ISBSA continued this trend and dealt with leaf identification using Fourier and Bayes (Kadir); analyzing leaf outlines with neural networks (Clark); use of machine vision in grain classification (Paliwal); sweet potato weevil distress sounds (Tatsuta); morphometric variation in elytal shape of the sweet potato weevil (Kumano): shape mapping and genome-wide association studies of human body shape genes (Wu); spiral shell growth in gastropods (Shimizu); pelvic variation in Hadrosaurid dinosaurs (Prieto-Marquez); color variation in ladybird beetles (Macleod); superimposition methods using 3D facial images (Daniels); effects of genomic ancestry and sex on facial morphology (Claes); the shape of the 3D human mandible seen in a clinical context (Khullar); surface mapping of craniofacial morphology (Moon); and finally, Fourier descriptors of the shape of whole body outlines (Lestrel).

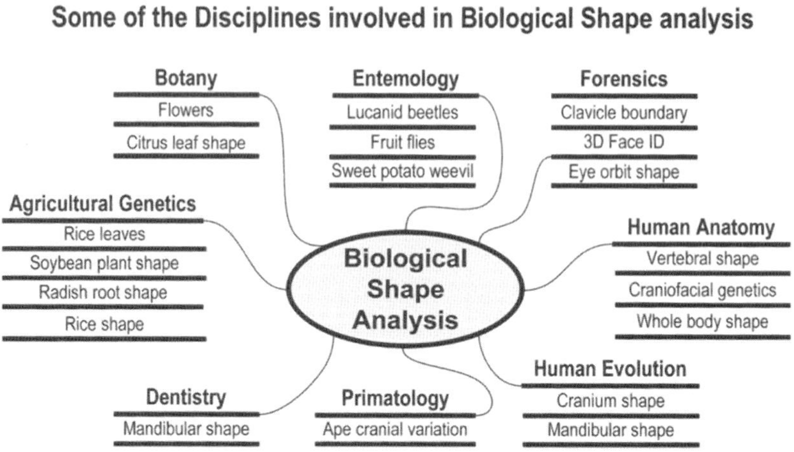

Fig. 1. Some of the various disciplines and their subject matter involved with biological shape analysis as based on the contributions to the 1st and 2nd ISBSA proceedings.

Special thanks again go to the co-organizers, Seishi Ninomiya, Hiro Iwata and Haruki Tatsuta for organizing the meetings at the University of Tokyo, their help insured that the meetings would be successful. Thanks also to my wife Dagmar for taking on the role of secretary at the symposium. Thanks again go to Charles Wolfe and Al Bodt for assistance with the reviewing process. Brian Bodt is to be again thanked for using Photoshop to improve the frontispiece photo of the participants. The programming experience of Charles Wolfe with the EFF and MLmetrics software packages is gratefully acknowledged. Finally, special thanks are again order to Sook Lim, Scientific Editor, World Scientific, Singapore, for agreeing to publish the 3rd proceedings of the ISBSA.

It is the fervent hope of the editor that the papers in this 3rd ISBSA proceedings, as well as those published in the 1st and 2nd ISBSA proceedings, will be as useful for those interested in the quantitative analysis of the biological form and especially its shape.

Pete E. Lestrel
Editor

Van Nuys, California, USA, March, 2015

REFERENCES

[1] Lestrel PE (2011) Preface **In**: PE Lestrel (Ed.) *Biological Shape Analysis: Proceedings of the First International Symposium of Biological Shape Analysis.* World Scientific, Singapore, pp. vi-xi.

[2] Lestrel PE (2013) Preface **In**: PE Lestrel (Ed.) *Biological Shape Analysis: Proceedings of the Second International Symposium of Biological Shape Analysis.* World Scientific, Singapore, pp. vii-ix.

[3] Lestrel PE (1997) *Fourier Descriptors and their Applications in Biology.* Cambridge University Press, New York.

CONTENTS

LIST OF SYMPOSIUM PARTICIPANTS

* ANZAI, W. Department of Evolutionary Biology, The University of Tokyo, Graduate School of Science, Japan, watanzai@um.u-tokyo.ac.jp

* CLAES, P. Medical Image Computing, ESAT/PSI, Department of Electrical Engineering, KU Leuven Medical Imaging Research Center, KU Leuven & UZ Leuven iMinds-KU Leuven Future Health Department Herestraat 49 bus 7003, 3000 Leuven, Belgium, peter.claes@esat.kuleuven.be

* CLARK, J. Y. Faculty of Engineering and Physical Sciences, Department of Computing, University of Surrey, Guildford, GU2 7XH, UK, j.y.clark@surrey.ac.uk

* DANIELS, K. Medical Image Computing, ESAT/PSI, Department of Electrical Engineering, KU Leuven Medical Imaging Research Center, KU Leuven & UZ Leuven iMinds-KU Leuven Future Health Department Herestraat 49 bus 7003, 3000 Leuven, Belgium, katleen.daniels@esat.kuleuven.be

ENDO, K. Department of Earth and Planetary Science, The University of Tokyo 7-3-1 Hongo Tokyo 113-0033 Japan, endo@eps.s.u-tokyo.ac.jp

* GUO, W. Institute for Sustainable Agro-ecosystem Services, Graduate School of Agricultural and Life Sciences, The University of Tokyo, 188-0002 Japan, guowei.net@hotmail.com

HIRONAKA, K. Laboratory for Developmental Morphogeometry, RIKEN Center for Developmental Biology (CDB). 2-2-3 Minatojima-minamimachi, Chuo-ku, Kobe, Hyogo 650-0047, Japan, hironaka@cdb.riken.jp

IMAIZUMI, K. 2nd Forensic Biology Section, National Research Institute of Police Science, 6-3-1, Kashiwanoha, Kashiwa, Chiba 277-0882, Japan, imaizumi@nrips.go.jp

IRITANI, R. Graduate School of Systems Life Sciences, Mathematical biology Laboratory, Kyushu University, 6-10-1, Hakozaki, Higashi-ku, Fukuoka, Japan, Lambtani@gmail.com

ISHIDA, F. Ishida Dental Clinic, 28-16 Inarimae, Tsukuba, Ibaraki 305-0061, Japan

* IWATA, H. Graduate School of Agricultural Life Sciences, The University of Tokyo, Bunkyo-ku, Tokyo 113-8657 Japan, aiwata@mail.ecc.u-tokyo.ac.jp

* KADIR, A. Department of Informatics Engineering, School of Computing Jenderal Achmad Yani, Indonesia and Department of Information Technology and Electrical Engineering Department, Gadjah Mada University, Indonesia, akadir@mti.ugm.ac.id

KANAZAWA, E. Professor Emeritus, Nihon University School of Dentistry at Matsudo 2-870-1, Sakaecho-Nishi, Matsudo, Chiba, Japan, mailing code 271-8587, kanazawa@koshin.or.jp

* KAWABE, S. Gifu Prefectural Museum, Gifu, Japan, kawabe_soichiro@yahoo.co.jp

* KHULLAR, R. Section of Orthodontics, UCLA School of Dentistry, Los Angeles, California, 90095-1668, USA. Now in private practice. drrkhullar@gmail.com

* KOMORI, M. Osaka Electro-Communication University Faculty of Information and Communication Engineering, 18-8 Hatsucho, Neyagawa, Osaka, Japan, komori@oecu.jp

KONDO, O. Department of Anthropology, Graduate School of Science, The University of Tokyo, Hongo 7-3-1, 113-0033, Tokyo, Japan, kondo-o@biol.s.u-tokyo.ac.jp

* KOYABU, D. University Museum, University of Tokyo, Hongo 7-3-1, Bunkyo-ku, 113-0033, Tokyo, Japan, koyabu@um.u-tokyo.ac.jp

* KUMANO, N. Okinawa Prefectural Plant Protection Center and Ryukyu Sankei Co. Ltd. Naha, Okinawa 902-0072, Japan, nrkumano@gmail.com

* LESTREL, P. E. Formerly School of Dentistry, University of California at Los Angeles (UCLA), USA, plestrel@earthlink.net or peter.lestrel@gmail.com

* MACLEOD, N. The Natural History Museum, Cromwell Road, London, SW7 5BD, UK, N.MacLeod@nhm.ac.uk

MASANORI, M. Medical Corporation Sanshinkai Hikari Dental Clinic, kqyhb723@ybb.ne.jp

MINAKA, N. National Institute for Agro-Environmental Sciences, Kannondai 3-1-3, Tsukuba, Ibaraka, Japan, minaka@affrc.go.jp

MITANI, Y. Medical Corporation Sanshinkai Hikari Dental Clinic, mitani@trust.ocn.ne.jp

* MOON, W. Section of Orthodontics, UCLA School of Dentistry, Los Angeles, California 90095-1668, USA, wmoonl@dentistry.ucla.edu

MORISHITA, Y. 2-2-3 Minami-machi, Minatojima, Chuo-ku, Kobe, Japan, morishita@cdb.riken.jp

NINOMIYA, S. Institute of Sustainable Agro-ecosystem Services, Graduate School of Agriculture and Life Sciences, The University of Tokyo, Tokyo 188-0002 Japan, snino@isas.a.u-tokyo.ac.jp

NOSHITA, K. Graduate school of Systems Life Sciences, Kyushu University, Hakozaki 6-10-1, Higashi-ku, Fukuoka 812-8581, Japan, noshita@bio-math10.biology.kyushu-u.ac.jp

* PALIWAL, J. Dept. of Biosystems Engineering, E1-342 EITC, University of Manitoba, Winnipeg, MB, Canada, R3T 5V6, J.Paliwal@umanitoba.ca

* PRIETO-MÁRQUEZ, A. Bayerische Staatssammlung für Paläontologie und Geologie, Richard-Wagner-Strasse 10, D-80333 Munich, Germany, redshore@gmail.com

* SHIMIZU, K. The University of Tokyo, Department of Earth and Planetary Science, Japan, shimizu@eps.s.u-tokyo.ac.jp

SHINTAKU, Y. 26 Kanrin, Inuyama, Aichi 484-0081, Japan, Shintaku.jmc@gmail.com

* TAKAHASHI, K. Graduate School of Environmental and Life Science, Okayama University, kaz_tak@cc.okayama-u.ac.jp

* TATSUTA, H. Department of Ecology and Environmental Science Faculty of Agriculture, University of the Ryukyus, 1, Sembaru, Nishihara, Okinawa 903-0213, Japan, htatsuta@agr.u-ryukyu.ac.jp

TRIWAHYUNI, T. C. JI. No. 1-B, Yogyakarta, Indonesia, terra-yuni@yahoo.co.id

* WU, R. Center for Computational Biology, Beijing Forestry University, Beijing 100083, China and Center for Statistical Genetics, Pennsylvania State University, Hershey, PA 17033, USA, rwu@phs.psu.edu

* YAMAMOTO, K. The University of Tokyo, Graduate School of Agricultural and Life Sciences, Japan, 8244012644@mail.ecc.u-tokyo.ac.jp

YONEHARA, Y. Atmosphere and Ocean Research Institute. The University of Tokyo 〒277-8564 5-1-5 Kashiwanoha, Kashiwa-shi, Chiba, yonehara@aori.u-tokyo.ac.jp

* YOSHIOKA, Y. Faculty of Life and Environmental Sciences, University of Tsukuba, Ibaraki, 305-8572, Japan, yoshioka.yosuke.fw@u.tsukuba.ac.jp

*** Active symposium participants**

Boundary Morphometrics: Some Theoretical Issues and Some Data Based Applications

Keynote Address

P. E. LESTREL[1, 2]

Abstract

Of the five senses, hearing, tasting, smelling, touching and seeing; vision is the predominant sense among primates, including humans. Humans rely almost exclusively on sight. It is via the visual system that we interact and make sense of the world around us. With the vision system as a starting point, the focus here will be primarily on three issues: (1) The role that visual perception plays in recognizing the biological **form**, (2) the function that form plays in biological processes, and (3) models as representations of form. The visual system can be considered as a part of the central nervous system (CNS) and acts as a major influence on other systems such as locomotion. However, more important, is the fact that it allows the brain to interact with the external environment. The human visual system is endowed with special qualities such as binocular vision allowing both eyes to look straight ahead facilitating 3D vision. Moreover, the vision system provides the ability to distinguish between different, often closely related forms, as well as allowing for the tracking of movement. All visual information entering the eyes is focused by the eye lens onto the retina. From the retina, it is transferred along the optic nerve to the visual cortex. The visual cortex lies at the posterior aspect of the brain. Sensory signals of objects are only recognized because similar experiences are stored in the brain as memories. Biological forms as perceived by the visual system are quite complex and attempts to characterize them can be accomplished by using models. Three such models are: (1) physical models, (2) heuristic models, and (3) mathematical models. For purposes here, the emphasis is on mathematical models; that is, to generate a mathematical representation of the biological form. More specifically, the focus here is on **shape**. Shape is defined as the outline of a biological form. It can be viewed in 2D or 3D terms. A mathematical model termed **Computerized Shape Analysis** is briefly presented here. It is composed of two aspects. *Elliptical Fourier Descriptors* (EFFs) to assess the boundary outline of a form and the *Continuous wavelet transform* (CWT) to extract localized features. This Fourier-wavelet model is intended as a model that generates substantially more of the visual information that is always present in the biological form.

---

[1] Formerly, Sections of Orthodontics and Oral Biology, UCLA School of Dentistry, Los Angeles, CA 90024-1668, USA, plestrel@earthlink.net or peter.lestrel@gmail.com.

[2] This chapter is a greatly enlarged version of the keynote address delivered at the 3rd ISBSA at the University of Tokyo, Yayoi campus.

INTRODUCTION

Mathematics is the door and key to the sciences. *Roger Bacon (ca. 1214-1294)*

No human investigation can be called a real science if it cannot
be demonstrated mathematically. *Leonardo da Vinci (1452-1519)*

...the universe...cannot be understood unless one learns to
comprehend...the language of mathematics. *Galileo Galilei (1564-1642)*

I often say that when you can measure what you are speaking about,
and express it in numbers, you know something about it; but when you
cannot measure it, when you cannot express it in numbers, your knowledge
is of a meagre and unsatisfactory kind. *Lord Kelvin (1824-1907)*

The purpose of this address delivered at the opening of the third ISBSA held at the University of Tokyo, Yayoi Campus on June 14-17, 2013, is two-fold. It is intended to: (1) introduce the participants and guests to the symposium and (2) stress the fundamental importance of the biological form, which is considered central to any assessment of biological processes such as growth and development or evolution.

First, I want to extend a warm welcome to you all for coming to the **Third International Symposium of Biological Shape Analysis** being held here at the University of Tokyo. This is the third time that we are meeting and this is the largest group so far. Speaking for the organizers, I want to especially thank those of you who have had to travel a long way to get here. This symposium is truly international. The initial idea of the organizers was to try to keep the size fairly small so that everyone can get to know everyone else in an informal setting. The ISBSA is a very unusual group in that it is composed of researchers from very diverse disciplines. These include agricultural genetics, botany, entomology, dentistry, physical anthropology, paleontology, zoology to name a few. What makes this symposium possible is a common interest in the biological form and how best describe it in quantitative terms.

In very general terms, this then leads to two aspects, which are: (1) how is the biological form perceived and once perceived (2) how is the biological form best described. The first part deals with the act of seeing and the second part deals with how to quantify what is being perceived. In an attempt to speak briefly to these two major issues, the following topics, considered of particular relevance, will be addressed:

Part one consists of:

(1) Visual perception and the role it plays in recognizing forms,
(2) Evolution of vision and the structure of the eye,
(3) Some recent quantitative applications of vision,

Part two deals with:

(4) The role that form plays in biological processes,
(5) Models viewed as representations of form,
(6) Computational Shape Analysis

Because the literature on vision, visual perception, structures of the eye, and more recently on the evolutionary and molecular physiology of the eye is extensive, it is only possible here to touch briefly on some these issues. The material here is divided into two parts. Part 1, deals with visual perception; that is, how the visual system is constructed, and how is this visual information transferred from the retina to the brain. This section ends with the quantification of aspects of the retina/brain relationship. The second part, Part 2, shifts the discussion from the visual system per-se to the analysis of forms as apprehended by the visual system. Discussion of the latter material will be followed by some examples of models of shape. Finally, the discussion concludes with the presentation of a Fourier-wavelet model. This last model joins the shape or boundary (global aspects) of a form with local changes (localized features) *along* the boundary contour. It is termed Computational Shape Analysis.

Vision has been considered as an information processor, analogous to a computer with a central processing unit (CPU) and memory and in that sense, the vision system can be viewed as consisting of an input that leads to a cascade of events that eventually produces a behaviorally appropriate response as an output [1].

PART 1: THE VISUAL SYSTEM

Vision as an Information System

With respect to humans, it can be argued that the most important basis for acquiring knowledge about the external world is the visual system [2]. The question that then arises (even if it cannot be satisfactorily answered in all aspects) is as follows. How can the normal brain process visual information from cues in the external environment to produce precise behaviors?

For example, consider the skull specimen displayed in Fig. 1 [3], viewed from the side (lateral view). Most reasonably educated nonprofessionals, would in all probability, identify it a human-looking skull based on their experience and education. However, if prompted to provide further details, they would probably be unable to do so. A physical anthropologist, on the other hand, would be able to readily identify it and describe it with a wealth of details. For example, that it is a Neanderthal specimen called La Ferrassie 1, discovered in 1909 by Denis Peyrony and Louis Capitan in a cave near the Dordogne region of France. It is an adult male of roughly 40-55 years of age. It is dated to roughly between 50,000 and 70,000 years ago. Associated faunal evidence indicates the presence of glacial cold conditions associated with the last glacial period. Archeological evidence suggests that La Ferrassie was an intentional burial. La Ferrassie had excessively worn teeth, giving hints as to the nature of the diet of Neanderthals.

This specimen also raises the interesting question of what, if any, is its relationship is to modern man (*Homo sapiens*)? This continues to be a hotly debated issue. Neanderthals are currently considered as a separate lineage from modern humans due to its distinct morphology. Whether it is a totally separate species has become increasingly unlikely as interbreeding with *Homo sapiens* appears to have occurred prior to *Homo neanderthalensis* disappearing from the fossil record. We will briefly return to this debate and the relationship between *Homo neanderthalensis* and *Homo sapiens* later in this address.

Returning to the issue here of vision as an information system, it can be reasonably presumed that both the educated individual and the anthropologist are viewing the same skull image. Further, neither observation (of the educated individual or the anthropologist), can be considered right or wrong. That is, one can assume that both visual assessments are closely similar, although perhaps not exact. What is distinctly different is the experience and memory that the observers bring to their assessment of the visual image. That is, their responses to the visual image are radically different due to their different experiences and memories. Thus, it should be apparent that it is impossible to separate the content of the visual image from the involvement of the brain's memory. In fact, one can argue that any observed image remains devoid of meaning to that observer in the absence of experience and memory.

One can then view the visual-brain relationship as an information system in that the visual image is transformed into a stream of signals that eventually become stored in the brain's memory. This allows for later recall leading to spoken and written descriptions of the visual image under consideration, even in the absence of the visual image. This leads to the challenging question of how the human visual system apprehends the external environment; and of specific interest here, the biological form. That is, how this process is accomplished is the focus of the first part of this address.

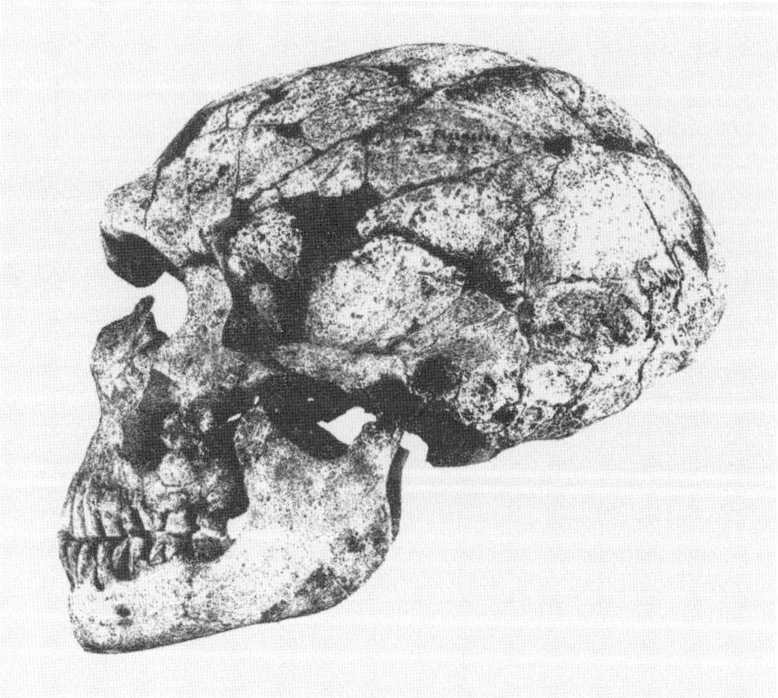

Fig. 1. La Ferrassie 1, a classic Neanderthal specimen discovered in 1909 at a cave site in the Dordogne of France in 1909 [3].

Visual Perception

The visual system perception involves much more than just the act of seeing. Visual perception acts to interpret or analyze the visual image. That is, it is much more than just the ability of seeing the external environment. It represents a complex interaction process of continuously combining the act of seeing with interpreting what is being seen, which involves the central nervous system (CNS). The visual system is involved in countless daily activities, some quite complex, such as reading this page, which involves a combination of visual and language skills. Another complex activity, walking, also involves visual and motor skills [1]. Thus, it is the visual system that allows the brain to interact with the external world. For primates in general and humans in particular, the information of the external world mediated and acted on by the brain, is overwhelmingly visual in character. While the other senses, touch, hearing, tasting, and smell have become less important, vision remains the primary sense.

However, it is important to realize that despite the enormous progress made in last 30 years or so of continuing brain research, questions remain regarding how brain function affects the activity of nerve cells (neurons) that affect specific behaviors. Now, there are still large gaps in the knowledge of how the firing of neurons affects cognition. That is, how do neurons affect perception and ultimately behavior? One current approach is to create a map of "anatomical connections" or synapses among neurons. Nevertheless, this is only a beginning and by itself still inadequate to document the continually varying electric signals that are responsible for cognitive behavior [4].

Some of the characteristics of the visual system include the ability: (1) for the CNS, in particular the brain, to interact with the external world, (2) to distinguish different forms from each other, (3) of pattern recognition and feature extraction (segmentation), (4) to recognize depth (3D) with binocular vision, (5) to track movement and (6) to distinguish color differences. All these capabilities raise the interesting question of how this very complex visual system developed in the first place.

Evolution of the Visual System

It is perhaps of interest to mention that the idea of biological complexity was recognized and appreciated by some scholars in the 19th century, notably Darwin; however, others such as William Paley, a theologian, drew entirely different conclusions from this apparent biological complexity. With respect to the human eye, Paley indicated that the eye was simply too complicated an organ to have arisen by natural processes, and as a consequence, had to be created by a designer, namely, God. This argument is also called "intelligent design" or ID and it is still being heard and advocated by creationists, especially in the more rural areas of the United States.

Paley, using the image of a watchmaker indicated:

> ... when we come to inspect the watch, we perceive ... that its several parts are framed and put together for a purpose, e.g. that they are so formed and adjusted as to produce motion ... the inference we think is inevitable, that the watch must have had a maker -- that there must have existed ... who comprehended its construction and designed its use ... [5].

Thus, as an intelligent watchmaker is needed to make a watch, only an intelligent designer could have created the universe and everything in it, and that this intelligent designer according to Paley, had to be the supreme deity:

> The marks of *design* are too strong to be gotten over. Design must have had a designer. That designer must have been a person. That person is God [5].

Leaving Paley's theological conclusions aside, Darwin, having read Paley did not take his arguments seriously but did admit, with respect the evolution of the eye, that the details were largely absent. In Darwin's words:

> To suppose that the eye, with all its inimitable contrivances for adjusting the focus to different distances, for admitting different amount of light, and for correction of spherical and chromatic aberration, could have been formed by natural selection, seems, I freely confess, absurd in the highest possible degree [6].

Nevertheless, Darwin correctly deduced that the development of the eye involved gradual stages over evolutionary time. It is worth quoting his words at some length:

> Yet reason tells me, that if numerous graduations from a perfect and complex eye to one very imperfect and simple, each grade being useful to it possessor, can be shown to exist; if further, the eye does vary ever so slightly and the variations be inherited, which is certainly the case; and if any variation or modification in the organ be ever useful to an animal under changing conditions of life, then the difficulty of believing that a perfect and complex eye could be formed by natural selection, though insuperable by our imagination, can hardly be considered real [6].

In other words, if natural selection operates over a long time and causes small changes leading to improvements of the eye in terms of survival, then the evolution of such a complex structure as the eye becomes quite conceivable in natural terms. It will be demonstrated that recent research has borne out the validity of Darwin's views.

The evolution of the eye developed independently in a number of different lines in different animal orders [7]. The earliest organisms developed photoreceptors, light-sensitive cells that determine the presence or absence of light. Eventually these photoreceptors develop into a cup-like structure that allows directionality of where the light is coming from. Next, the cup edges grow together producing a small opening (pinhole camera). The light-sensitive spot at the bottom of the cup develops into a retina. A protective lens develops over the hole at the front. The eyecup structure becomes sealed containing liquid. All this proceeds in gradual steps over a long time period under natural selection. This process is outlined in Table 1 and displayed in Fig. 2.

An example of primitive or proto-eye development is *Planaria*, one of many non-parasitic flatworms. The planaria have depressions or cup-like eyespots called *ocelli* the primary function of which is to detect the absence or presence of light. Once they detect light, they move away from it. They can also very roughly distinguish light direction. Planaria can roughly differentiate the direction and intensity of light because of their cup-shaped, heavily pigmented retina cells, which shield the light-sensitive cells from exposure in all directions except for the single opening for the light. These ‘pit eyes’ are also found in other invertebrates such as snails living today.

As stated earlier, the evolution of the eye proceeded independently in a number of evolutionary directions. The development of compound eyes called *Ommatidia* (containing many lenses) as seen in flies, for example, represents one of at least three independent evolutionary developments of the eye. While these compound eyes cannot make out details, they can effectively recognize movement. They also allow for almost 360 degrees of vision. Thus, flies and other insects represent one way in which the eye has evolved.

Another independent development of the eye is found in mollusks of which the cephalopods (octopus, squid) are a sub-group. The development of the octopus eye closely resembles the vertebrate eye (human) but with some notable differences. One of these is that these invertebrate animals contain what is called a “verted” retina. This verted retina is in contrast to the “inverted” one found in all vertebrates including humans. This retinal structure will be described in more detail later. The general pattern or stages, as seen in existing living species, are diagrammatically displayed in Fig. 2. It should be emphasized that Fig. 2 refers to the stages of eye development leading to and ending with mollusks not vertebrates. While the mollusk eye is close to that of vertebrates in many ways, there are significant differences. Vertebrates and mollusks represent independently evolved developments. The vertebrae eye in contrast to the mollusk one developed a blind spot were the nerves pass through the retina [8].

However, recent work has suggested that the idea of numerous separate independent evolutionary developments of the eye may not be correct because of the newly found presence of a control gene, Pax6 that is present in both insects and vertebrates. Pax6 codes for highly conserved transcription factors that are thought to control eye development. This suggests the possibility of a more monophyletic origin of the eye rather than the presence of numerous (three or more) independent developments current held. Clearly, the story of the evolution of the eye is an on-going one and a complete picture remains to be drawn [9].

Table 1. The Stages in the Evolution of Vision. The presence of light-sensitive photoreceptors can be traced to 600 million years ago. The first eyes are presumed to have appeared in the Cambrian around 550 million year ago.

[1]	Development of light sensitive patch of photoreceptors (cells sensitive to light).
[2]	Depression of a patch or pit, which allows for directional sensitivity to light.
[3]	The pit edges develop into a cup by growing together (pin-hole camera effect).
[4]	The light-sensitive spot at the bottom of the cup develops into a retina.
[5]	Development of a “protective” lens at the opening of the cup margin.
[6]	The eye-cup eventually becomes a sealed structure containing a liquid.

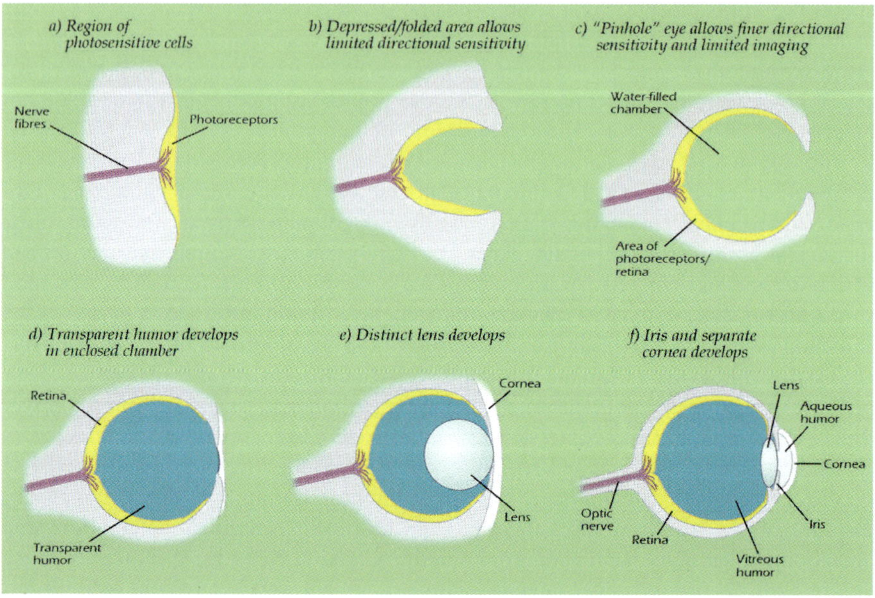

Fig. 2. Major stages in the development of the evolution of the mollusk eye: [a] shows the development of **photoreceptive or light-sensitive** cells. These cells can sense light or darkness only. These are already seen in some single-celled organisms; [b] shows a depression or **cup** development leading to limited directionality toward light, example is **planaria**; [c] shows the narrowing of the cup to produce the small hole (**pin-hole camera**); [d] displays the development of a **retina** from the photoreceptive cells and the sealing an enclosed chamber; [e] the **lens** develops as well as the beginning of a protective **cornea**; [f] shows the eventual mollusk and vertebrate eye with an **iris** and a separate cornea. Note: **Vitreous humor** is found inside the eye and **aqueous humor** is found between the cornea and the iris (Adapted from [8]).

Finally, it should be noted that neurons (see discussion in the next section) probably evolved earlier than actual eye structures like the early "eye spots" containing photoreceptors. These primitive photoreceptors were capable of recognizing light and thereby mediating phototaxis (locomotion toward/away from light). Presumably, these neurons appeared in the Pre-Cambrian around 600 million years ago in a group of organisms called *Cnidaria* suggesting that neurons are of ancient origin [10, 11]. However, image-forming eye structures did not develop until the Cambrian beginning around 550 million years ago.

Structures of the Eye

Due partially to space constraints and to the complexity of the visual pathways involved, it is important to recognize that some of the material discussed in the next five sections represent a simplification of the actual details involved. The sections are intended more as an overview than an emphasis on details. Finally, it should be noted that ongoing research will undoubtedly modify some of the results involved in the vision-brain-interrelationship as reported here.

At this point, it may be useful to take a brief look at some of the structures of the mammalian eye. In many ways, the eye is like a camera in that both the eye and the camera focus on objects and control the amount of light entering. There are several major structures of the eye. The iris (the colored part of the eye) controls the amount of light entering the eye. The pupil is the dark center opening of the iris through which light passes through the lens. The lens is the structure that focuses the image onto the retina. Both the lens and the cornea (a transparent protective covering in front of the eye) act in a joint fashion to focus the light onto the retina. The retina is the light sensitive tissue at the posterior aspect of the eye. The liquid found between the cornea and the iris is called the *aqueous humor*. The liquid that is internal to the eye is called *vitreous humor* (Fig. 3).

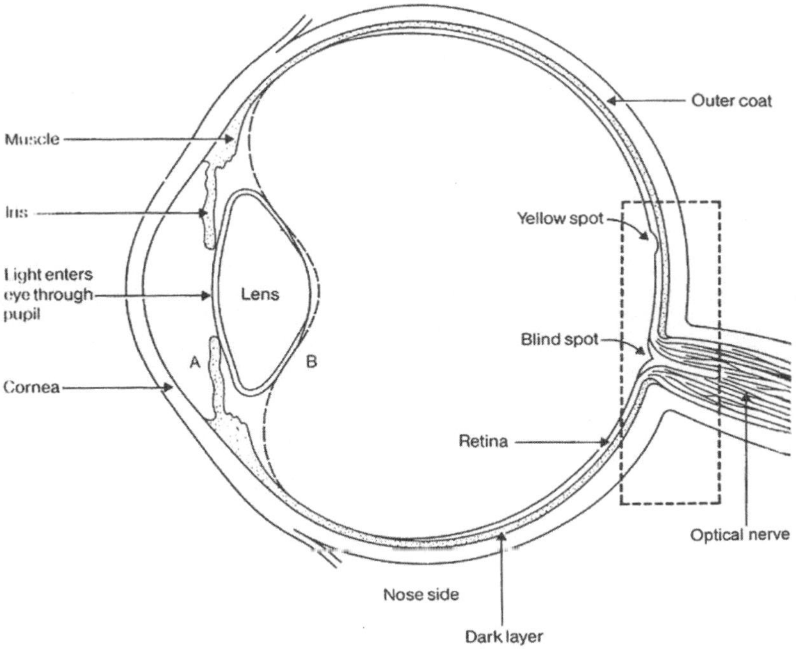

Fig. 3. A schematic representation of the eyeball (refer to text).

The retina is the structure at the back of the eyes that is the conduit by which visual images, received and focused by the lens, is transferred to the optical cortex of the brain. The retina is a crucially important structure in that it converts the light (photons) into electrochemical energy in order for the brain to interact with external world [2]. It is has been argued that the retina is actually part of the brain [12].

The retina is composed of photoreceptors called rods and cones. There are roughly 120 million rods and 7 million cones. These can be distinguished by their shape. The pigment in the smaller rods is more efficient at converting the electrochemical impulses so they are more useful in low light levels, but not for color. The larger cones, on the other hand, are more efficient at the ability of recognizing different wavelengths, or colors. Cones are located in the center of the retina (the fovea), while rods are distributed more peripherally. Thus, this central area, the fovea, is a region that is especially sensitive to detail and is the only one that provides a clear image of a visual scene [13].

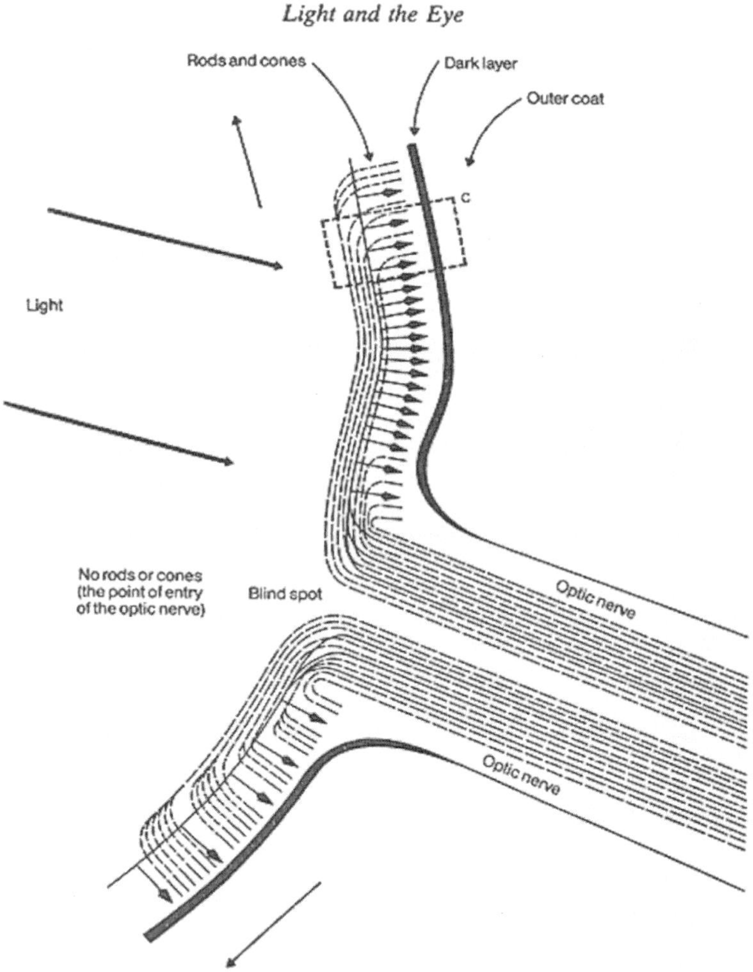

Fig. 4. The detailed structure of the retina (refer to text). It represents an enlargement of the back of the eye where the retina adjoins the optic nerve shown as a "box" in Fig. 3.

In technical terms the retinal structure in all vertebrates including humans, is "inverted". This arises, because, somewhat surprisingly, light passes through the ganglion cells first (on the left in Fig. 5) before reaching the rods and cones and then travels along the axons of the ganglion cells to a centrally-located optic nerve bundle (composed of axons) and then to the brain; specifically, the visual cortex. The retinal structure, found in invertebrates such as the squid, octopus and in mollusks, has a "verted" retina. This means that the light entering the eye travels through the verted retina to reach the rods and cones first and then directly to the ganglion cells and then on to a series of separate optic nerves (rather than a single optic nerve bundle found in the human eye) (Figs. 4 and 5). In humans, the retina has a thickness of about a quarter of mm [12].

Photons of light initially pass through the ganglion and bipolar cells to reach the photoreceptors (rods and cones). The bipolar cells mediate the signal from the photoreceptors to the ganglion cells. Each ganglion cell "sees" a small view of the visual scene, termed the receptive field. It is a small roughly circular region of the retina, which is stimulated by the light in the scene (the receptive field), which in turn transmit the visual information as nerve impulses to the axons that form the optic nerve connecting the eye to the brain [13].

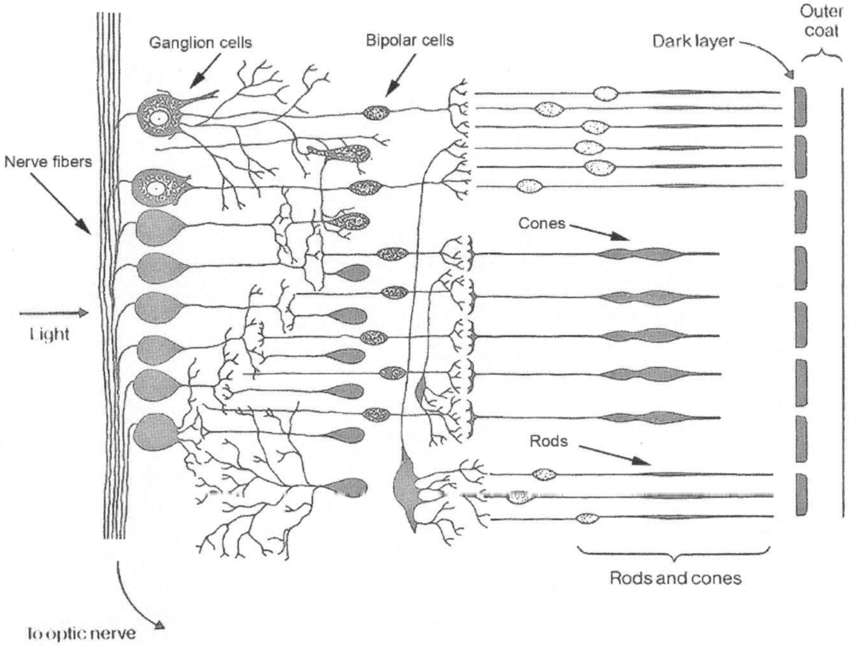

Fig. 5. Displayed here is an enlargement of section "c" in Fig. 4 to better elucidate the structures of the retina (refer to text for details).

The information along the fibers of ganglion cells crosses at the optic chiasm and travel to the opposite side of the brain to end up at the visual cortex. The temporal retina side (closest to the temporal bone), on the other hand, does not cross the midline but remains on the same side of the brain. It is this "dual" structural aspect, not present in most other vertebrates that allow primates and humans to see in stereoscopic vision or in 3D (Fig. 6).

The visual pigment molecules of the photoreceptor eye cells can be considered as the inter-phase between light and the organism. Opsins form a family of light-sensitive proteins found in photoreceptor cells of the retina. They are responsible for the conversion of photons of light into electrochemical signals. The visual pigment *rhodopsin* is found in the rods of the retina, while *iodopsin* is involved with the cones of the retina. These opsins (proteins) play a key role in all regulatory processes of living organisms. The signaling pathways regulated by these receptor proteins regulate numerous biological processes including sensory perception of primary interest here [11].

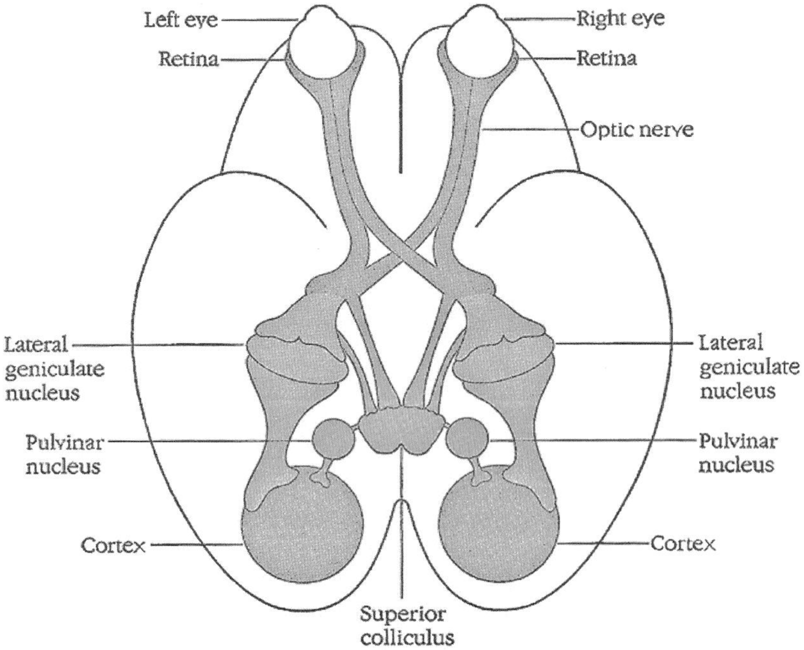

Fig. 6. Schematic of how the visual information is transferred from the retina through the lateral geniculate nucleus on to the visual cortex (refer to text) [1].

Eye-Brain Pathways

Now that the structures of the eye have been briefly elucidated, the next step is to describe how the visual information travels from the retina to the brain. Visual information from the retina to the brain travels along neurons or specialized nerve cells. Neurons contain an outer membrane that encloses the cell nucleus (containing DNA). The size of a neuron is about 20 microns but this can be quite variable (4-100 microns). Five neurons, side by side, can easily fit in the width of a human hair. Each neuron contains several branches or spines called dendrites. Additionally, a thin long tendril called an axon extends from the neuron body (Fig. 7). These axons can be as long as a meter or more.

Electrical impulses (spikes or action potentials) flow from the axon to the dendrite in a one-way direction. The connection or junction point between the axon and the dendrite is called a synapse. Synapses are the connections where the electrical impulse signal from the visual image along the axon is converted into a chemical signal (a neurotransmitter) before being re-converted back into an electrical signal in the dendrite [14, 15].

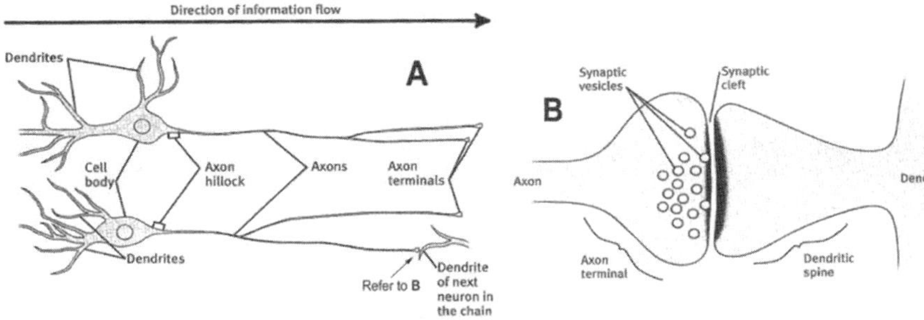

Fig. 7. Schematic structure of neurons and synapses (refer to text). Adapted from [14].

At this point, it may be asked how these electrical impulses (action potentials) operate and how visual information is actually transmitted along axons. Simply put, it turns out that there is a continuously present difference in voltage (–70 millivolts) called the resting potential, between the inside and the outside of the cell membrane of the nerve cells (neurons). This electrical difference is a consequence of an ion gradient, an imbalance between the concentrations of ions inside and outside the cell membrane. For this electrical charge to actually flow requires openings in the cell membrane so that potassium (K^+) ions can cross the cell membrane causing the ion gradient or resting potential [15].

However, this resting potential is insufficient to generate a spike or action potential, which is responsible for conveying the visual information of interest here. It turns out that when visually stimulated, the sodium (Na+) ion enters the cell changing the internal voltage to +40 millivolts (generating a spike), this is almost immediately countered by other potassium (K^+) ions, which are leaving the cell resulting in a dissipation of the spike and a quick return to the resting potential of –70 millivolts [15].

Once an action potential has occurred in one region of the axon, it triggers activity in a neighboring region, which triggers another action potential. In this fashion there is a cascade of action potentials along the entire axon until it reaches the dendrite of the adjoining neuron and passes the, in our case, visual signal across the synapse. Nevertheless, it is important to note that this is a very brief and simplistic discussion of an extremely complex process. For more details, the interested reader is encouraged to consult the following references [2, 11, 12, 15, 16, 17, and 18].

How then, exactly, do these action potentials travel from the retina to the brain? It turns out that this involves two pathways rather than one. These pathways are represented by two types of ganglion cells in the retina. These are the P cells and the M cells. The P stands for *Parvocellular* or "Parvi" for short (meaning small) and the M stands for *Magnocellular* or "Magni" for short (meaning large). P cells are particularly sensitive to color and fine detail and allow for recognition of objects (faces) but they are limited to a small center part of the visual field (fovea). These cells are generated from the cones in the retina. M cells, on the other hand, respond to dim light and motion, cover a larger visual field, and are insensitive to color or detail. These M cells are involved with location and depth of field of visual objects. Their information is derived from the rods in the retina [2].

These different neurons are activated (the firing of action potentials) depending on their initial ganglion source in the retina (the P and M cells), and then travel along the optic nerve (actually a bundle of axons) to the lateral geniculate nucleus (dLGN), a sub-structure of the thalamus. On their arrival at the dLGN, the axon signals continue on to the primary visual cortex (V1) located at the posterior aspect of the brain or occipital lobe, but in reduced number. That is, only a minority of the selective V1 region cells show activity. These now code for *specific* aspects of the visual information initially projected onto the retina. It is in this sense that only restricted elements of the visual image of interest are actually acted on.

Continuing, the signals from these two cell types travel in parallel along axons from the retina to the (dLGN) and then on to the visual cortex. Once arriving at the primary visual cortex, region V1, the two signals separate with the motion or M cells signal following multiple pathways leading to various regions of the parietal lobe while the color or P cells signal are sent via multiple pathways to various regions of the temporal lobe. There is also experimental evidence of interactions between the P and M pathways [2].

These two electrical signal streams eventually merge seamlessly and converge to integrate the visual message originating in the retina. To the observer, this integration of the two signals of the visual image is largely unconscious and occurs with great rapidity. Travel from the neurons along the axons to the next dendrite is roughly of the order of 45 meters/sec (Fig. 7, [14]). This speed is clearly necessary to avoid injury and can be seen as vital for survival in evolutionary terms. This is especially apparent in the autonomic nervous system (fight or flight response).

Computational Neuroscience: Image Discretization

The perception of a visual image is selective in that one focuses on certain aspects of the image and ignores other aspects that are not of central interest. This is an efficient way of avoiding image overload leading to confusion and even danger to the individual. That is, aspects of the visual image that are of peripheral interest are generally ignored or treated briefly as largely noise in the system. Moreover, with this selective process, humans recognize visual images effortlessly with great rapidity allowing many cognitive decisions to be made almost unconsciously [14, 16].

As stated earlier, the visual image focused by the lens and cornea is projected onto the retina. The retina captures this image composed of photons of light, which are in turn projected onto a set of photoreceptors composed of rods and cones that are discrete in character. Consider: (1) a computer screen composed of small elements called pixels or (2) the letters on a printed page, which if magnified can be shown to be composed of dots. Thus, as image magnification increases in both instances, the actual image deteriorates until only various small rectangles of color, brightness, shade, etc., (pixels) exist in the computer image. Likewise, for the black and white printed page, as the magnification increases, the words and letters are reduced to small black dots. Therefore, for both of these examples prior to magnification, the visual image, or text is readily recognizable as a visual image in spite of being discretinized. This is conceptually analogous to how the eye/brain structures apprehend visual information.

Each of these visual areas of the brain (more than 30 have been identified) responds to and separately analyzes specific *discrete* aspects of the image projected onto the retina thereby facilitating the process of recognizing different features of the visual image allowing the observer to interact with external world. This clearly suggests complexity in that other areas of the neocortex are involved in vision, not just the visual cortex at the occipital aspect of the brain. Note the emphasis on the word 'discrete'. The importance of this discretization will be made clearer subsequently.

The retina is initially composed of 126 million photoreceptors; however, the optic nerve that transmits the image to the brain contains only 1 million photoreceptors, a 126-fold reduction in image information [17]. Thus, this step can be considered as a sort of a pre-processing of the retinal image [18]. Why is this loss mostly inconsequential? The reason has to do with relevance. It turns out that much of the information in a visual scene is largely irrelevant or redundant (Fig. 8B). While the pixilated image is only a rough approximation of the eye in the actual photo (Fig. 8A), it is still recognizable. This is precisely analogous to the compression techniques used in digital cameras to reduce the size of an image to manageable levels for storage. Since the algorithm used for compression is known, it is a comparably simple matter to uncompress the image for display on a computer screen [17]. Thus with respect to the eye, an analogous "compression" process occurs. Despite the brain only receiving a fraction of the visual information that falls on the retina, most of this information is largely redundant and ignored by the brain. One can therefore view the transfer process of the visual information from the retina to the brain as a filter, which rejects the redundant information and passes only the critically relevant information.

In summary then, vision involves the translation of the retinal image (composed of a pattern of various shades of light and color at various intensities) to neurons (composed of cells that electro-chemically transmit this visual information as neuronal firings) into dendrites as input and along axons as output. This translation is, in general, a faithful reproduction; that is, with very little loss of information [17].

A photo can be viewed as composed of pixels (Fig. 8) and the photoreceptors in the retina can be equivalently visualized as composed of specific output voltages. However, this description is partially misleading as it suggests a one-to-one correspondence of the pixel/photoreceptor analogy. While perhaps, appropriate at the retinal level with its ganglion cells (which act as filters at different spatial scales [17]), this "picture" changes noticeably as the visual information is transferred along the optic nerve to the structures of brain. One can ask if this information transfer (the output from the ganglion cells) is a "good" representation of the retinal image. The answer is yes in that it is an "information preserving" representation in the sense that only non-redundant information (such as edge detection, color, motion, etc.) is transferred from the retina to the brain [17].

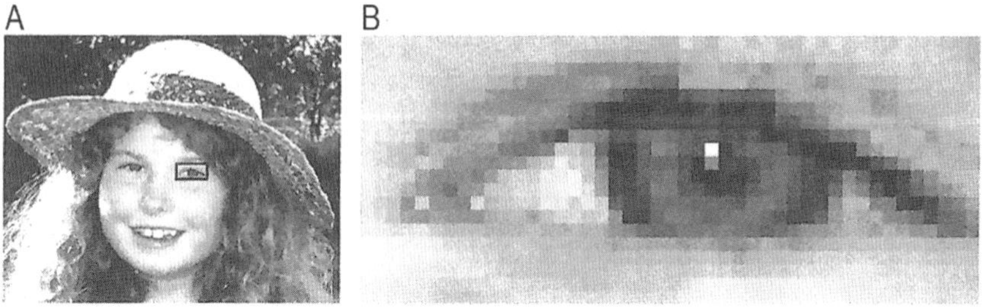

Fig. 8. Fig. 8A illustrates an image of the original photo. Fig 8B is a zoom of the boxed eye in Fig. 8A where each pixel is based on a gray level (0-255). This is analogous to each photoreceptor (rod or cone) corresponding to a small region of the retinal image (modified from [17]).

The earlier sections have focused on some of the actual structures involved in vision; that is, the neuronal organization allowing for the firing of chemical messages across synapses along pathways leading to various brain structures implicated in vision and memory. While this information is essential as an explanation and understanding of vision, it nevertheless remains incomplete. In fact, one can argue that it is not an explanation of visual perception as such, as it does not completely answer the central question of exactly how one apprehends form.

David Marr [19] stated that "Trying to understand vision by only studying the neurons is like trying the understand bird flight by studying only feathers. It cannot be done". A fuller understanding of the "computational analysis of vision", leads to the application of information theory to visual perception. This is, in effect, an attempt to quantify aspects of the process of vision.

Information Theory Applied to Vision

The theory of information flow has had widespread application in many fields since it was initially proposed by Shannon in 1948 [20, 21]. Information theory implied that the efficient flow of information is present in many systems including biological ones such as vision. This approach will be briefly pursued here, as it is one of the major theoretical approaches, termed the efficient coding hypothesis, currently in vogue [17].

Information theory here is concerned with the efficient coding of different image properties such as edge detection, line orientation, color, etc. The unit of information is defined as a *bit*. It is composed of only two elements, a zero (0) or a one (1). Nevertheless, while computer programmers are familiar with this binary code consisting of 0's and 1's as it forms the basis of machine language, there is a crucial difference between a bit (an amount of information) and a binary digit (composed of a 0 and a 1).

To illustrate this further, consider standing at a fork in the road, without prior information as to whether to take the left road or the right road. That is, one is faced with two equally likely choices. If one chooses the left branch instead of the right one, one bit of information (left = 1) has been involved. The right turn (right = 0) would represent the other binary bit [17]. Having one bit implies two equally likely probabilities (1 bit = 2^1), with three bits there are eight equally likely probabilities (3 bits = 2^8). The example of forks in the road is displayed in Fig. 9. So, for n bits of information (2^n) there are m = 2^n equally likely possibilities. For ease in computation, this is usually re-stated in logs to the base of two or as $n = log_2m$.

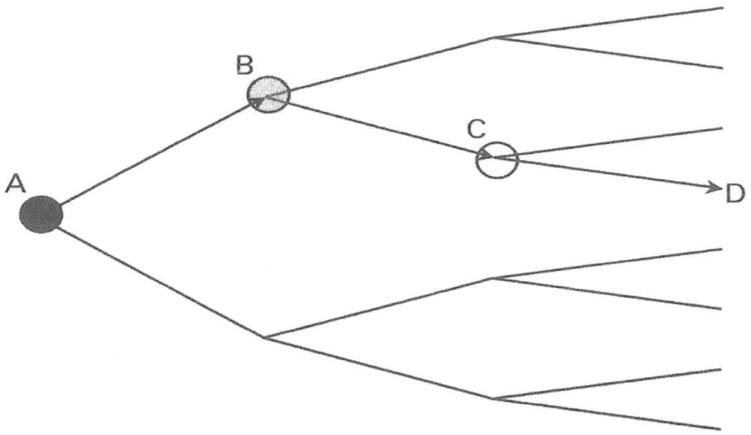

Fig. 9. Each fork in the road requires one bit of information. Assign the fork from A to B as left = 1 and the right fork = 0. The first fork, from A to B, eliminates half of the eight equally possible roads to take. The second fork, from B to C, eliminates another half leaving two roads. After the third fork, from C to D, the elimination leaves only one road to the destination. Thus, three equally likely choices (forks in the road) require three bits of information to choose one out of eight (2^3) possible destinations (from [17]).

Now rather than assuming equal probabilities (50% for each of the left and right paths), consider the fork going to the left (left = 1) as correct with a 90% probability. This means that one would "expect" the left fork to be the correct choice given the high probability and the right fork to be "unexpected" (right = 0) since it only has a 10% probability. This leads to the interesting question of whether this "unexpectedness" can be measured. Shannon [20, 21] suggested the following equation: $H = log_2 1/(p(s)$ where s refers to a binary digit (0 or 1), $p(s)$ refers to the probability of a specific value of s and H is now known as the "Shannon information". If it is assumed that if s = 1 and it occurs with a probability of 90%, one can compute the Shannon information as $H = log_2 1/0.9 = 0.152$ bits. That is, if the probability of choice is high then the information to be gained is low, as here, in effect an expected result. In contrast, if the right fork is chosen as the correct path with s = 0, then $H = log_2 0/0.1 = 3.322$ bits, a much larger value reflecting an unexpected result [17]. If these values were computed over a long time to generate an average value of H for the two alternatives, it would be 0.469 bits. This occurs because one of the alternatives is nine times more probable than the other one. This average in Shannon information is now known as the entropy of s.

This information approach can now be applied to the retinal photoreceptor structure. To do this one needs to divide the possible photoreceptor output voltages into discrete intervals or bins to use Stone's terminology [17]. These eight levels

(Fig. 10) generate three bits of information. The actual number of intervals here is arbitrarily set to eight, a number that is certainly too low, but will allow comparisons with Fig. 9. In actuality, this number of levels can be increased with each binary digit (0's and 1's) contributing one additional bit, each additional bit cuts the uncertainly of the photoreceptor voltage output value in half. Alternatively, each additional bit doubles the number of levels or bins and halves the voltage range between the levels. The use of eight bits to describe the photoreceptor voltage output generates 256 (2^8) possible values. This would be exactly equivalent, information wise, to the pixel array displayed in Fig. 9 [17].

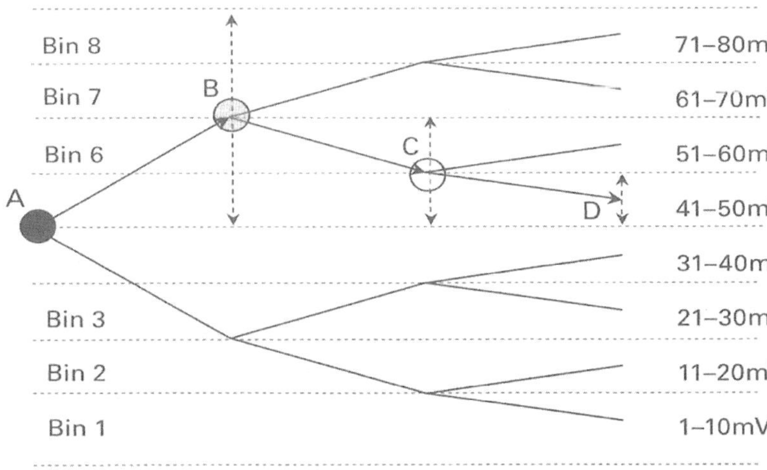

Fig. 10. Arbitrarily dividing the output voltage of a cone into eight levels or bins (clearly too low in actuality), yields three bits of information. In an equivalent fashion to Fig. 9, three steps (A to B, B to C, and C to D) are required to reach the desired photoreceptor output voltage (from [17]).

The use of information theory here has enabled the quantification of one aspect of the vision system; namely, the flow of photoreceptor voltage output as an information flow from the rods and cones. The same information theory process can also be applied to the ganglion cells. The data here consists of action potentials that also carry information. The firing rates of these action potentials can also be modeled using the bit approach outlined earlier.

Finally, one other quantitative approach is briefly mentioned but not detailed here. This refers to another aspect or level of the vision/brain interaction. It is apparent that much of the visual information that impinges on the retina is greatly compressed along its optical nerve pathway to the visual cortex. One of the results of this is that the amount of visual information that the brain receives is greatly reduced; nevertheless, the brain is still able, so to speak, to connect the dots, and re-create a perceived image. This re-enforces the concept that the brain relies on previous experience to reach visually mediated decisions. As Stone indicated:

> So, what you perceive is based partially on the data that is the retinal images and partly on your expectations about what the retinal image should contain, which can be summarized as follows: perceived image = image data + prior expectations [17].

This relationship leads to a numerical approach that is particularly suitable for problems of perceptual processing. This is *Bayes, rule* or *Bayes' theorem*. The basic equation of Bayes' theorem is $p(\mathbf{x}, c) = p(c \mid \mathbf{x}) \bullet p(\mathbf{x})$ where $p(\mathbf{x}, \mathrm{c})$ is the *joint probability* that both events **x** and *c* will occur, which is equal to the probability that *c* will occur *given that* **x** has already occurred times the probability that **x** will occur. Note that $p(c \mid \mathbf{x})$ is a *conditional probability* in that it expresses the probability that that *c* will occur conditional upon **x** having already occurred. For purposes here, the letters *c* and **x** refer to a perceived image data value and to an actual retinal image value respectively [17]. Since the alternate $p(\mathbf{x}, c) = p(\mathbf{x} \mid c) \bullet p(c)$ also obtains, the two equations can be equated leading to $p(c \mid \mathbf{x}) = p(\mathbf{x} \mid c) \bullet p(c) / p(\mathbf{x})$. Stone suggests that the following formulation is easier to recall as posterior = *k* x likelihood x prior. For further details of Bayes' theorem, refer to [17, 22, and 23]).

Viewed in more technical terms, $p(c \mid \mathbf{x})$ is the posterior probability density function, $p(\mathbf{x} \mid c)$ refers to the likelihood function (the maximum likelihood estimate MLE), $p(c)$ is the prior probability density function. This prior probability function is divided by $p(\mathbf{x})$. In sum then, Bayes' theorem allows for a more accurate estimate of the probability of a specific image being correctly identified if prior knowledge is also utilized.

The two quantitative illustrations here, using information theory and Bayes' theorem, are just two examples intended to demonstrate the increasing sophistication of recent studies of vision. Research in computational neuroscience is ongoing and will undoubtedly produce a more complete picture in the future. Nevertheless, the visual process is a complex and multilayered process suggesting that a single theory of vision is unlikely in the near future. Finally, it needs to be emphasized that the material here has been simplified for clarity and brevity. The interested reader is urged to consult the relevant sources for more details [12, 17, 20, 21, 22, and 23].

Now that we have briefly examined the mechanics involved in vision with all its complexities, it is now time to shift the discussion to how best represent the actual biological form; that is, to measure it, the subject of primary interest here.

The first section of this keynote address has dealt with vision as an "input"; namely, how the biological image enters the eye and how it travels to brain structures. The second section, below, focuses on what might be termed as "output"; that is, how the brain interprets the biological image allowing one, in both qualitative and quantitative terms, to arrive at meaningful conclusions about the biological form.

PART 2: THE ANALYSIS OF FORM

It is hardly too much to say that the whole science of biology has its origin in the study of form. *C. H. Waddington (1968: p. 43)*

From Vision to Form

Now that one has gained some clarity of how the form as a visual image is obtained with the eye/brain system, it is time to shift the discussion to the form itself, the subject matter of this symposium. The visual system facilities the apprehension of form and thereby allows for organismal interaction with the environment. This form (or object) information is routinely used in numerous ways for cognition and generates behavioral responses. The central question here is how to best characterize, in this specific case, the *biological* form (with its numerous attributes such as size, shape, color, structure, etc.) so that most of it (as visualized) can be retrieved without a significant loss of information.

The very act of recognition of the visual image requires interpretation, which by necessity, involves the brain. As stated earlier, without brain involvement no understanding of the visual image, as projected onto the retina, is even possible. The involvement of the brain as a crucial element in the interpretation of the meaning of the visual image and has been termed visual intelligence; in contrast, the eye (in of itself) contains no visual intelligence [14]. The eye does not 'see' in the conventional meaning of the word. What the eye actually, 'sees' consists solely of photoreceptor activity activated by photons of light falling on the retina; that is, consisting of physiochemical activity of the *rhodopsin* in the rods and the *photopsin* in the cones of the retina prior to being transferred as action potentials via the optic nerve to various regions of the brain.

Consequently, it is also important to realize that the eyes do not project visual images as such onto the retina and then on to the visual cortex as a precise representation of the real world. Instead, only certain aspects of that visual image are stored in various regions of the neocortex allowing for later recall [10]. That is, the representation of the actual image being observed becomes, in effect, an abstraction when it reaches the visual cortex.

In spite of the last three decades of visual perception research, the central question remains; namely, how precisely does this visual intelligence construct reflect ones visual experience? That is, how can the visual image (or essential elements of it) be retrieved later from memory leading to reliable cognitive behaviors? A partial, although not complete answer, to this question can now be given based on recent research. Part of this answer is based on the process of discretization of the visual image as discussed earlier [14, 17]. Thus, visual information (our biological form of interest) is composed of discrete elements and these are transmitted as electrical signals are sent via the optic nerve as axons to the lateral dLGN and then on to the primary visual cortex from where it radiates to other regions of the neocortex as discussed in previous sections.

Ontogenetic and Phylogenetic Diversity

As the above quote from Waddington implies, the study of biology can be largely reduced to the study of form. This, in turn, leads to two of the major issues in biology: (1) how does the form of the individual organism develop over time and (2) how does the form change over the course of evolution.

With respect to the visual system, one can effortlessly and often instantaneously recognize the differences between, for example, gourds, and squashes (Fig. 11). While one can apprehend the biological form and aspects of that form effortlessly with the visual system, it is an entirely different matter to describe those differences quantitatively. In fact, one can argue that only a small fraction of the visual information so readily perceived by the eye has been adequately measured [24, 25].

In fact, considerable discussion ensued by the current author and his co-workers in the early 1990's on how to develop a generalized model that quantifies a majority of this visual information [25]. This model is briefly mentioned in this presentation in the discussion at the end of the section on Biological Shape Analysis.

Fig. 11. Example of form variation: The variety of shapes of gourds and squashes [24].

Another illustration (Fig. 12) depicts that biological organisms do not only display variability in shape, but that this variability can take on many aspects. In the example shown here, it is apparent that these lady beetles do not differ very much in size or shape but they do display widely different coloration as well as spot markings on the carapace [26]. This variability is exhibited in colors (largely black and red) as well distinctly different patterning. This suggests that alternative techniques are required to assess this biological variation. A pioneering study, of how to analyze color variability as displayed by a sample of lady beetles, was independently carried by N. MacLeod and this research is presented in the current proceedings [34].

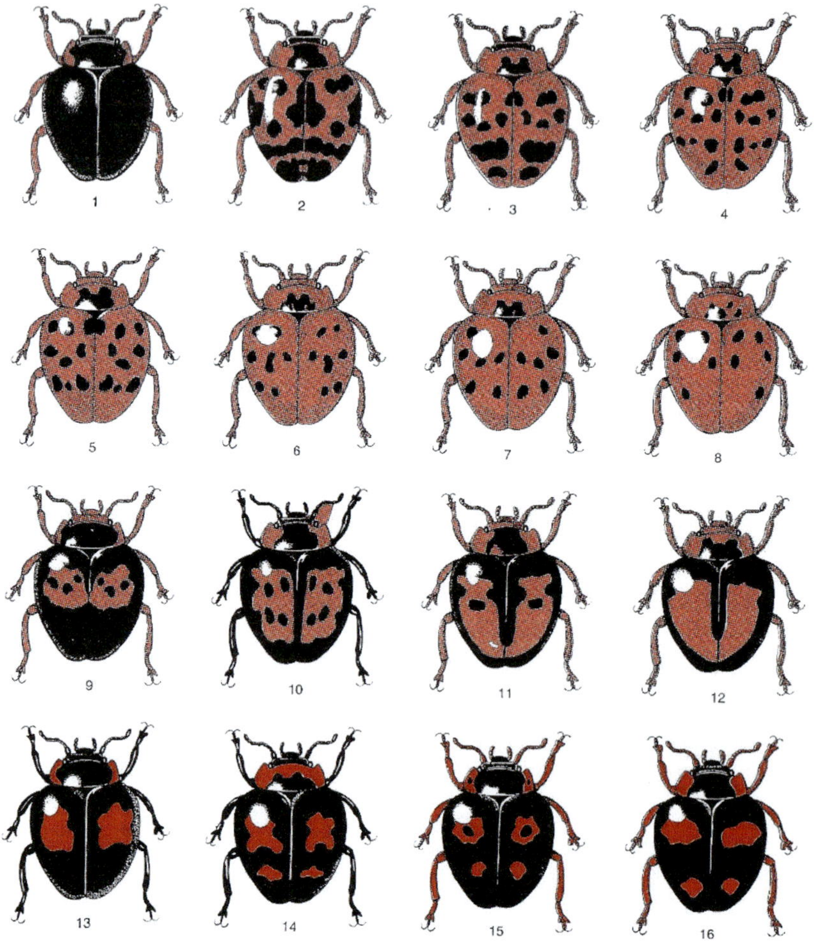

Fig. 12. Lady beetle morphological variation that is independent of form or shape [26].

Here is another example of this variability, this time closer to the area of interest of the author (Fig. 13). Visually, the differences between these skull specimens seen from the side (lateral view) are readily apparent. However, the question that arises is of how to describe these form differences (for example the shape changes in the outline of the skull/mandible) in a quantitatively meaningful way that captures most of the visual information. Clearly, this is not a trivial issue but continues to represent considerable challenges and will be taken up subsequently [28].

The last example in this section focuses on differential growth as a change in structure over time (Fig. 14). Clearly, this is the norm for most if not all ontogenetic processes. Again, the question that arises is how to best quantify these form changes over time [24].

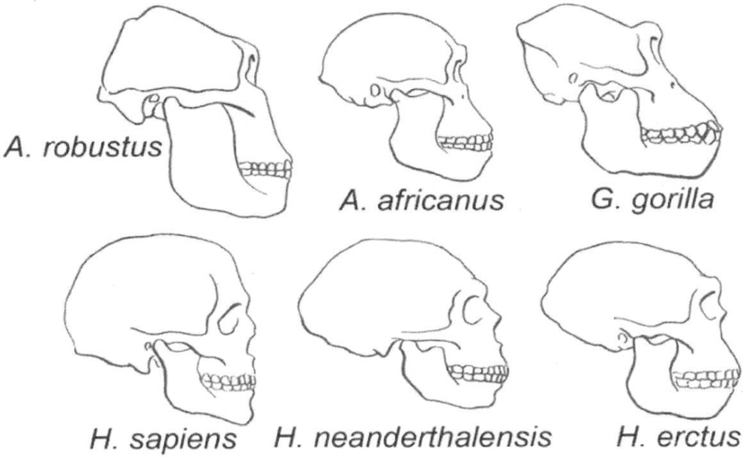

Fig. 13. Hominoid skull diversity (modified from [27]).

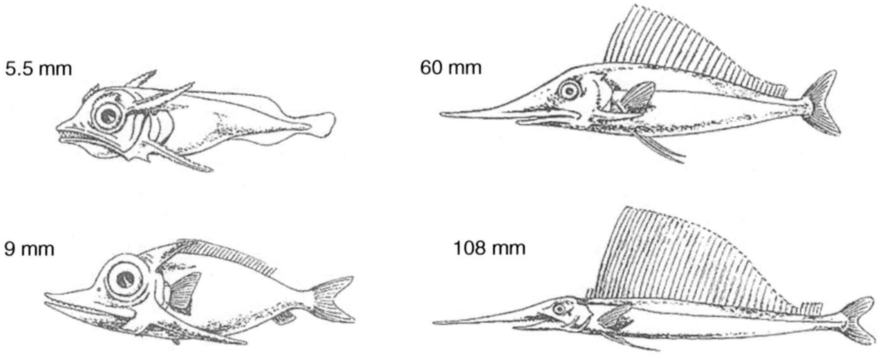

Fig. 14. Illustrated here is an example of upper jaw differential growth in the sailfish. Shown are the four stages in the growth of a young sailfish. Note that the snout (upper jaw) grows differentially faster than the lower jaw with the result that it extends much further in the adult specimen (modified from [24]).

Biological Processes

Richard Goldschmidt in the 1960's, made a very cogent observation in that: "The prime action of the genes is to govern the rate of biological processes" [24]. This rather prescient observation has turned out to be largely correct.

It is now known that genes control enzymes, which as catalysts provide the mechanisms to control growth rates. Although many of the details still remain unclear. However, it is not only the rate of growth that is important, but also the direction that this growth takes. If we accept that the form, and especially the shape of an organism, is determined by the rate of growth and the direction of growth over time, this raises the question of how are biological forms determined.

Outlined here are some of the major biological processes involved over the lifespan of an organism. Although, it is focused on humans, it applies equally as well to most other organisms. Each of the six steps listed in Table 2, others could also have been listed, signifies a process that leads to changes in the biological form, some minor others major.

Table 2. Biological Processes from Genes to the Development of the Phenotype

[1]	**Sexual reproduction leading to zygote development.**
[2]	**Cellular division and differentiation leading to morphogenesis.**
[3]	**Process of birth resulting in a change of internal to external environment.**
[4]	**Organismal growth and development.**
[5]	**Reproduction, which leads back to [1] above.**
[6]	**Aging, senility, and death.**

Models as Representations

The central issue of this address is concerned with the measurement of the biological form or more precisely its shape, in such a way that a majority of the variability being expressed is captured. This has not been very successful in the past because of the approach used. The universally used methodology consisting of distances, angles, and ratios or conventional metrical approach (CMA) is inefficient when applied to the largely irregular morphologies exhibited by most biological organisms; thus, what is required is a numerical representation of the biological form, one that closely maps (ideally in a one-to-one relationship) to what is actually observed. Such a mathematical representation can be considered as a *model*. This will be made clearer below.

Models are always simplifications of the real world. All such models attempt to simplify the behaviors that are involved in what are often very complex systems. This is especially the case in the biological sciences where even the simplest biological organism is a quite complex system. Therefore, ways have to be found that reduce this complexity to more manageable terms. Note that models apply in a general way to a whole host of disciplines from engineering, the physical sciences, as well as the biological and social sciences.

There are three specific kinds of models that will be mentioned here. These are: (1) physical models, which are perhaps the easiest to understand and visualize because they are used in everyday life; (2) heuristic models, these are bit more complicated in that they are more abstract. They often consist of block diagrams to illustrate the flow of particular processes; and (3) mathematical models. These can be quite simple or very complicated.

The first model, a physical model, is easy to comprehend. It is a representation of an actual physical object. This could be any object including biological ones. It is a simplification in that it is often a smaller, more manageable representation, and necessarily lacks the details of the actual object. Such models may be more or less detailed. Shown here is a Spitfire, a famous aircraft of World War II. This is a photo of one of the very few Spitfires still in flying condition. It needs to be emphasized that this is NOT a model but the actual object, which is to be modeled (Fig. 15).

Here is the model (Fig. 16). It is a physical representation of the actual object but in this case at 1/72 scale. Clearly, it is not an exact representation; for one thing, this model cannot fly. However, there are Spitfire models that can fly. Nevertheless, this Spitfire model displayed here is a good, if rough approximation, of the real object. Everyone can recognize what it is intended to represent.

Fig. 15. A Spitfire of World War II fame. This is a photo of an actual flying airplane.

The second model to be discussed is called, for the lack of a better term, a heuristic one. The idea behind a heuristic model is that it is more abstract. It is intended to model the flow of a complex process in simple terms. It is often composed of block diagrams. It is, of course, a simplification of the process and, in general, does not show the details. Although, one can develop more fine-grained heuristic diagrams by expanding each of the block diagrams to show more details. Generally, however, the purpose of such a heuristic model is to outline the major features of a complex process, not so much the details.

Here is an example of a heuristic diagram that models how one does science as a flow diagram. It displays in simple terms a complex process (Fig. 17). Starting with a biological system (upper left), one then carries out observations and/or experiments under controlled conditions that look at just some variables since the system is too complicated, in the process you may build mathematical models, which are in themselves, simplifications (upper right). The process then continues with the derivation of analytical conclusions based on calculations, models, etc. (lower right). These analytical conclusions should provide information about the biological process (lower left) and finally, the conclusions will need to be tested or verified against the real world to see if they are indeed valid [24].

Fig. 16. Shown here is a photo of a model of a Spitfire at 1/72 scale.

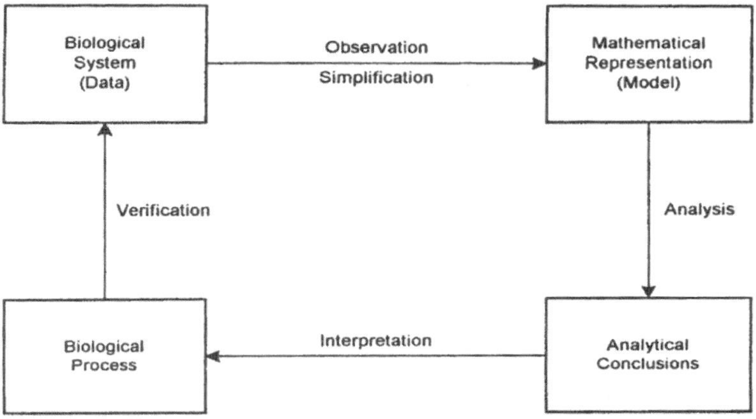

Fig. 17. An example of a heuristic model set up as a flow diagram to illustrate the path of scientific research in the biological sciences (refer to text) [24].

The third model to be covered here refers to a mathematical model. A mathematical model can be quite simple, as will be illustrated or quite complex. It often is a representation of a process over time. Eventually, a mathematical model will be presented here that is a representation of the biological form; specifically, one that focuses on the shape of organisms.

An example of a very simple model is the tooth *crown shape index* (Fig. 18). While this has been widely used, it is really a description of a rectangle (or a diamond) and has little to do with the shape of an actual tooth. The *actual shape* of the crown, its outline or boundary, is not being measured. While the MD/BL or MD x BL indices may be useful as rough measures, there is the very legitimate question of what is really being measured. As indicated at the beginning of this section, the use of CMA is inefficient and this example based on ratios clearly shows why that is the case. It is in this sense, that CMA is a poor model.

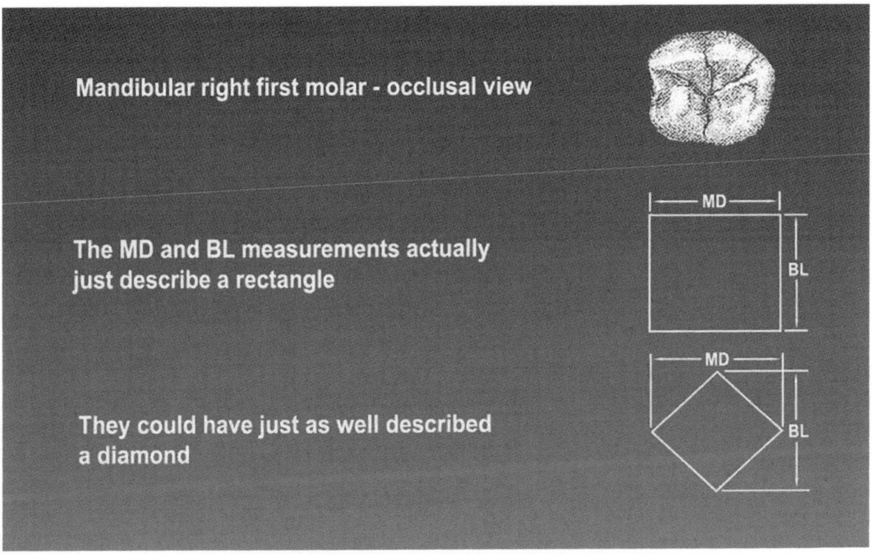

Fig. 18. An example of a simple biological model.

The next example illustrated here is a more complex mathematical model. It is the comparison of two hominin groups, *H. sapiens* on the left and *H. neanderthalensis* on the right (Fig. 19). The differences in skull shape are obvious and distinct. Each has been fitted with a conventional Fourier function, the coefficients of which are displayed as columns next to the skulls. The numbers clearly describe the differences seen visually. Specifically, the second harmonic, A2, measures the height of the cranial vault, which is 18.14 for *H. sapiens* and 9.92 for the *H. neanderthalensis* case. Moreover, the third harmonic, A3, is 7.97 for *H. neaderthalensis* and 2.23 for *H. sapiens*. This harmonic measures the pronounced brow ridges and occipital bun shape of *H. neaderthalensis* in contrast to *H. sapiens*. Details can be found in [27]. In this sense, the Fourier function can be viewed as a representation or model of the skull outline.

The emphasis now focuses on the biological form from a shape perspective. Three mathematical representations will be described. Their pros and cons discussed. While the examples here are in 2D, their extension to 3D can also be inferred.

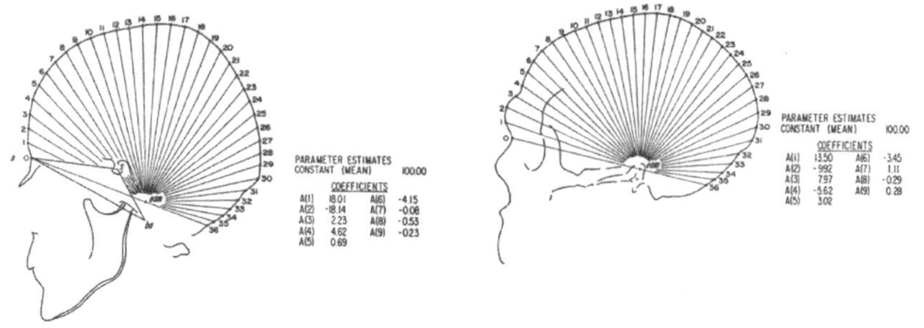

Fig. 19. An example of a more complex biological model (refer to text [27]).

Three Representations of Form

The first representation to be discussed is the CMA. This approach is generally limited to homologous points. Homologous points are those can stand in a one-to-one correspondence or mapping across forms. Displayed here is the lateral view of a human cranial base in 2D (Fig. 20). It is being measured with three homologous landmarks. Then using the CMA, the three distances forming a triangle, are drawn connecting the three homologous landmarks. The three angles here have been traditionally considered as measures of shape. However, it is important to note that these three linear/angular measurements do not even begin to capture the actual cranial base as seen visually. This illustration is not just to emphasize the issue here but rather represents real data as traditionally used by anthropologists. (Fig. 20).

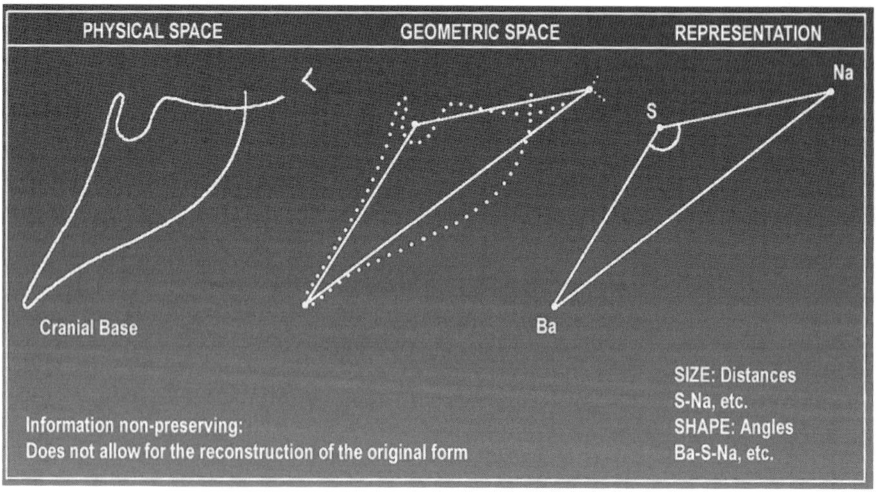

Fig. 20. The universally used model based on homologous points. The conventional metrical approach (CMA) composed of distances, angles and ratios used to measure a cranial base (*Norma lateralis*) (modified from a figure originally developed by Read [33]).

Fig. 21 represents a summary figure of the CMA listing its pro and cons. CMA has been used for the last two centuries and continues to be used today in spite of its drawbacks. The one advantage, of course, of the CMA is that it easy and convenient to use. Nevertheless, the CMA is a poor model because it does not provide a close representation of the biological form or its shape from a visual perspective. CMA continues to be *the* major measurement system in the biological sciences in spite of a number of limitations such as the sampling of points being too sparse to capture adequately the shape. Moreover, the choice of landmarks selected is subjective. Finally, one cannot re-construct the form from the measurements.

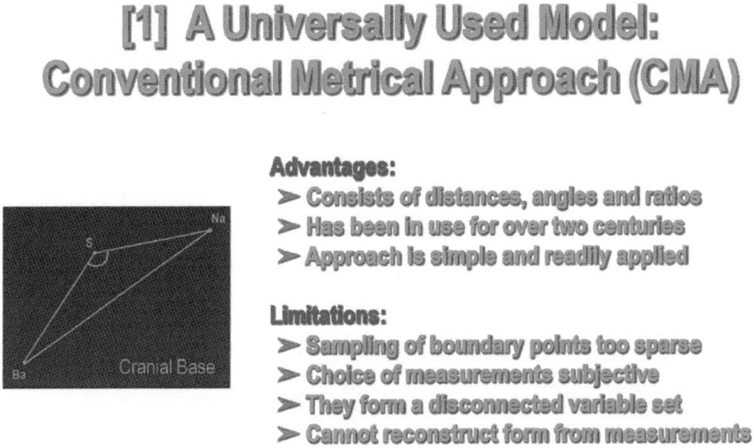

Fig. 21. Pros and cons of the conventional metrical approach (CMA).

The second representation is the use of elliptical Fourier functions (EFF). EFFs provide a precise mathematical analog or representation of the form, as well as allowing for the re-creation of the original form at any time in the future. EFFs transform the spatial domain into the frequency domain. The spatial domain is composed of the set of closely spaced points describing the form and the frequency domain contains amplitudes, phase, and power (a new set of variables).

Rather than using integrals, as required with the Fourier transform, Kuhl and Giardina in 1982 devised an algebraic approach (Fig. 22). That is, a parametric solution in which the x and y coordinates of the points along the boundary outline are set up as functions of a third variable, *t*. Note that the EFF is periodic; that is, composed of the trigonometric sines and cosines [21]. This approach confers a major advantage in that the spacing between points can now be unequal in contrast to conventional Fourier descriptors (FDs). Fig. 23 illustrates the pros and cons of the EFF approach Refer to [24] for an extended discussion of EFFs.

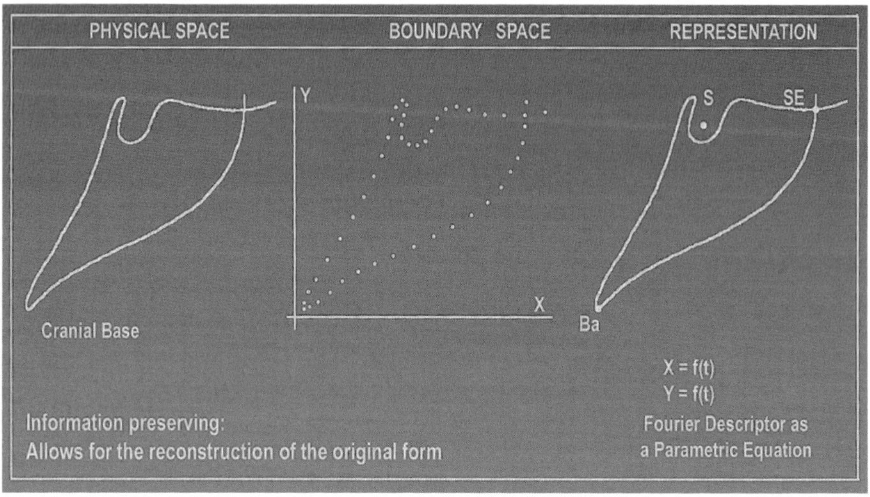

Fig. 22. An application of elliptical Fourier functions (EFF) to the lateral cranial base is shown here. While the example here is in 2D, the EFF can be extended to 3D as a curve in space. (This figure was modified from one originally developed by Read [33]).

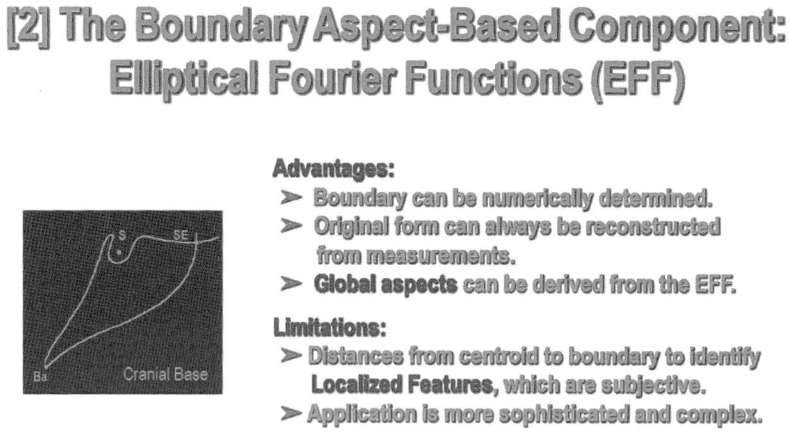

Fig. 23. A diagram of the pros and cons of Elliptical Fourier functions (EFF).

Further, there are two other attributes of shape that need to be considered. These are global aspects and localized features. While global aspects can be characterized with EFFs, the identification of localized features remains subjective. However, this latter issue can be now objectively handled by another method, wavelets, to be described subsequently. One advantage of EFFs is that one can re-construct the form at any time from the coefficients; in effect a Fourier inverse-transform.

The third representation is the continuous wavelet transform (CWT). The CWT, as with EFFs, transforms the spatial domain into the frequency domain. This produces a set of wavelet components at different frequencies. Wavelets are based on multiresolution analysis, which consists of two aspects: position along the boundary outline of the form (b) and scale (a). In this way, it contrasts with EFFs. Points along the boundary, s, must be sampled as a power of (2^n) and the intervals between points must be equal. While these two restrictions do not apply to EFFs, they must be observed here. The inverse CWT allows for the original form to be reconstructed. The great advantage of wavelets is that localized boundary aspects (singularities) can now be objectively identified. This resolves the subjectivity issue inherent with EFFs and all Fourier Descriptors with respect to locality. Fig. 24 illustrates the pros and cons of the CWT approach.

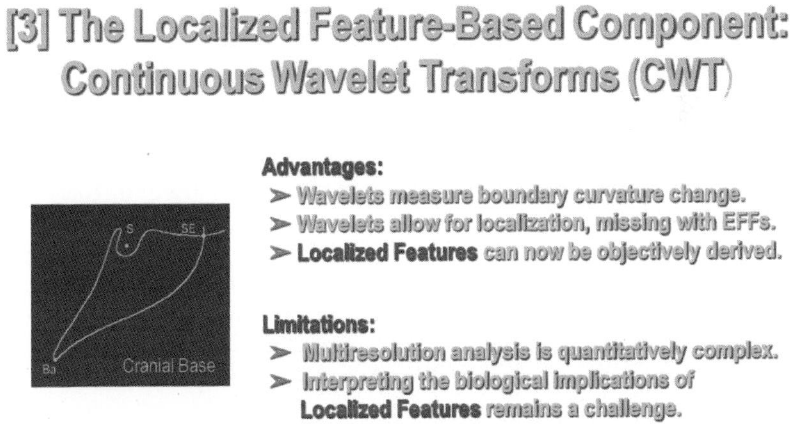

Fig. 24. Localized feature-based component: Continuous wavelet transforms (CWT). Shown here are the pros and cons of this approach.

The CWT can also be viewed as a transformation from the spatial domain into the frequency domain. However, the functions are now *not* periodic as with the EFFs or other Fourier-based methods. Thus, the wavelet transform represents an alternative decomposition of the spatial domain into *two* wavelet components, one determining the *scale* or *level* and the other the *position* or *location* along the boundary of the form. Each of the levels refers to a different resolution or scale, which represents a doubling of the frequency.

In a similar way to the Fourier transform, the closely spaced points on the boundary outline must again be even divisions of a power of two. In addition, in contrast to EFFs, which ranges over the 2π interval $[-\pi,\pi]$, wavelets are based on "compact support". This means that they range over a very narrow interval, this results in a small wave, hence the name wavelets. However, as mentioned earlier and must be emphasized, localization with EFFs is still subjective.

Wavelets, on the other hand, allow for the precise identification of these localized features in a totally objective fashion. This is accomplished by computing the change in curvature along the boundary outline. Thus, it is the fact that wavelets use a very narrow interval (compact support) that allows for the identification of localized features. It is called a multiresolution decomposition because it can focus on low frequency components (high scale or level values) and high frequency ones (low scale or level values). The reader is highly recommended to peruse the following references dealing with wavelets [28, 29, and 30].

Two Examples based on Craniofacial Data

The first example here is based on lateral cephalometric cranial radiographs that were available through the courtesy of Prof. Eisaku Kanazawa from the Nihon University School of Dentistry (Matsudo, Chiba Prefecture, Japan). The total cranial base data for this study consisted of 297 specimens divided into five archaeological age groups. The age groups utilized are from the Yayoi, Kofun, Kamakura, Edo, and Modern eras. The skulls of Yayoi, Kofun, and Kamakura specimens are from Kyushu University (Fukuoka, Japan). The Edo specimens are from the National Science Museum (Tokyo, Japan), while the Modern samples consisted of students from the Nihon University School of Dentistry.

The example here is limited to the Yayoi specimens (n = 32 ♀, n = 32 ♂) and involves the superimposition of the mean ***female*** Yayoi cranial base viewed in the lateral view on the mean ***male*** Yayoi (Fig. 25). The male trace, the one with the ***dotted lines,*** shows that the region around the pituitary fossa; that is, around points 25 to 30, is displaced anteriorly, that is to the right, in comparison to the female trace. This anterior displacement was present for all archeological comparisons [30]. This pattern of an anterior male displacement at the pituitary fossa is an ancient primate pattern, at least 30 my old as it was also seen in a previous study of sexual dimorphism in *Macaca nemestrina*, where the same pattern was obtained [31].

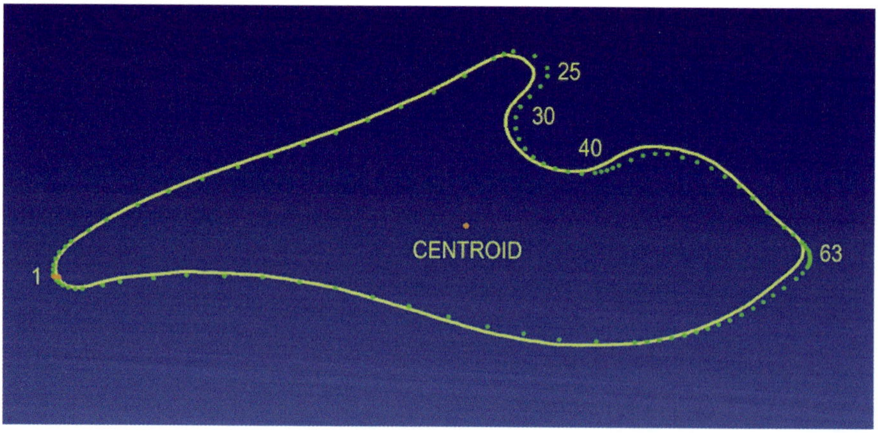

Fig. 25. Size standardized superimposition on the centroid of the Yayoi females (solid) and the Yayoi males (dotted).

Since there were statistically significant differences between the sexes, this difference in sex, was further examined by subtracting the coefficient wavelet matrices of the *male* data from the *female* data. Each matrix is composed of 96 levels by 512 points; or, 49,152 individual coefficients. This sex difference is illustrated in

Fig. 26. The major area of change between the Yayoi *females* and *males* is around points 25 to 30; that is, at the pituitary fossa. Other changes can also be discerned, but they do not exhibit as strong a signal. It is important to emphasize that the sexual dimorphic differences seen in the Yayoi cranial base at the pituitary fossa and displayed as wavelet differences in Fig. 25, were identified in a completely objective manner by simply running the CWT.

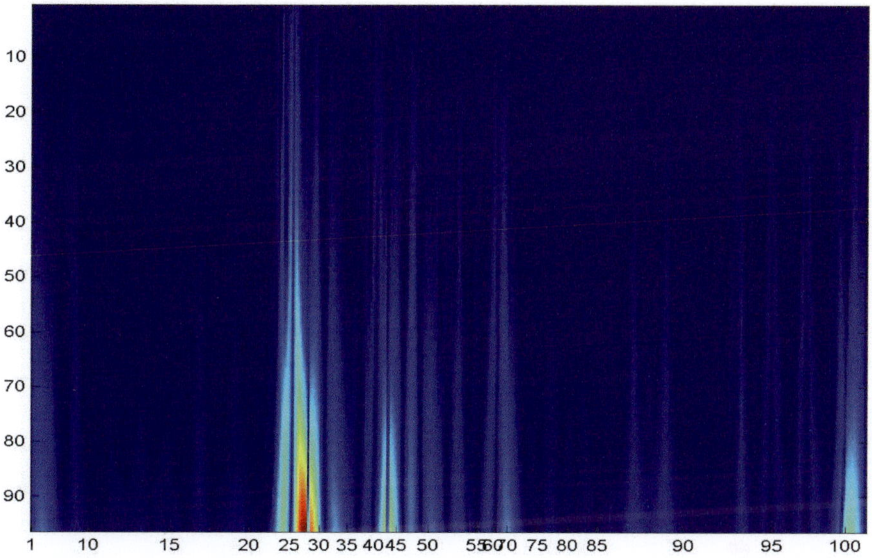

Fig. 26. Shown here is the wavelet displaying sexual dimorphism of the cranial base among Yayoi males and females. The numbers along the horizontal axis are point numbers along the cranial base outline (refer to text).

The second cranial example here is based on standardized photographs of the cranial vault in the lateral view [32]. Because numerous specimens consisted of only the skull vault with the facial aspect missing, a geometry was devised that solely described the cranial vault, otherwise too many specimens, especially *H. erectus* ones, would have had to be excluded. The 104 specimens included the following samples: *H. erectus* (n = 30) covering a span of approximately 2.0Ma to 40,000 years; *H. neanderthalensis* (n = 22) from roughly 300,000 years to perhaps 30,000 years and *H. sapiens* (n = 52). Of the 52 *H. sapiens* specimens, 41 were archaic *H. sapiens* (~150,000 ybp to 11,000 ybp) and 11 represented modern *H. sapiens* (less than 125 years ago).

Distinct visual differences (statistically significant) were found between *H. sapiens* and both *H. neaderthalensis* and *H. erectus*. In contrast, the differences in cranial vault between *H. erectus* and *H. neaderthalensis* were small and statistically insignificant. A conclusion drawn from this result is that there is a reasonable probability that *H. erectus* and *H. neaderthalensis* form a single evolutionary lineage, and *H. sapiens* represents a separate and later developing lineage. The example here displays the superimposition on the centroid of the cranial vault means of the *H. sapiens* and *H. neanderthalensis* samples to illustrate their differences. These distinct visual differences between the two samples are depicted with the arrows (Fig. 26).

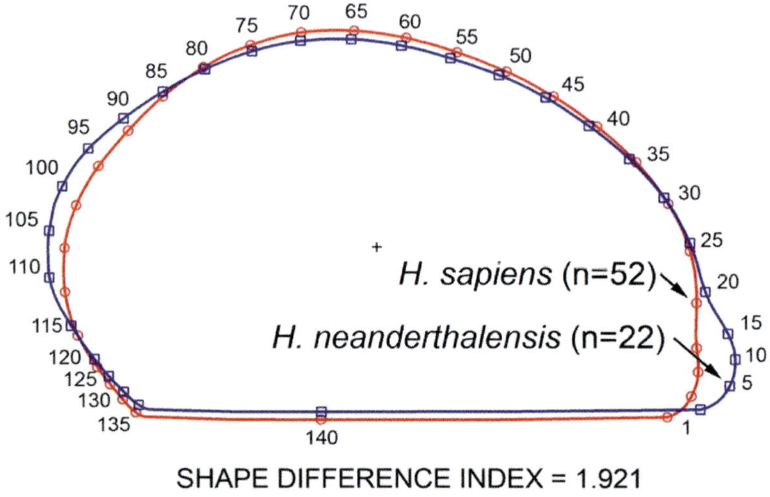

Fig. 27. Centroid superimposition of the mean trace of the cranial vault of *Homo neaderthalensis* versus *Homo sapiens*. These samples are area-standardized to remove differences in size.

Fig. 28 shows the subtraction of the *H. sapiens* wavelet matrix from the *H. neanderthalensis* matrix displaying the differences between the two cranial outlines. Most of the difference is visually displayed at the *Nasion* area around point 15 and at the occipital aspect around point 100. The difference around 135 represents a discontinuity to be ignored (Fig. 28).

According to the wavelet data, the major change is in the region of points 5-15 and a more moderate change around points 95 to 110 (Fig. 28). Again, it needs to be mentioned that these differences in shape at *Nasion* and the occipital aspect were objectively and independently derived with the use of the CWT.

Biological Shape Analysis

As this symposium is on biological shape analysis, it raises the natural question of what exactly is biological shape analysis. So what is meant by *Biological Shape Analysis* and why is it important? Note that this does not mean Biological Form Analysis; the emphasis here is on Shape. The difference between Form and Shape will be explored subsequently.

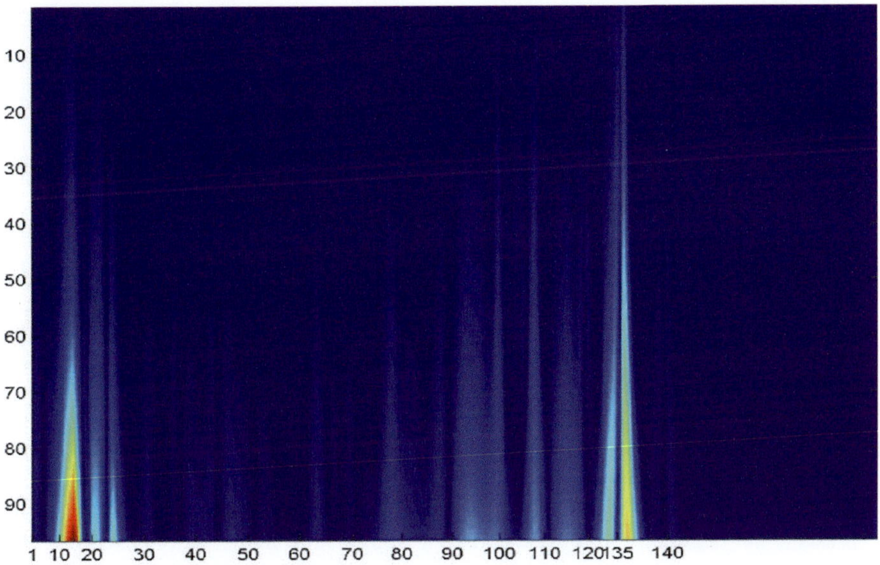

Fig. 28. Boundary differences utilizing the wavelet between the mean cranial vaults of *Homo sapiens* and *Homo neanderthalensis.* The numbers along the horizontal axis are point numbers along the cranial base outline (refer to text).

Morphometrics represents the mapping of visual information into a mathematical representation. However, morphometrics as currently used, generally focuses on aspects such as primarily size and shape. Little attempt has been made to date to develop a more complete model of form. It is almost as if just to take a few size measurements is enough to qualify one to be doing "morphometrics". It is argued here that that is not sufficient. There is much more to morphometrics than just size and shape.

Returning to the question of what is biological shape analysis. What this means is that one has to first distinguish between form and shape. These two concepts are often conflated. It is especially important in science that definitions be precise. Form is defined as the totally of the morphology in question. It can also be a part or a whole. Shape is defined here as only the outline of a form or part of a form. It can be viewed in 2D or 3D. Thus, it is important to distinguish between Form and Shape otherwise confusion ensues. Having defined what shape is (the boundary outline of a form), is just the beginning. One then needs develop a mathematical model that is a precise representation of the shape of a form. More importantly, the measurement of shape is not a trivial issue but continues to be a difficult and challenging problem.

As discussed earlier, humans are especially aware of visual objects and rely on vision quite heavily in ones daily life. So also in scientific research, there is a major emphasis on how we see objects. So although there are various other attributes of form such as colors and texture, shape is one aspect that the visual system tends to focus on.

A number of theoretical approaches have been suggested to model this visual information. Three of these are: (1) Form = Size + Shape; (2) Form = Size + Shape + Structure; and Form = f < Size, Shape, Orientation, Surface, Interior, etc.>. This last formulation treats form as function but makes no assumptions about linearity in contrast to the first two models. However, to date, no model is available that includes a significant amount of the visual information that is observed in biological organisms [24].

To re-emphasize the visual context of biological forms, shown here (Fig. 29) is an idealized drawing or sketch if you will, of the lateral view of an idealized *H. erectus* skull. Here are shown six attributes, such as the outline, surface texture, etc. While this visual information is readily observed, its integration into a quantitative model has been less than successful to date. The problem is how to incorporate the *H. erectus* visual information displayed into a numerical representation or model.

It is also of considerable interest that the work of Norman MacLeod as presented at these proceedings marks a notable initial step in focusing on other aspects of form such as structure in terms of surface patterning and color utilizing the surface configuration of lady beetles [34]. It is therefore hoped that this represents the beginning of a more complete approach to numerically describing the biological form in all its beauty and complexity.

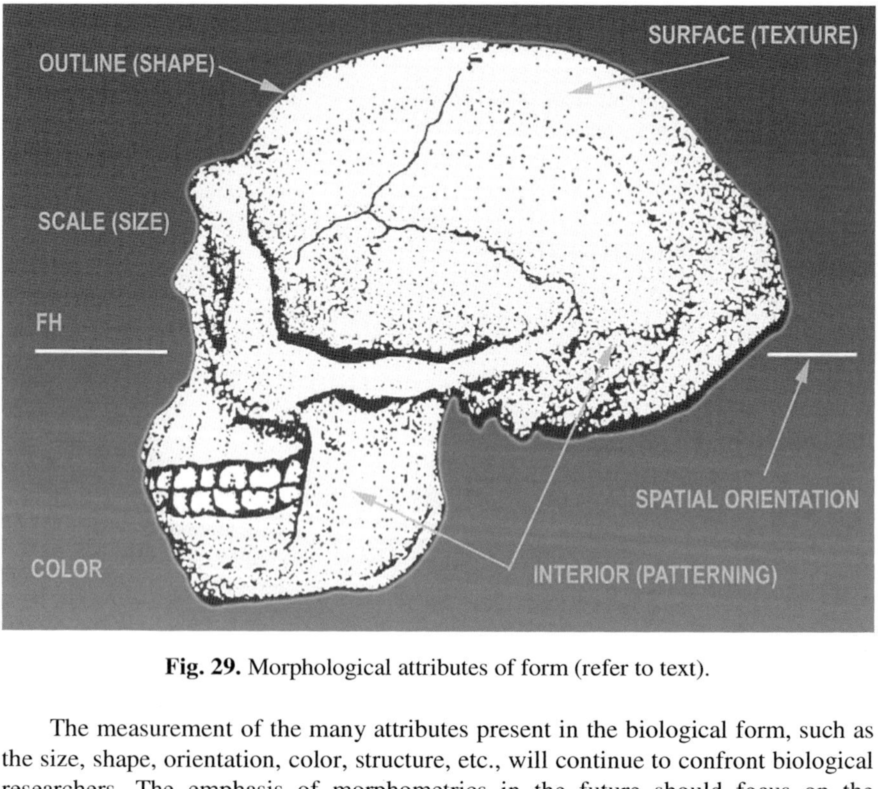

Fig. 29. Morphological attributes of form (refer to text).

The measurement of the many attributes present in the biological form, such as the size, shape, orientation, color, structure, etc., will continue to confront biological researchers. The emphasis of morphometrics in the future should focus on the development of more sophisticated models to allow for the extraction of more of the visual information that is always present in the biological form.

Some Conclusions

In sum then, this keynote address has focused on two main topics: (1) how the eye "sees" the biological form in visual terms and (2) how this visual information once apprehended, can be numerically described. While considerable progress has been made, knowledge in both areas remains incomplete. Numerous issues remain to be resolved. On the visual front, there is still a paucity of data to exactly how the brain is able to interpret the visual image with all its aspects, such as color, movement, details, etc., as it is received by the retina/brain. That is, there is still "no single view or equation that explains everything" to quote Marr [18]. While various details of the vision system are now understood, there is no general model of vision. Equivalently, on the numerical front, there is no general model yet that incorporates a majority of the visual information into a single coherent quantitative model.

REFERENCES

[1] Spoer KT and Lehmkuhle ST (1982) *Visual Information Processing*. W. H. Freeman and Co., San Francisco, CA.

[2] Zeki S (1993) *A Vision of the Brain*. Blackwell Scientific Pub. Oxford, UK.

[3] Schwartz JH and Tattersall I (2002) *The Human Fossil Record*, Vol. 1. John Wiley and Sons, Inc. New York, NY.

[4] Lein E and Hawrylycz (2014) The genetic geography of the brain. *Sci. Amer.* **310**: 70-77.

[5] Paley W (1809) Natural Theology: or, Evidences of the Existence and Attributes of the Deity. 12th edition, Printed for J. Faulder, London.

[6] Darwin C (1859) *On the Origin of Species*. John Murray, London, p. 186-187.

[7] Nilsson E and Pelger S (1994) A pessimistic estimate of the time required for an eye to develop. *Proc. Roy. Soc.* B. **256**: 53-58.

[8] Evolution of the Eye. http://en.wikipedia.org/wiki/Evolution_of_the_eye. Accessed on 11-23-2013.

[9] Gehring WJ (2005) New Perspectives on eye development and the evolution of eyes and photoreceptors. *J. Heredity* **96**: 171-184.

[10] Purves D Monson BB Sundararajan J and Wotjtch (2014) How biological vision succeeds in the physical world. *Proc. Natl. Acad. Sci. USA*.

[11] Ostrovsky M (2008) Evolution of vision. **In** *Evolution from cellular to social scales.* (Skjeltorp AT and Belushkin AV (Eds.) Springer Science, Dordrecht, The Netherlands.

[12] Hubel DH (1988) *Eye, Brain, and Vision.* Scientific American Library, W.H. Freeman & Co., New York, NY.

[13] O'Shea M (2007) *The Brain.* Oxford University Press, Oxford, U. K.

[14] Linden D (2007) *The Accidental Mind.* Belknap Press. Cambridge, Mass.

[15] Kandel ER (2006) *In Search of Memory.* W. W. Norton and Co., New York, NY.

[16] Hoffman DD (1998) *Visual Intelligence.* W. W. Norton and Co., Inc. New York, NY.

[17] Stone JV (2012) *Vision and the Brain: How We Perceive the World.* The MIT Press. Cambridge, Mass.

[18] Gregory RL (1997) *Eye and Brain: The Psychology of Seeing.* Princeton (5th Edition) University Press. Princeton NJ.

[19] Marr D (1982) Vision: A computational investigation into the human representation and processing of visual information. W. H. Freeman and Co. San Francisco, CA.

[20] Shannon CE (1948) A mathematical theory of communication. *Bell Systems Tech. J.* **27**: 379-423, 623-656

[21] Shannon CE and Weaver W (1949) *The mathematical theory of communication.* University of Illinois Press, Urbana, Illinois.

[22] Davis JC (1986) *Statistics and Data Analysis in Geology* (2nd Ed.). John Wiley & Sons, New York, NY.

[23] Nadler M and Smith EP (1993) *Pattern Recognition Engineering.* John Wiley and Sons, Inc., New York, NY.

[24] Bonner JT (2002) *The Ideas of Biology.* Dover Publications, Inc. New York, NY.

[25] Lestrel PE (1997) *Fourier Descriptors and their Applications in Biology.* Cambridge University Press, Cambridge, UK.

[26] Ayala F (1988) *Genetic variation and Evolution.* Carolina Biological Supply Co. Burlington, NC.

[26] Radinsky LB (1987) *The Evolution of Vertebrate Design.* The University of Chicago Press. Chicago, IL.

[27] Lestrel PE (1980) A quantitative approach to skeletal morphology: Fourier analysis. *Proc. Soc. Photo-Optical Inst. Engrs.* (SPIE) **166**: 80-93.

[28] Hubbard BB (1998) *The World According to Wavelets* (2nd Ed): A. K. Peters, Wesley, Mass.

[29] Costa LF and Cesar Jr. RM (2001) Shape Analysis and Classification: Theory and Practice. CRC Press, Boca Raton, Florida.

[30] Lestrel PE Cesar Jr. RM Takahashi O and Kanazawa E (2005) Sexual dimorphism in the Japanese cranial base: A Fourier-Wavelet representation. *Am. J. Phys. Anthrop.* **128**: 608-622.

[31] Lestrel PE and Moore RN (1978) *Macaca nemestrina*: A quantitative analysis of size and shape. *J. Dent. Res.* **57**: 395401; erratum, **57**: 947.

[32] Lestrel PE, Ohtsuki F and Wolfe CA (2010) Cranial vault shape in fossil hominids: Fourier descriptors in *Norma lateralis. J Comp Hum Biol* (HOMO) **61**: 287-313.

[33] Read DW (1990) From multivariate to qualitative measurement: representation of shape. *Hum Evol* **5**: 417-429.

[34] MacLeod (2015) The Direct Analysis of Digital Images (eigenimage) with a Comment on the Use of Discriminant Analysis in Morphometrics **In**: PE Lestrel (Ed.) *Biological Shape Analysis*: *Proc 3rd Int Symp Biol Shape Anal* (ISBSA) World Scientific Pub, Singapore.

Leaf Identification Using Polar Fourier Transform and Linear Bayes Normal Classifier

ABDUL KADIR[1]

Abstract

A novel leaf identification system that incorporates shape, color, texture and vein features has been developed. The Polar Fourier Transform (PFT) was used to handle the shapes of leaves and lacunarity was used to describe texture on the leaves. Color was also added to improve the performance of the leaf identification system. To classify a leaf, a linear classifier called Linear Bayes Normal classifier was applied. Several experiments were conducted using the Flavia dataset, a common dataset that contains 32 kinds of green leaves. The Foliage dataset, a dataset that consists of 60 kinds of leaves with various colors and shapes was also utilized. The results show that the system yielded an accuracy rate of 95.94% when using the Flavia dataset and 93.25% when using the Foliage dataset.

INTRODUCTION

Researchers are continuing to develop leaf identification systems using various methods and features. The goal is to create systems that can identify a leaf automatically. Researchers have proposed different methods for recognizing leaves. A method called Partial Dynamic Time Warping (PDTW) was used to compare two curves based on the Dynamic Time Warping (DTW) to match curves [8], Radial Basis Probabilistic Neural network (RBPNN) was incorporated in [1] to recognize leaves and General Regression Neural Network (GRNN) was used in [9]. To implement these identification systems, they involved various leaf features, from just 5 features to more than 30 features. Five features derived from shape and vein of leaves were applied in [5], while 73 features derived from Zernike moments and others were used in [4].

Choosing what features to use for leaf identification is one of the important aspects in leaf identification systems. Features used to represent shapes have been incorporated in such systems. Moment invariants were used in [9], Zernike moments were used in [1, 4]. Several researches [4, 5] applied vein features to improve the system's performance. Others [4, 6] included color features. Moreover, texture features were also used to obtain better performance of the leaf identification systems, such as GLCM in [4] and Gabor filter in [10].

The goal of the current research was to develop a leaf identification system with a high accuracy rate. To accomplish this objective, several features were examined and applied using a linear classifier. The Flavia dataset was tested to compare the results against previous research results. However, the current research also used the Foliage dataset in order to show its results in recognizing leaves with various colors.

This chapter is organized as follows. The initial section reviews all features that were used in this research and explains the classifier used in the identification system. The next section describes the proposed system. A further section shows the experimental results. Finally, the last section concludes with the results.

[1] Engineering Technology Faculty, Technical University of Malaysia, Malacca, Malaysia, akadir64@gmail.com.

MATERIALS AND METHODS

Several researchers [1, 2, 3, and 4] have used the Flavia dataset (http://flavia.sourceforge.net), a dataset prepared by Wu and colleagues [5]. Another public dataset called Foliage (http://rnd.akakom.ac.id/foliage) was also used in [4]. However, others have prepared their own datasets [6, 7].

Shape Features

The current research incorporated three kinds of shape features: (1) solidity, (2) convexity, and (3) the Polar Fourier Transform (PFT) as a shape descriptor. All features here are invariant under translation, scaling, and rotation.

Solidity is a ratio between the area of the leaf and the area of its convex hull (see Fig. 1) [11]. The formula for calculating solidity is:

$$solidity = \frac{area\ of\ leaf}{area\ of\ convex}. \tag{1}$$

Convexity is defined as ratio between the convex hull perimeter of the leaf and the perimeter of the leaf [11]. Mathematically, it is notated as:

$$convexity = \frac{convex\ perimeter}{perimeter}. \tag{2}$$

Fig. 1 shows an example of a convex hull (represented by the straight lines bounding the figure). The convex hull was calculated by using the "Graham Scan" algorithm [12].

Fig. 1. A convex hull (solid grey lines) bounding a leaf.

The third shape feature, PFT [13], is defined as:

$$PFT(\rho,\phi) = \sum_r \sum_i f(\rho,\phi_i)\exp[j2\pi(\frac{r}{R}\rho + \frac{2\pi}{T}\phi)], \tag{3}$$

where

$0 \leq r < R$ then $\theta_I = I(2\pi/T)$ $(0 \leq I < T)$; $0 \leq \rho < R$, $0 \leq \phi < T$,
R is the radial frequency resolution,
T is the angular frequency resolution.

How to obtain the descriptors of the PFT is described as follows. Consider that there is an image such that $I = \{f(x, y); 0 \leq x < M, 0 \leq y < N\}$. Initially, the image is converted from the Cartesian space to the polar space so that $I_p = \{f(r,\theta); 0 \leq r < R, 0 \leq \theta < 2\pi\}$, where R is the maximum radius from the center of the shape. The origin of the polar space becomes the center of the space in order to be translation invariant. The centroid (x_c, y_c) is calculated by using formulas:

$$x_c = \frac{1}{M}\sum_{x=0}^{M-1} x, \qquad y_c = \frac{1}{N}\sum_{y=0}^{N-1} y, \tag{4}$$

The values, (r, θ), are computed by using:

$$r = \sqrt{(x - x_c)^2 + (y - y_c)^2}, \qquad \theta = \arctan\frac{y - y_c}{x - x_c}. \tag{5}$$

Rotation invariance is obtained by ignoring the phase information in the coefficients. Therefore, only the magnitudes of coefficients are retained. To get the scale invariance, the first magnitude value is normalized by the area of the circle and all the magnitude values are normalized by the magnitude of the first coefficient. Lastly, the Fourier descriptors are computed as follows:

$$FDs = \{\frac{PF(0,0)}{2\pi r^2}, \frac{PF(0,1)}{PF(0,0)}, \ldots, \frac{PF(0,n)}{PF(0,0)}, \ldots, \frac{PF(m,0)}{PF(0,0)}, \ldots, \frac{PF(m,n)}{PF(0,0)}\} \tag{6}$$

In the above formula, m is the maximum number of the radial frequencies and n is the maximum number of angular frequencies.

Color Features

For describing leaf features based on their colors, these color features were obtained by applying statistical calculations on each leaf in the color channel R, G and B. Table 1 lists the color features that were used in this research. In Table 1, M and N refer to the height and width of the leaf images used.

Table 1. The Features Representing the Colors in a Leaf

Feature	Formula	Description
Mean	$\mu = \frac{1}{MN}\sum_{i=1}^{M}\sum_{j=1}^{N} P_{ij}$	The average color
Standard deviation	$\sigma = \sqrt{\frac{1}{MN}\sum_{i=1}^{M}\sum_{j=1}^{N}(P_{ij} - \mu)^2}$	Shows variation or dispersion compared to the average color
Skewness	$\theta = \frac{\sum_{i=1}^{M}\sum_{j=1}^{N}(P_{ij} - \mu)^3}{MN\sigma^3}$	A measure of symmetry around the mean of colors
Kurtosis	$\gamma = \frac{\sum_{i=1}^{M}\sum_{j=1}^{N}(P_{ij} - \mu)^4}{MN\sigma^4} - 3$	A descriptor of the shape of the probability distribution

Vein Features

Vein features are features derived from the veins of the leaf. There were four features, defined as follows:

$$V_1 = A_1/A \tag{7a}$$

$$V_2 = A_2/A \tag{7b}$$

$$V_3 = A_3/A \tag{7c}$$

$$V_4 = A_4/A \tag{7d}$$

A_1, A_2, A_3 and A_4 are number of pixels that construct the vein and A is the area of the leaf. The vein of the leaf was constructed by using morphological operation called opening [5]. That operation was performed on the gray scale image with flat, disk-shaped structuring elements of radius 1, 2, 3, and 4, respectively, and then subtracted from the remaining image to the margin. The difference between A_1, A_2, A_3 and A_4 is in leaf vein width derived from the morphological operation.

Texture Features

A fractal refers to a feature based on a repeating pattern. A form with a perfect fractal is a form that can be generated by repeating the pattern. Although the pattern of the leaves is actually not a repeated pattern or more perfectly, by several parts, especially the veins, it can be viewed as composed of a fractal. Therefore, the current research attempted to apply a fractal approach. A fractal measurement called lacunarity was described by Petrou and Sevilla [14].This fractal measurement can be used to distinguish between two fractals with the same fractal dimension. The lacunarity are defined as follows:

$$L_s = \frac{\frac{1}{MN}\sum_{m=1}^{M}\sum_{n=1}^{N} P_{mn}^{\ 2}}{\left(\frac{1}{MN}\sum_{k=1}^{M}\sum_{l=1}^{N} P_{kl}\right)^2} - 1, \tag{8}$$

$$L_a = \frac{1}{MN}\sum_{m=1}^{M}\sum_{n=1}^{N}\left|\frac{P_{mn}}{\frac{1}{MN}\sum_{k=1}^{M}\sum_{l=1}^{N} P_{kl}} - 1\right| \tag{9}$$

$$L_p = \left(\frac{1}{MN}\sum_{m=1}^{M}\sum_{n=1}^{N}\left(\frac{P_{mn}}{\frac{1}{MN}\sum_{k=1}^{M}\sum_{l=1}^{N} P_{kl}} - 1\right)^p\right)^{1/p} \tag{10}$$

Originally, those formulas are applied to gray scale images, where P_{mn} is a gray value at coordinate (m, n). However, in the current research, L_s, L_a, and L_p were used and applied to the color components R, G, and B and the intensity in gray scale image as well. The value of p in Eq. 10 is an even number (2, 4, 6,…., n).

Linear Bayes Normal Classifier

The Linear Bayes Normal Classifier (LBNC) is a linear classifier based on a Gaussian or normal probability density with an equal covariance matrix, which is included in the PRTools package [15] and defined as ldc. This approach is used to calculate all probabilities between the reference objects and the target objects to be identified. Then, the greatest probability produced by a reference object and the target object means that the target object has a close similarity to the reference object. This classifier follows the rule of Bayes decision theory as:

$$P(\omega_i \mid x) = \frac{p(x \mid \omega_i)P(\omega_i)}{p(x)} \tag{11}$$

where

$$p(x) = \sum_{i=1}^{c} p(x \mid \omega_i)P(\omega_i) \tag{12}$$

$P(\omega_i)$ is an *a priori* probability of class ω_i; $P(\omega_i \mid x)$ is a *posteriori* probability density function (pdf) of x_i; $p(x \mid \omega_i)$ is the class conditional pdf of x given ω_i where i = 1,2, …, c and c is number of classes. Then, x is assigned to the class ω_i if:

$$P(\omega_i \mid x) > P(\omega_j \mid x), \forall j \neq i \tag{13}$$

In this particular case, the data in each class are distributed as a Gaussian distribution, $N(m, S)$, where m_i is the mean of the class ω_i and S_i is the covariance matrix of the class ω_i.

The original LBNC supports two regulation parameters, named R and S as described in [15], but both parameters were not used in the current research because they did not provide any contributions. Therefore, the computation of the covariance matrices and other values for calculating $P(\omega_i \mid x)$ become simpler and faster.

Since the covariance matrices are equal, the joint variance matrix can be calculated by averaging those matrices by considering *a priori* probabilities. The pseudo code for obtaining the covariance matrix based on dataset A with c classes is as follows:

```
U ←mean(A)
P ←get_a_priori(A)
G ←covariance(A1)
G ←G * Pi

fori← 2 to c
   tmpG← covariance(Ai)
   G ←G + Pi * tmpG
end-for
```

The outcomes are the vector mean (U), *a priori* probabilities (P), and the covariance matrix (G), that could be used to calculate $P(\omega_i \mid x)$.

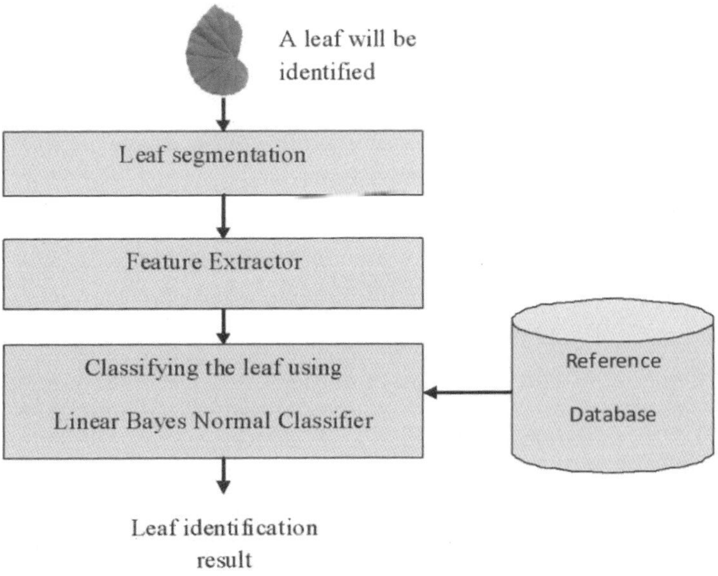

Fig. 2. The architecture of the proposed system.

The Proposed System

Fig. 2 shows the architecture of the proposed leaf identification system. The characteristics for recognizing a leaf were pre-calculated and all features were saved in a database. The image of the leaf is segmented to obtain the area of the leaf. Then, all features of the leaf are computed. Subsequently, the features are processed by the LBNC based on all features in the reference database and the leaf classified according to the type of leaves.

All mentioned features were examined for classifying a leaf. Descriptors of the PFT were calculated by using the maximum number of the radial frequencies = 4 and the maximum number of the angular frequencies = 6. Meanwhile, L_p (Eq. 10) were calculated for p = 2, 4, and 6. Therefore, there are 20 features to represent texture features (L_s, L_a, L_2, L_4 and L_6, each with four features).

In order to calculate the system's performance, the following formula was used [16]:

$$Performance = \frac{n_r}{n_t} \tag{14}$$

where n_r is the relevant number of images and n_t is the total number of queries. The meaning of the relevant number of images here is the number of images matched between its label and the result label. This performance represents the accuracy rate of the leaf identification system.

RESULTS

In the experiments, the Flavia and the Foliage datasets were tested separately. For the Flavia, there were 1280 leaves for reference. Each type of leaf was represented by 40 leaves. Samples of leaves in the Flavia dataset are shown in Fig. 3. For the Foliage dataset, there were 6000 leaves for reference. Each type of leaf was represented by 20 leaves. Samples of leaves in the Foliage dataset are shown in Fig. 4. Table 2 shows the experimental results using the Flavia and the Foliage datasets. With both datasets, combination of descriptors of the PFT, vein features, color features, and 20 features of the lacunarity, solidity and convexity yielded the maximum performance. The experiments also showed that shape features derived from the PFT were still not enough to give an optimal accuracy rate (Table 2, Number 3). Therefore, the solidity and convexity variables were added.

Table 2. Results of Several Experiments Using the Flavia and the Foliage Datasets

No.	Features	Foliage	Flavia
1	PFT + color + vein features	90.58%	94.06%
2	PFT + color + vein features + La + Ls	91.25%	93.13%
3	PFT + color + vein features + 20 features of lacunarity	92.00%	95.31%
4	PFT + color + vein features + 12 features of L2, L4, and L6	91.00%	94.37%
5	PFT + color + vein features + L2 and L4	90.58%	94.69%
6	PFT + color + vein features + 20 features of lacunarity + solidity and convexity	93.25%	95.94%

Fig. 3. Samples of the Flavia dataset.

Fig. 4. Samples of the Foliage dataset.

Table 3. Comparison to the Performance of Other Systems that Used the Flavia Dataset

Method	Accuracy Rate	Notes
Fourier moment [2]	46.30%	32 classes
PNN [5]	90.31%	32 classes
PNN-PCNN [2]	91.25%	32 classes
Zernike moments + PNN [4]	93.44%	32 classes
Zernike moments + City block distance [4]	93.75%	32 classes
LDA [3]	94.30%	20 classes
Pseudo Zernike Moment + RBPNN [1]	94.52%	32 classes
Zernike moments + Euclidean distance [4]	94.69%	32 classes
PFT + GLCM + PNN [17]	94.69%	32 classes
Proposed system	**95.94%**	32 classes

Table 4 shows the performance of the proposed system and the performance of other researches that used the Foliage dataset. It shows that the proposed system has a better accuracy rate. Compared to [4] that used the Foliage dataset, the proposed system is more efficient. The proposed system involved 54 features; meanwhile the system in [4] used 73 features. Zernike moments are the mappings of an image onto a set of complex Zernike polynomials which are orthogonal. However, it should be noted that the Zernike moments data were taken from other sources and are not part of this research. They were just added for comparison.

Table 4. Comparison to the Performance of Other Systems that Used the Foliage Dataset

Method	Accuracy Rate	Notes
Zernike moments + City block distance [4]	90.58%	60 classes
Zernike moments + Euclidean distance [4]	91.75%	60 classes
Zernike Moment + PNN [4]	92.92%	60 classes
PFT + GLCM + PNN [17]	93.08%	60 classes
Proposed system	**93.25%**	60 classes

DISCUSSION AND CONCLUSIONS

The proposed leaf identification system using a combination of shape, color, vein and texture feature was developed in the current study. This system gave better accuracy rates compared to other published research results, based on using the Flavia or the Foliage datasets. The current results show that the system yielded an accuracy rate of 95.94% when using the Flavia dataset and 93.25% when using the Foliage dataset. However, more research is still needed to improve the performance of the leaf identification system especially for handling multicolored leaves.

REFERENCES

[1] Kulkarni AH, Rai HM, Jahagirdar KA, and Kadkol, RJ (2011) A Leaf Recognition System for Classifying Plants Using RBPNN and Pseudo Zernike Moments. *Int J Latest Trends in Engr Tech (IJLTET)*, **2**: 6-11.

[2] Singh K, Gupta I, and Gupta S (2010) SVM-BDT PNN and Fourier Moment Technique for Classification of Leaf Shape. *Int J Sig Proc, Image Proc Patt Recog* **3**: 67-78.

[3] Shabanzade M, Zahedi M, and Aghvami SA (2011) Combination of Local Descriptors and Global Features for Leaf Recognition. *Signal and Image Proc: An Int J (SIPIJ)*, **2**: 23-31.

[4] Kadir A, Nugroho LE, Susanto A, and Santosa P. (2012) Experiments of Zernike Moments for Leaf Identification. *J Theoret Applied Info Tech* **41**: 82-93.

[5] Wu SG, Bao FS, Xu EY, Wang YX, Chang YF, and Xiang QL (2007) A Leaf Recognition Algorithm for Plant Classification Using Probabilistic Neural Network. *Proc IEEE 7th Int Symp Sig Proc Info Tech* **7**: 11-16.

[6] Man QK, Zheng CH, Wang XF, and Lin FY (2008) Recognition of Plant Leaves Using Support Vector. *Int Conf on Intell Comput* **8**: 192-199.

[7] Ehsanirad A (2010) Plant Classification Based on Leaf Recognition. *Int J Comput Sci Info Security,* **8**: 78-81.

[8] Tak YS and Hwang E (2007) A Leaf Image Retrieval Scheme Based On Partial Dynamic Time Warping and Two-Level Filtering. *7th Int Conf Comput Info Tech* **7**: 633-638.

[9] Zulkifli Z (2009). *Plant Leaf Identification Using Moment Invariants and General Regression Neural Network.* Master Thesis, Faculty of Computer Science and Information Systems, Universiti Teknologi Malaysia, Malasia.

[10] Chaki J and Parekh R (2012) Plant Leaf Recognition using Gabor Filter. *Int J Comput Applications* **56**: 26-29.

[11] Russ JC (2011) *The Image Processing Handbook.* CRC Press, Boca Raton, Florida.

[12] Goodrich MT and Tammasia R (2002). *Algorithm Design.* John Wiley and Sons, New York.

[13] Zhang D (2002) *Image Retrieval Based on Shape.* Unpublished Dissertation, Faculty of Information Technology, Monash University, Australia.

[14] Petrou M and Sevilla PG (2006) *Image Processing Dealing with Texture.* John Wiley and Sons, Ltd., Chichester, UK.

[15] Duin RPW, Juszczak P, Paclik P, Pekalska E., de Ridder P, and Tax DMJ, and Verzakov S (2007) *PRTools 4.1, A Matlab Toolbox for Pattern Recognition*, Delft University of Technology, Netherlands.

[16] Jyothi B, Latha YM, and Reddy V (2010) Medical Image Retrieval using Multiple Features. *Adv Comput Sci Tech* **3**: 387-396.

[17] Kadir A, Nugroho LE, Susanto A, Santosa PI. (2011) Neural Network Application on Foliage Plant Identification. *Int J Comput Applications* **29**: 15-22.

Image Processing and Artificial Neural Networks for Automated Plant Species Identification from Leaf Outlines

JONATHAN Y. CLARK[1], DAVID P. A. CORNEY[2], SCOTT NOTLEY[1] and PAUL WILKIN[3]

Abstract

Rapid, accurate identification of plant species is urgently needed for surveys of species diversity in the light of the global crisis in biodiversity driven by factors including climate change. Such identification systems are often most effective when they use vegetative parts alone, given the ephemeral nature of plant reproductive organs. This chapter describes a system that can be used as a practical method for identification of botanical herbarium specimens that have been digitized as images and has potential for creating large morphological datasets. The system is composed of an almost completely automated leaf shape extraction component and an artificial neural network, specifically a multilayer perceptron (MLP). First the leaves are found using deformable templates and evolutionary algorithms, and then a level set method is used to separate the leaf outline from the background. The length and width and other shape measurements are then extracted automatically, together with numerous measurements of the marginal teeth. The neural network is then able to identify plants from leaf shape alone. Furthermore, the system is also able to automatically refer specimens to a botanist for expert examination, in cases of uncertainty. Thus a methodology is presented here to provide a practical way for taxonomists to use a combination of neural networks and image processing as a tool for automated plant identification. A case study is provided using data extracted from specimens of four species of the tree genus *Tilia* in the herbarium of the Royal Botanic Gardens, Kew, UK. Over half the test specimens were identified correctly to species level using 41 automatically extracted leaf shape parameters. The remaining specimens were referred for further botanical study, together with suggestions as to their identity.

INTRODUCTION

Plant identification is important for those who need to be certain which species they are dealing with, e.g. if some species contain important compounds of medical interest, and others who are interested in establishing levels of biodiversity, for instance to investigate changes in geographic distributions due to climate change. Biological identification is still usually carried out using a printed "taxonomic key", though there is a growing trend towards computer-based or computer-aided identification systems [1, 2]. Such traditional keys are followed manually, the user making sequential choices from usually pairs of contrasting statements, usually concerning morphology, and eventually terminating in a name. These statements contain facts, values, or states of one or more characters or attributes, and the user simply chooses the option that fits the specimen to be identified. The identification performance thus depends largely on the experience of the author of the key, and how it is interpreted by the user.

Artificial neural networks (ANNs) are computer programs that, like humans, are able to learn from examples and can thus similarly perform recognition of previously

---

[1] Department of Computing, University of Surrey, Guildford, GU2 7XH, UK, j.y.clark@surrey.ac.uk.
[2] IDEAS Research Institute, Robert Gordon University, Aberdeen, AB10 7QB, UK.
[3] Royal Botanic Gardens, Kew, Richmond, Surrey, TW9 3AB, UK.
Partly funded by the Leverhulme Trust (grant F/00 242/H): Morphological Herbarium Image Data Analysis Project (MORPHIDAS).

unseen data. A multilayer perceptron (MLP) is a so-called supervised artificial neural network (ANN) and is most suitable for identification, because it is trained using data where the class (e.g. species) of each data record is known. Such training is achieved by presenting a group of data records (known as the *training* dataset) to the network, each record containing data from a specimen or record of known identity. Periodically during training the generalization ability of the network, i.e. its ability to recognize previously unseen patterns, is tested using a similar, but independent *validation* dataset, also containing data records with known classes. This independent testing using the validation set is important to enable training to be terminated before over-training occurs. A completely independent *test* dataset containing the actual unknown data records to be identified is then presented to the network. Information derived from this test set should never be used to optimize the parameters of the network, as that would introduce an unscientific bias to recognize the data in that particular test set. Instead the performance of the network using the validation set (which is already involved in preventing over-fitting) can be used to optimize network parameters. Further information about ANNs is available in [3] and [4].

Here a case study is presented regarding ANN-based automated identification of species of the genus *Tilia* (Tiliaceae/Malvaceae) from leaf shape features extracted using image processing methods. This comprises around 23 species of woody trees, widely distributed in north temperate regions, many being commonly cultivated in gardens or as street trees. These deciduous trees, with heart-shaped, pointed leaves, are often known as limes, lindens or basswoods, though they are unrelated to the citrus tree known as lime. The images referred to are photographs of herbarium specimens (e.g. Fig. 1: *Tilia cordata* COR463) - actual physical dried and pressed plant specimens, labeled and mounted on stiff white cartridge paper and kept in paper folders stored in cabinets to use as voucher specimens for identification.

A review of other plant species identification using image processing methods is given in [5]. This includes leaf and flower shape analysis, leaf texture analysis and vein analysis, as well as other species identification methodologies. The most similar effort to that presented here uses a MLP to distinguish between species and varieties of the genus *Banksia* [6]. This work used software to automatically extract characters from leaves, such as area and roundness, for use in identification. However, the characters obtained were extracted from images of single undamaged leaves as opposed to whole herbarium specimens, making the character extraction relatively straightforward.

Although a number of classical printed taxonomic keys have already been created for the identification of *Tilia* species (recent examples are those in Pigott's recent monograph [7]), the only computer-based identification systems relating to the genus *Tilia* alone are those by Clark [8, 9, 10, and 11], and Clark *et al.* [12]. This kind of identification task, as also presented here, is especially challenging to automate as one is considering species within a single genus, which are similar by definition. It is usually much simpler to distinguish between species in different genera (or other higher level taxa, for that matter).

Fig. 1. *Tilia cordata* herbarium specimen (COR0463).

In earlier similar work involving identification of *Tilia* [10] 19 species were considered. However, character states were manually recorded in the traditional way by observation and measurement. In more recent work [12], and that presented here, information is extracted almost totally automatically from images of complete herbarium specimens, and therefore a large number of specimens of each species are needed. The scope is therefore restricted to 4 species of *Tilia*, as these 4 species were the only ones present in sufficient numbers in the Kew herbarium. The work presented here is the first to use the combination of artificial neural networks, data extracted automatically from herbarium specimen images, plus logical referral to botanists, where appropriate.

MATERIALS AND METHODS

Datasets and Feature Extraction

Training of the neural network was carried out using data derived from 177 different *Tilia* specimens in the Herbarium of the Royal Botanic Gardens, Kew (K). Preliminary studies suggested that large numbers of named leaves were needed for this kind of study, and therefore it was decided to concentrate work on the four species of *Tilia* that had the most leaves available in the Kew Herbarium (*Tilia cordata* Mill., *T. platyphyllos* Scop., *T. americana* L. and *T. tomentosa* Moench, referred to here as COR, PLA, AME and TOM respectively). For the purposes of this study, all specimens in folders labeled with these species names were initially included, and all names were assumed to be correct, in order to validate the principle that herbarium specimens alone can be used in this kind of study without additional specialist knowledge.

Full details are given in Corney *et al.* [13], but briefly these are as follows. The software finds leaves in three stages. First it identifies a set of all regions in the image that might be leaves using a deformable templates approach [14], optimized with a simple evolutionary algorithm [15]. The edges of the image are found using a Canny edge detector [16], and approximate matches to leaf shapes found using trigonometric deformations [14] of initial templates. Secondly, the boundary of each candidate leaf is iteratively adjusted using a level set method [17] until it corresponds closely to the high-contrast edges in the image. In this way, the leaf boundaries can be found precisely. The third stage is to filter the set of candidate leaves to remove non-leaf objects. Here, in the only non-automated step of the process, a user is presented with a number of candidate leaves, and they simply have to click on some actual leaves (about 100, ideally) whose outlines have been correctly determined, in order to provide information to the system about the appearance of a leaf. The user does not have to identify the species, nor carry out time-consuming tasks such as drawing the outline. This stage allows some flexibility in the technique, rather than being restricted to a particular genus. Non-leaf objects are then rejected by comparing centroid-contour distances [18] with those of the chosen leaves. The remaining objects, presumed to be leaves, are used for subsequent character extraction.

In this case study, this method resulted in 441 data records, each containing data from a single leaf on a named specimen. Data for 41 morphological characters (see Table 1), comprising some conventional and some new, were extracted automatically and algorithmically from the leaf outlines. These include 17 of the 22 characters used in earlier studies [12] plus 24 new characters, these mostly pertaining to asymmetric morphometric measurements of the leaf base (with the two lobes referred to as A and B). There were different numbers of specimens for each species in the herbarium, due to the historical development of the herbarium and the general availability of specimens. Also, the software resulted in different numbers of leaves being extracted from each specimen image. Since training ANNs with very unbalanced class numbers is difficult [20], the over-represented classes were randomly discarded until all classes had roughly the same number of examples, namely 119 to 126. The exception was the data from *Tilia americana* (AME), where data from only 77 leaves were available at this stage. The (single leaf) data records were divided into independent training, validation and test sets in a ratio of approximately 70:20:10 respectively. Care was taken, however, to ensure that leaf data from the same specimen was not present in more than one set.

These data were converted to a text format suitable for input to the neural network. Unlike earlier, similar studies [12], preliminary work showed that additional Gaussian noise did not improve the overall classification accuracy, so none was added here.

Table 1. List of Characters (Attributes) Extracted

#	Character	Description - List of characters (attributes) extracted
01	Length of leaf blade	From petiole (leaf stalk) insertion point to tip
02	Width of leaf blade	Width at widest point, perpendicular to length-axis
03	Relative position, widest point	Distance along main axis: 0 = at insertion point, 1= at tip
04	Width-25	Leaf blade width, 25% from insertion point to tip
05	Width-50	Leaf blade width, 50% from insertion point to tip
06	Width-75	Leaf blade width, 75% from insertion point to tip
07	Total perimeter, leaf blade	Including teeth
08	Total area of leaf blade	Including teeth
09	Width A	Largest width of one half of leaf blade
10	Width B	Largest width on other half of leaf blade
11	Width A-25	Width at 25% from tip to petiole (leaf blade half A)
12	Width A-50	Width at 50% from tip to petiole (leaf blade half A)
13	Width A-75	Width at 75% from tip to petiole (leaf blade half A)
14	Width B-25	Width at 25% from tip to petiole (leaf blade half B)
15	Width B-50	Width at 50% from tip to petiole (leaf blade half B)
16	Width B-75	Width at 75% from tip to petiole (leaf blade half B)
17	Lobe length A	Vertical depth from petiole insertion, base of blade half A
18	Lobe length B	Vertical depth from petiole insertion, base of blade half B
19	Area A	Area of leaf blade half A
20	Area B	Area of leaf blade half B
21	Max lobe length	maximum of lobe length A and lobe length B
22	Max Area	maximum of areas A and B
23	Lobe asymmetry ratio	Min. length (lobe A, lobe B) / max length (lobe A, lobe B)
24	Area asymmetry ratio	Min. area (lobe A, lobe B) / max. area (lobe A, lobe B)
25	Width asymmetry ratio	Min. width (lobe A, lobe B) / max. width (lobe A, lobe B)
26	Width asymmetry ratio-25	As above at 25% from tip to petiole
27	Width asymmetry ratio-50	As above at 50% from tip to petiole
28	Width asymmetry ratio-75	As above at 75% from tip to petiole
29	Shape ratio 1	Width 25% / width 50%
30	Shape ratio 2	Width 25% / width 75%
31	Shape ratio 3	Width 50% / width 75%
32	Aspect ratio	Width/length
33	Total number of teeth	Total count around the leaf blade outline, excluding the tip
34	Total area of teeth	Total area in mm^2
35	Mean angle	Angle at tip of tooth, averaged over all teeth
36	Tooth frequency	Number of teeth / inner length
37	Total outer edge length	Total length of outer edges of all teeth
38	Tooth edge ratio	Average ratio of lengths of two outer edges of each tooth
39	Tooth area / blade ratio	Total tooth area / leaf blade area excluding teeth
40	Tooth number / blade ratio	Number of teeth / total leaf blade perimeter
41	Average tooth area	Total area of teeth / number of teeth

Neural Network

Here, a simple feed-forward Multi-Layer Perceptron (MLP) with one input layer, one hidden layer, and one output layer was used. One input node corresponded to each character (attribute) and one output node was assigned to represent each of the four species. Thus, there were 41 input nodes and 4 output nodes. The number of hidden nodes was varied to help optimize the network performance. There were no connections between nodes in the same layer, and no recurrent connections. The network architecture is shown in Fig. 2, though the actual number of nodes in the input and hidden layers differed from that shown. The input vectors were normalized independently for each character over all training records between ±0.9 to reduce the training time required, and to prevent character weighting. The minimum and maximum values for each character over the entire training set were used during similar normalizations of the validation and test data sets to make sure that scaling was comparable. The network weights were initialized to small random values between ±0.5 [3], and the presentation order of input vectors (leaf data records) was randomized between epochs (each epoch being one complete run through the training set). A bias input of 1.0 was used. Further details on parameters and training algorithms used are given in [9]. The error value reported in the results is the Squared Error Percentage (SEP) [19], with corrections [8], given by:

$$E = 100\frac{1}{NP(o_{max} - o_{min})^2}\sum_{p=1}^{P}\sum_{i=1}^{N}(o_{pi} - t_{pi})^2 \tag{1}$$

where o_{max} and o_{min} are the maximum and minimum output values used in training, here 0.9 and 0.1 respectively. N is the number of output nodes (equal to the number of species), and P is the number of records (in this case leaf data records) in the data set under consideration. o_{pi} is the actual output at output node i when input pattern p is presented. t_{pi} is the target (desired) output at output node i when pattern p is presented.

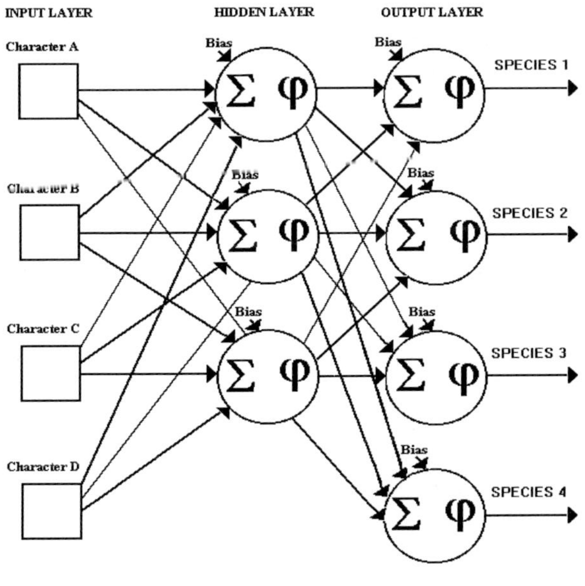

Fig. 2. MLP network architecture for species identification.

Training was initially carried out with a constant learning rate of 0.1, without momentum and with a fixed random seed, with a variable number of nodes in the hidden layer. Neural networks are known to have a tendency to become 'overfitted', if too much training is carried out, and that means that they then perform badly when presented with previously-unseen data. That is, their ability to generalize is reduced. To mitigate this problem, a validation dataset was used to test the generalization ability of the network periodically every 10 training epochs. Thus, the principle of early stopping was used. When the validation set error began to rise, training was terminated and the previous state was restored. The principle here is that in order to test generalization properly, information from the test set cannot be used to optimize network parameters, However, the validation set is partially involved in training, and is used periodically to test generalization, so the performance with the validation set can be used to help decide such parameters. Thus, the Percentage Error (E_{val}) on the validation set can be used to decide which networks are best.

The number of nodes in the hidden layer was thus optimized by performing a number of trials with different numbers of hidden nodes. The configuration that resulted in the lowest Squared Error Percentage (E) on the validation set was considered to have an optimized number of hidden nodes. After fixing the number of hidden nodes, a number of runs were carried out with variable learning rate, in order to determine an optimized value. After the optimized number of hidden nodes and learning rate had been fixed, the random seed itself was varied, in order to establish a population of results to collate, to reduce the effect of the random seed itself on the results. This variation of the random seed had the effect of changing the initial random weights before training. Then, finally, the generalization ability of the network was tested with the previously unseen test data set in order to calculate the final accuracy likely to be representative of actual identifications of unknown specimens. A winner-takes-all principle was used to determine the identification provided by the trained network, the output node with the highest output being used to determine the species identification corresponding to the chosen species (see Fig. 1).

Expert Referral

In this chapter a novel and practical method of collating results from this kind of study was introduced, in order to automatically and logically decide the species identification of specimens in the test set, or refer to a botanist those specimens that cannot be identified. Here, after determining the optimized number of hidden nodes and learning rate, and the best random seed, the results from the test set are examined more closely, and the concept of a 'trusted' leaf identification is introduced.

In order to establish this level of trust, a special test is carried out, where the fully trained network with the best performance is tested again, but this time, using the original training set as a dummy test set:

$$MSE_{correct} = \frac{1}{NC} \sum_{1}^{N} \left(\frac{(o_{pi} - t_{pi})^2}{(o_{max} - t_{max})^2} \right), \tag{2}$$

$$MSE_{wrong} = \frac{1}{NW} \sum_{1}^{N} \left(\frac{(o_{pi} - t_{pi})^2}{(o_{max} - t_{max})^2} \right), \quad (3)$$

and

$$threshold\ of\ trust = \frac{MSE_{correct} + MSE_{wrong}}{2} \quad (4)$$

Then the Mean Squared Error (Eq. 2) is calculated for those leaves the network identified 'correctly', and similarly the Mean Squared Error (Eq. 3) for those leaves that were 'wrongly' identified. The mean of these two values is then used as a *threshold of trust* (Eq. 4). The principle here is that if the system cannot identify a leaf well even if it is in the training set, then it cannot be expected to give a trustworthy identification of a similar leaf in the independent test set either. These thresholds are determined separately for each species, because some species will be easier to identify than others, and then are considered in conjunction with the network results when using the real test set. A leaf record in the real test dataset identified by the system to be a given species is then only considered to be trusted (regarded with high confidence) if that identification has an MSE less than, or equal to, the respective threshold for *that* species. It is then possible to label each leaf identification as 'Good' or 'Bad', even when the identity of the specimens in the test set is unknown (as in a practical situation).

However, these results could still be a little impractical, as they represent the overall identification on a leaf by leaf basis. Since the practical purpose of such a system is primarily to identify specimens (it is clearly not usually sensible to consider two leaves on the same specimen as belonging to different species), one can now consider leaf identifications on a specimen basis. In practice, it is usually specimens that need identifying, not individual leaves. Furthermore, in a real situation, a user would not know the real named identity of a specimen, but instead would want to use the system as an advisory tool to help identify unknowns. It is therefore necessary to introduce some rules that can be applied algorithmically to provide such advice, without any prior knowledge of the identity of the specimen to be identified. The rules proposed here, and demonstrated in this case study are:

1. Consider the final test run (from the random seed test population) with the *minimum* Error (E_{val}), using the validation set, as providing the identification decisions of the system all leaves in the test dataset.

2. Firstly, *refer specimens whose leaves are all 'Bad' to an expert botanist for further study.* These are assumed to be too poor (from a detectable leaf outline point of view) for the system to attempt, and are likely to be difficult for an expert botanist to identify. A low-confidence suggestion can be passed to the expert by simply choosing the leaf on that specimen whose decision has the lowest MSE.

3. For the remaining specimens (those with at least one 'Good' leaf), *use a majority rule voting principle* using decisions for all 'Good' leaves to make a species identification decision for that specimen.. That is, *a specimen can be considered to be identified 'correctly' if at least one leaf is 'Good'*, and the identification decision is based on 'Good' leaves only. Although this does

not happen in this case study, in the case of a disagreement between an equal number of 'Good' leaves, it would be possible to base the decision on the one with the lowest MSE.

4. Then *consider identifications of all leaves in all 10 of the random seed test population.* The *'Confidence'* of each specimen's identification is then defined as the proportion of identifications for leaves on that specimen that agree with the decision made by rules 2 or 3.

RESULTS

The results are presented for different numbers of nodes in the single hidden layer, varied between 4 and 116 (Table 2). In each case, the error (E_{val}) and recognition accuracy (R_{val}) produced by the network on presentation of the validation set at the point of the termination of training is provided. The number of hidden nodes which resulted in the lowest mean validation error (E_{val}) was found to be 36. Table 3 shows results produced using networks with 36 hidden nodes, with the learning rate varied between 0.001 and 0.149. Similarly, having fixed the number of hidden nodes to 36, the optimized learning rate was determined to be 0.065.

A summary of the results from tests using the above network parameters with the 10 different random seeds is shown here in Table 4. Each *Image ID* represents a different herbarium specimen image; *Leaf ID* is a unique number for a given leaf on that specimen. *Leaves 'Ident as'* is the majority rule identification for that leaf; *Quality* is *Good* if the identification for that leaf was less than or equal to the *threshold of trust* for that species; quality is *Bad* if identification for that leaf was greater than the *threshold of trust* for that species. Regarding the whole *Specimens 'Ident as'* is the majority rule species identification decision, considering all the leaves on that specimen, from the best (lowest E_{val}) test run only; *Trust/Refer* relates to whether to trust the decision of the system, or to refer the final decision to a botanist, based on the *threshold of trust*; *Confidence* is the *confidence rating* of each specimen identification as described above. Species identifications in bold are those which are considered 'correct', that is, they agree with the label of the herbarium folder in which the specimen is kept.

The misidentification matrix for all the test set identifications in the population of tests with 10 different random seeds is shown in Table 5 for individual leaves, and on a specimen basis with a majority decision using all extracted leaves from each specimen. The rows refer to the 4 species to be identified in the test set. Similarly, the columns show the species leaf or specimen that is identified by the system. Percentages are shown of the total samples of the row test species that are identified as belonging to the corresponding column species. Ideal (correct) identifications are shown in bold and the winner (highest %) underlined. The best result (lowest E_{val}) is given by the network with the random seed 239. This gave an R_{tst} (recognition percentage for the test set, on a single leaf basis) of 56.1%.

In general terms, if we simply consider all the leaf identifications individually, then more than half (56.1%) are identified correctly, with an average of 84.4% confidence. If we only consider the 'trusted' leaves, then 13 out of 23 leaves were identified correctly, giving an accuracy of 56.5%. This may not seem high, but this is achieved with minimum (non-expert) human intervention, and is only operating using leaf outlines obtained almost completely automatically from images of whole herbarium specimens. Furthermore, a botanist would use features from many other parts of the plant.

Table 2. Determination of the Optimized Number of Hidden Nodes (H)

H	4		12		20		28		36	
	E_{val}	R_{val}	E_{val}	R_{val}	E_{val}	R_{val}	E_{val}	R_{val}	E_{val}	R_{val}
	15.5346	56.67	15.5166	55.56	15.5570	52.22	15.8637	47.78	15.3192	53.33
H	44		52		60		68		76	
	E_{val}	R_{val}	E_{val}	R_{val}	E_{val}	R_{val}	E_{val}	R_{val}	E_{val}	R_{val}
	16.4549	42.22	15.5424	52.22	15.6085	50.00	15.9363	47.78	17.3011	38.89
H	84		92		100		108		116	
	E_{val}	R_{val}	E_{val}	R_{val}	E_{val}	R_{val}	E_{val}	R_{val}	E_{val}	R_{val}
	15.3419	57.78	16.2216	51.11	16.2938	52.22	15.9859	50.00	15.8192	52.22

Table 3. Determination of Optimized Learning Rate

LR	0.001		0.005		0.009		0.013		0.017	
	E_{val}	R_{val}	E_{val}	R_{val}	E_{val}	R_{val}	E_{val}	R_{val}	E_{val}	R_{val}
	16.0843	45.56	15.5974	54.44	15.4740	54.44	15.3077	54.44	15.2974	51.11
LR	0.021		0.025		0.029		0.033		0.037	
	E_{val}	R_{val}	E_{val}	R_{val}	E_{val}	R_{val}	E_{val}	R_{val}	E_{val}	R_{val}
	15.3607	51.11	15.4631	55.56	15.4279	55.56	15.4167	55.56	15.4233	56.67
LR	0.041		0.045		0.049		0.053		0.057	
	E_{val}	R_{val}	E_{val}	R_{val}	E_{val}	R_{val}	E_{val}	R_{val}	E_{val}	R_{val}
	15.3942	53.33	15.3480	53.33	15.3146	54.44	15.2913	53.33	15.2761	53.33
LR	0.061		0.065		0.069		0.073		0.077	
	E_{val}	R_{val}	E_{val}	R_{val}	E_{val}	R_{val}	E_{val}	R_{val}	E_{val}	R_{val}
	15.2672	54.44	15.2633	54.44	15.2635	54.44	15.2666	53.33	15.2718	53.33
LR	0.081		0.085		0.089		0.093		0.097	
	E_{val}	R_{val}	E_{val}	R_{val}	E_{val}	R_{val}	E_{val}	R_{val}	E_{val}	R_{val}
	15.2786	53.33	15.2863	54.44	15.2947	54.44	15.3034	54.44	15.3124	54.44
LR	0.101		0.105		0.109		0.113		0.117	
	E_{val}	R_{val}	E_{val}	R_{val}	E_{val}	R_{val}	E_{val}	R_{val}	E_{val}	R_{val}
	15.3215	53.33	15.3307	53.33	15.6598	51.11	15.6587	51.11	15.6598	51.11
LR	0.121		0.125		0.129		0.133		0.137	
	E_{val}	R_{val}	E_{val}	R_{val}	E_{val}	R_{val}	E_{val}	R_{val}	E_{val}	R_{val}
	15.6631	48.89	15.3807	50.00	15.3918	48.89	15.4035	50.00	15.4156	51.11
LR	0.141		0.145		0.149					
	E_{val}	R_{val}	E_{val}	R_{val}	E_{val}	R_{val}				
	15.4282	51.11	15.4415	51.11	15.4554	51.11				

In this study, 8 of the 24 test specimens have no trusted leaves and are referred to a botanist for further study. These specimens are assumed to be too poor (from a detectable leaf outline point of view) for the system to attempt to identify, and are likely to be difficult for an expert botanist. For the case study here, 13 of the 24 specimens have trustworthy species identifications with 100% confidence rating. However, if we look at the actual (presumed) identity of these, we see that 9 are correct and 4 wrong, giving an identification performance of 69.2%. The mean confidence rating of all 13 trustworthy specimen identifications is 84.4%.

Table 4. Leaf and Specimen Identifications

Leaves				Specimens		
Image ID	**Leaf ID**	**Ident as**	**Quality**	**Ident as**	**Trust/Refer**	**Confidence**
COR0449	1	**COR**	Good	**COR**	Trusted	100
COR0434	1	AME	Good	AME	Trusted	100
COR0421	1	**COR**	Bad	**COR**	Trusted	93.3
COR0421	2	**COR**	Good			
COR0421	3	**COR**	Good			
COR0426	1	PLA	Good	PLA	Trusted	100
COR0463	1	**COR**	Good	**COR**	Trusted	100
COR0463	2	**COR**	Good			
COR0463	3	**COR**	Good			
COR0427	1	**COR**	Good	**COR**	Trusted	100
PLA0151	1	**PLA**	Bad	COR	Referred	50
PLA0151	2	COR	Bad			
PLA0134	1	COR	Bad	COR	Referred	100
PLA0136	1	**PLA**	Good	**PLA**	Trusted	96.7
PLA0136	2	**PLA**	Good			
PLA0136	3	**PLA**	Bad			
PLA0169	1	**PLA**	Bad	TOM	Trusted	66.6
PLA0169	2	TOM	Good			
PLA0169	3	**PLA**	Bad			
PLA0146	1	**PLA**	Bad	**PLA**	Trusted	95
PLA0146	2	**PLA**	Good			
PLA0171	1	COR	Good	COR	Trusted	100
AME1150	1	**AME**	Bad	**AME**	Referred	80
AME1148	1	TOM	Bad	TOM	Referred	80
AME1156	1	PLA	Good	TOM	Trusted	66.6
AME1156	2	TOM	Good			
AME1156	3	TOM	Good			
AME1129	1	PLA	Good	PLA	Trusted	95
AME1129	2	PLA	Good			
AME1136	1	PLA	Bad	PLA	Referred	90
AME1128	1	PLA	Bad	COR	Trusted	100
AME1128	2	PLA	Bad			
AME1128	3	COR	Good			
TOM0060	1	PLA	Bad	PLA	Referred	60
TOM0713	1	AME	Bad	AME	Referred	70
TOM0101	1	**TOM**	Good	**TOM**	Trusted	100
TOM0014	1	**TOM**	Good	**TOM**	Trusted	100
TOM0097	1	**TOM**	Bad	**TOM**	Referred	55
TOM0097	2	**TOM**	Bad			
TOM0039	1	**TOM**	Bad	**TOM**	Trusted	80
TOM0039	2	**TOM**	Good			

Note: Leaf identifications in bold are "correct" identifications

Table 5. 10-run Miss-identification Matrices

LEAVES	**Identification % (rounded up to nearest %)**			
Label	COR	PLA	AME	TOM
COR	**78**	12	10	0
PLA	20	**65**	5	10
AME	15	48	**14**	24
TOM	3	19	16	**63**
SPECIMENS	**Identification % (rounded up to nearest %)**			
Label	COR	PLA	AME	TOM
COR	**67**	17	17	0
PLA	50	**33**	0	17
AME	17	33	**17**	33
TOM	0	17	17	**67**

COR = *Tilia cordata* Mill. PLA = *Tilia platyphyllos* Scop.
AME = *Tilia americana* L. TOM = *Tilia tomentosa* Moench

DISCUSSION AND CONCLUSIONS

This chapter builds on earlier studies that used neural networks for identification of *Tilia* herbarium specimens from images. Here, the concept of automated referral is introduced, in which it is accepted that some specimens are not identifiable by the system and are instead referred to an expert botanist. This is reasonable since here one is attempting to identify specimens automatically from images, with minimal human intervention, and also only using leaf outlines. The quality of herbarium specimens can vary considerably, and if there are many leaves overlapping, it is difficult to separate individual leaves by image processing. Furthermore, one aim of this study is to show the use of existing herbaria as repositories of information and accumulated expert knowledge resulting from many years of human experts studying and naming the specimens. Such a system is not perfect, but this study does indeed show that it is possible to use such data to help identify some unknown specimens, without the help of expert knowledge. For instance, a trainee, non-expert in the plant group concerned could be provided with help by this kind of system, and then refer difficult identifications to an expert on the group.

Referring to the Misclassification Matrices, individual leaves of *T. americana* (AME) leaves are largely (48%) misidentified as *T. platyphyllos*. It is possible that this might indicate that the system is mostly influenced by general leaf size, as the majority of mature *T. americana* leaves have a similar (large) size to those of *T. platyphyllos*. Considering herbarium specimens as a whole, this also makes sense as the smaller, immature leaves of *T. americana* are more likely to be similar to the size of mature *T. tomentosa* leaves. The smaller, immature leaves of the otherwise fairly large-leaved *T. platyphyllos* resemble the somewhat smaller mature leaves of *T. cordata*, so on a specimen basis, many *T. platyphyllos* specimens that have a predominance of immature leaves, or just those with naturally small mature leaves are misidentified as *T. cordata*. Another influence is likely to be the number of leaves that the system can successfully extract from images of whole specimens - usually this is only about 2 or 3, and the decision of the system is thus likely to be heavily influenced by the proportion of immature to mature leaves extracted.

It is interesting; however, that some of the incorrect identifications by the system are understandable; for instance one specimen (COR0434) is of a stump sprout of *T. cordata*. It is extremely difficult, if not impossible, for a human to identify leaves on shoots sprouting from *Tilia* tree stumps, as their leaf morphology is often extremely different from the normal canopy leaves. Indeed, all existing keys to identify *Tilia* species state that they are for identifying shoots from the canopy, and also often specify flowering or fruiting shoots. Also, many test specimens referred to a botanist had problems that would make the decision of the system understandable, for example, PLA0151 and TOM0060 had paper strips crossing the leaves, as part of the specimen mounting, and PLA0134 is most likely to be *Tilia tomentosa* (though the system identifies it, tentatively, as *T. cordata*). The referral of TOM0097 is interesting because it is *Tilia tomentosa* 'Petiolaris', a distinctive cultivar. It is clear that *Tilia americana* (AME) is not identified well by the system, and that is the species with fewer data records. It is suggested that the number of records is below a critical threshold to compensate for the variation in leaf morphology, and the addition of more specimens/leaves would help in this instance - indeed the main limitation of this methodology may well be the number of available leaves for creating the identification model.

The performance of the system would clearly be much enhanced by the addition of other characters, also automatically extracted. For example, the presence or absence of staminodes in the flowers would consistently separate *T. americana* and *T. tomentosa* from *T. cordata* and *T. platyphyllos*. If that character could be extracted, it could greatly increase the performance of the system. Here we only work with leaf outlines, and it is not known how well a botanist, even one familiar with the genus, would perform if trying to identify *Tilia* species using only outlines.

An obvious limitation of the system is that it can currently only be used to help identify four species of *Tilia*, and the specimen to be identified must be known to be one of these four in order for the system to work. Extending the system to cover other *Tilia* species would be advantageous, but undoubtedly it would be necessary to extract other character information such as the floral structure mentioned earlier, or information regarding the hairs on the leaves, which are critically important in published keys to the genus [7].

ACKNOWLEDGEMENTS

Grateful thanks are due to the Trustees of the Royal Botanic Gardens, Kew for permission to study and photograph specimens in the Herbarium, and to include some photographs here. Also, gratitude is also due to Donald Pigott for helpful comments and encouragement regarding *Tilia* and especially for writing his wonderful monograph. Thanks also go to Lilian Tang, Y. Hu and J. Jin for help with extraction algorithms, and to Kulamagul Mahendrarajah for development of the Java-based shell for the neural network program.

REFERENCES

[1] Pankhurst RJ (1991) *Practical Taxonomic Computing*. Cambridge University Press, Cambridge, UK.

[2] MacLeod N (2007) *Automated Taxon Identification in Systematics* (Systematics Association Special Volume). CRC Press, Boca Raton, FL, USA.

[3] Freeman JA and Skapura DM (1992) *Neural networks: algorithms, applications, and programming techniques*. Addison-Wesley, Reading, Massachusetts, USA.

[4] Haykin S (1994) *Neural networks - a comprehensive foundation*. Macmillan College Publishing Company, New York.

[5] Cope JS, Corney DPA, Clark JY, Remagnino P, Wilkin P. (2012) Plant species identification using digital morphometrics: a review. *Expert Systems with Applications* **39**: 7562-7573.

[6] Messina G, Pandolfi C, Mugnai S, Azzarello E, Dixon K, and Mancuso S (2009) Phyllometric parameters and artificial neural networks for the identification of *Banksia* accessions. *Austral Syst Bot* **22**: 31-38.

[7] Pigott D (2012) *Lime Trees and Basswoods: A Biological Monograph of the genus* Tilia. Cambridge University Press, Cambridge, UK.

[8] Clark JY (2000) *Botanical identification and classification using artificial neural networks*, PhD Thesis, University of Reading, Reading, UK.

[9] Clark JY (2003) Artificial neural networks for species identification by taxonomists. *BioSystems* **72**: 131-147.

[10] Clark JY (2004) Identification of botanical specimens using artificial neural networks. *IEEE Sym Comput Intelligence Bioinformatics Comput Biol (CIBCB)*, San Diego, CA, USA.

[11] Clark JY (2007) Plant identification from characters and measurements using artificial neural networks. **In**: *Automated Taxon Identification in Systematics* (Systematics Association Special Volume). MacLeod N (Ed.). CRC Press, Boca Raton, FL, USA.

[12] Clark JY, Corney DPA, and Tang HL (2012) Automated plant identification using artificial neural networks. *IEEE Sym Comput Intelligence Bioinformatics Comput Biol (CIBCB)*. San Diego, CA, USA.

[13] Corney DPA, Clark JY, Tang HL, and Wilkin P. (2012) Automatic extraction of leaf characters from herbarium specimens. *Taxon* **61**: 231-244.

[14] Jain AK, Zhong Y, and Lakshmanan S. (1996) Object matching using deformable templates. *IEEE Trans Pattern Anal* **18**: 267-278.

[15] Fogel DB (2006) *Evolutionary Computation: Toward a new philosophy of machine intelligence*, 3rd Ed. Wiley-IEEE Press, Hoboken, NJ, USA.

[16] Canny JF (1986) A computational approach to edge detection. *IEEE Trans Pattern Anal* **8**: 679-698.

[17] Malladi R, Sethian JA, and Vemuri BC (1995) Shape modeling with front propagation: a level set approach. *IEEE Trans Pattern Anal* **17**: 158-175.

[18] Ye L, Keogh E (2009) Time series shapelets: a new primitive for data mining. *Proc of the 15th ACM SIGKDD Int Conf Knowledge Discovery Data Mining.* Paris.

[19] Prechelt L (1994) *Proben1 – A set of neural network benchmark problems and benchmarking rules.* Technical Report 21/94, Universität Karlrühe, Germany.

[20] He H and Garcia EA (2009) Learning from imbalanced data. *IEEE Trans of knowledge and data Engr* **21**: 1263-1284.

Towards Machine Vision Based Grain Classification: Challenges and Future Prospects

JITENDRA PALIWAL[1], CHYNGYZ ERKINBAEV[1] and HIBRU K. MEBATSION[2]

Abstract

Cereal grains are one of the major export agricultural commodities of Canada contributing sizable amounts to its annual GDP. Grain has to pass, a number of quality evaluation steps and grading before it comes to the global market. In the current Canadian grain grading system, grain visual appearance is the main component of a measure of quality. With significant research and advances in machine vision systems (MVS) and computer based grading and sorting the Canadian grain industry is on the verge of adopting an automated procedure for routine cereal grain grading. Most MVSs, however, rely on extracting visual appearance in terms of shape, size and colour features acquired from cereal grain digital images. These features are then used for automatic classification of the cereal grains into pre-determined categories. The challenge of automatic classification lies in the fact that grain kernels are not simple geometries that could be defined by simple mathematical equations and hence their visual features cannot easily be related to their shape features. Members of the Imaging and Spectroscopy Laboratory of the Department of Biosystems Engineering, have been developing various algorithms that extract morphological (shape and size), colour and textural features from grain kernels to characterize their visual appearance. This chapter discusses advances that have been made in computer-based classification algorithms and future prospects in this exciting and challenging research field. In this regard, mathematical techniques including morphological analysis, Fourier descriptors, colour quantification, and textural analysis (gray level run-length matrices and gray level co-occurrence matrices) coupled with non-linear data analysis techniques will be discussed. The chapter concludes with a look at future research that is believed to be essential in machine vision based classification before these systems are adopted across the grain industry worldwide.

INTRODUCTION

Grain is one of the most important foods in the world both as a source of nutrition and income. Determination of grain type and quality is a crucial aspect at every stage of grain postharvest handling. Traditional manual analysis and grading of grains has many limitations and certain drawbacks: (i) it is subjective as it is highly influenced by anthropogenic and environmental factors [1]; (ii) it is inefficient and time-consuming; and (iii) it is prone to grading inconsistencies caused by the influence of external factors on human perception [2]. Therefore, what is needed is a novel system that is capable of objectively and rapidly applying evaluation criteria and would be an advanced tool for grain handling and quality control. With increasing demand for high quality grains and high risks of grain contamination during post-harvest storage and handling, there is a need for a simple, accurate and robust grain classification system. The development of such a technique would lead to the advancement of an automated grain sorting, cleaning, and grading system that could be performed on-line. In this regard, machine vision systems

[1] Department of Biosystems Engineering, University of Manitoba, 75A Chancellor's Circle, Winnipeg, R3T 2N2, Canada, J_Paliwal@UManitoba.Ca.
[2] Color Ad Packaging Ltd., 200 Beghin Ave, Winnipeg, MB R2J 3W2, Canada.

(MVSs) have strong potential as a non-destructive and inexpensive technique to accomplish this task. MVS refers to an automated system comprising of charge coupled device (CCD), cameras, and computers to perform the routine task involving a human eye and brain [3].

The MVS technology has been extensively used to identify and classify cereal grains and various foreign materials using one or few morphological [4, 5, and 19]; colour [6]; texture [7], and wavelet features [8]. However, these approaches had certain drawbacks and limitations: (i) the classification required a higher number of grain features; (ii) the classification of grain type and variety using mentioned features showed poor consistency [8, 9] and (iii) grain separation was limited to manually presenting non-touching kernels to the imaging system.

The need for comprehensive and robust grain image segmentation algorithms that could enhance accuracy of classification has motivated researchers across the globe. In this respect, a separation and curvature calculation algorithm, such as the elliptic Fourier series approximation as proposed by Kuhl and Giardina [10], was later applied on buckwheat [11] and several grains [12, 13, 14, and 15]. Mebatsion and Paliwal [12] achieved a near-perfect classification of selected cereal grains by applying the elliptic Fourier approximation method with a limited number of shape features. Inspired by these achievements, this chapter discusses the classification of selected cereal grains using mathematical techniques: morphological analysis, colour quantification, Fourier descriptors, and textural analysis coupled with multivariate non-linear data analysis techniques. The chapter concludes with a discussion of future research on computer-based classification algorithms and application of MVS in the grain industry.

Grain Classification Models

Morphological features

Kernel size is one of morphological parameters that is directly and indirectly associated with quality of grains, which can be ascertained by perimeter, area, roundness and circularity of grains. Grain perimeter and area are important parameters that identify the grain geometry and used in determination of roundness and circularity of grains. Roundness is defined as $R = 4A/ \pi M^2_d$, where M_d – major diameter and A – area. Circularity is calculated as $C = 4A/\pi p^2$, where P – perimeter is of the evaluated cereal grains. Kernel shape is another important morphological parameter associated with grain quality. Grain genetic background [16] and agricultural type [11, 17] were studied using information on kernel shape. These geometrical parameters: kernel length, width, maximum diameter, contour length, area, aspect ratio as well as normalized central moments of grain kernels have been extensively studied in grain classification [5, 9, and 18]. However, these morphological parameters do not provide efficient grain classification and are less informative than grain shape [20]. A mathematical quantification of grain shape using Elliptic Fourier descriptors (EFDs) method was studied by Mebatsion and Paliwal [12] and Mebatsion *et al.* [14]. This method describes shapes by transforming contour coordinates into Fourier coefficients. Detailed description of elliptic Fourier series approximation for grain shape variability will be presented in Section. 4.1.

Color features

Color is one of the parameters that can be used for characterization of grains. In several studies on cereal grain classification using color [5, 6], the color features used in classification models were obtained from the values of values of red, blue and green (RGB) indexes of the grain image. These color parameters were then transformed into hue, saturation, and brightness value parameters of the HSV color system. The reason for this conversion is that the HSV system is closer to human eye vision and easy in interpretations of color sensation than the RGB system [21]. Hue index is defined as a degree of a pure color. The saturation shows the value of dilatation of a color with white light and brightness shows the values from total black to completely saturated image color. The RGB can also be converted into $L^*a^*b^*$ system, a uniform color space with the values, a and b, for the color-opponent dimensions, and L for lightness.

Elliptic Fourier Approximation

Elliptic Fourier series descriptors

The elliptic Fourier series approximation method is based on solving a set of parametric equations [10]. The pair of parametric equations describe the curve with (x, y) points as a function of the third (t) variable. The Fourier coefficients are calculated based on the discrete Fourier series approximation of chain-coded boundary contours. The chain code is then described by boundary pixels of an object [12]. The elliptic Fourier approximation series of a closed contour that is projected onto the x and y-axis can be defined as the following sets of equations [12, 22, and 23]:

$$y_N(t) = C_0 + \sum_{n=1}^{N} c_n \cos\left(\frac{2n\pi t}{T}\right) + d_n \sin\left(\frac{2n\pi t}{T}\right) \qquad (1)$$

$$y_N(t) = C_0 + \sum_{n=1}^{N} c_n \cos\left(\frac{2n\pi t}{T}\right) + d_n \sin\left(\frac{2n\pi t}{T}\right) \qquad (2)$$

Where t is the step of movement of pixel unit along the closed contour, which gives the following condition $t_{p-1} < t < t_p$ for $1 \leq p \leq K$. K is the total number of points of chain-code and N is the number of Fourier harmonics. A_0 and C_0 are the first coefficients at the frequency equal to 0. They define the average size of the shape contour. When the contour between the (i-1)th and the ith chain-coded points is linear and counter length from the start point to the pth point, the following equation shows estimation of the perimeter of the contour t_p:

$$t_p = \sum_{i=1}^{p} \Delta t_i \qquad (3)$$

T is the basic period of the chain code that defines step required to traverse the entire contour. $T = tk$, where Δt_i is the distance between the (i-1)th and the ith points.

The ending point is denoted as *K*th. If the x-coordinate of the *p*th point is give as x_p, and the y-coordinate as y_p then:

$$x_p = \sum_{i=1}^{p} \Delta x_i \tag{4}$$

and

$$y_p = \sum_{i=1}^{p} \Delta y_i \tag{5}$$

where Δx_i and Δy_i are the distances along the *x* and *y* axes between (*i*-1)th and the *i*th point. If the neighbouring points are defined by linear interpolation, then the elliptic Fourier coefficients (Eq. 1) of the *n*th harmonic (a_n; b_n; c_n and d_n) can be estimated using the following two pairs of equations (6, 7 and 8, 9) [12, 22, 23, and 24]:

$$a_n = \frac{T}{2n^2\pi^2}\sum_{p=1}^{K}\frac{\Delta x_p}{\Delta t_p}\left(\cos\left(\frac{2n\pi t_p}{T}\right) - \cos\left(\frac{2n\pi t_{p-1}}{T}\right)\right) \tag{6}$$

$$b_n = \frac{T}{2n^2\pi^2}\sum_{p=1}^{K}\frac{\Delta x_p}{\Delta t_p}\left(\sin\left(\frac{2n\pi t_p}{T}\right) - \sin\left(\frac{2n\pi t_{p-1}}{T}\right)\right) \tag{7}$$

$$c_n = \frac{T}{2n^2\pi^2}\sum_{p=1}^{K}\frac{\Delta y_p}{\Delta t_p}\left(\cos\left(\frac{2n\pi t_p}{T}\right) - \cos\left(\frac{2n\pi t_{p-1}}{T}\right)\right) \tag{8}$$

$$d_n = \frac{T}{2n^2\pi^2}\sum_{p=1}^{K}\frac{\Delta y_p}{\Delta t_p}\left(\sin\left(\frac{2n\pi t_p}{T}\right) - \sin\left(\frac{2n\pi t_{p-1}}{T}\right)\right) \tag{9}$$

The mean Fourier power spectrum is used for determination of the number of required harmonics. Fourier power is an amplitude based parameter that describes the shape information in terms of harmonics. The Fourier power for the *n*th harmonic is calculated by the following equation [25]:

$$Fourier\ power = \frac{\sum_{n=1}^{N}(a_n^2 + b_n^2 + c_n^2 + d_n^2)}{2} \tag{10}$$

Mebatsion and Paliwal [12] truncated the Fourier harmonics at the value of $N = 13$. At this N number the average cumulative power. N_{max} is number of the boundary points. The estimated Fourier coefficients varies in with size and orientation of grain kernels during image acquisition and therefore, cannot be used as powerful indicators of variability in grain kernel comparison [14, 16, and 26]. For this reason, the Fourier coefficients have to be normalized by first harmonic ellipse as follows. If the standardized coefficients of the *n*th harmonic are a_n^{**}, b_n^{**}, c_n^{**}, and d_n^{**}, then:

$$\begin{bmatrix} a_n^{**} & b_n^{**} \\ c_n^{**} & d_n^{**} \end{bmatrix} = \frac{1}{E^*} \begin{bmatrix} \cos\omega & \sin\omega \\ -\sin\omega & \cos\omega \end{bmatrix} \begin{bmatrix} a_n & c_n \\ b_n & d_n \end{bmatrix} \begin{bmatrix} \cos n\theta & -\sin n\theta \\ \sin n\theta & \cos n\theta \end{bmatrix} \quad (11)$$

where:

$$E^* = (a_1^{*2} + c_1^{*2})^{\frac{1}{2}}, \qquad \begin{bmatrix} a_1^* \\ c_1^* \end{bmatrix} = \begin{bmatrix} a_1 & b_1 \\ c_1 & d_1 \end{bmatrix} \begin{bmatrix} \cos\theta \\ \sin\theta \end{bmatrix} \quad (12)$$

$$\theta = \frac{1}{2} arctan \left[\frac{2(a_1 b_1 + c_1 d_1}{(a_1^2 + c_1^2 - b_1^2 + d_1^2} \right] (0 \leq \theta \leq \pi) \quad (13)$$

$$\omega = arctan \begin{bmatrix} c_1^* \\ a_1^* \end{bmatrix} (0 \leq \omega \leq 2\pi) \quad (14)$$

Fig. 1 displays the various grains used in the current study fitted with the boundary outline using the Fourier series approximation. These were: (a) Barley; (b) Wheat from the Canadian Western Amber Durum (CWAD); (c) Wheat from the Canadian Western Red Spring (CWRS); (d) Rye; and (e) Oats [14]. Fourier series approximation of selected cereal grains with Fourier power of 99% is presented here. It is shown that the Fourier series approximation (white contours) perfectly matched with boundary (red contours) of the grains.

Application of the standardization process results in decreasing the total number of coefficients by 3 and the terms a_n^{**}, b_n^{**}, c_n^{**} become independent of the kernel shape reducing to 49 coefficients and N-independent terms [14]:

$$\text{Independent terms} = N * 4(a_n; b_n; c_n; d_n) - 3(a_n^{**}; b_n^{**}; c_n^{**}) \quad (15)$$

The standardized Elliptic Fourier descriptors were divided into symmetrical and asymmetrical (based on variation for the central axis) groups. The coefficients a_n^{**}, and d_n^{**} represent symmetrical relationships while the b_n^{**}, and c_n^{**} coefficients describe asymmetrical variations [27].

Fig. 1. Boundary contours of five cereal grain kernels obtained by Fourier series approximation. The boundary contour is given in white, whereas the Fourier series approximation is in red. (a) Barley; (b) Wheat (CWAD); (c) Wheat (CWRS); (d) Rye; and (e) Oats [14]. Used with permission of Elsevier.

Invariant Fourier coefficients

The invariant Fourier coefficient is derived from normalized first coefficients of the Fourier descriptors and can be determined by following equations [15]:

$$A_x = sqrt(a_{nn}^{**2} + b_{nn}^{**2}), \qquad A_y = sqrt(c_{nn}^{**2} + d_{nn}^{**2}) \tag{16}$$

$$B_x = \frac{A_x}{a_0}, \qquad B_y = \frac{A_y}{c_0} \tag{17}$$

$$IFC = abs\left(\frac{B_x}{B_x(1)}\right) + abs\left(\frac{B_y}{B_y(1)}\right) \tag{18}$$

where a_0 and c_0 are the first coefficients of the Fourier descriptors in the x and y coordinate system.

Mebatsion *et al.*, [14] applied principal component analysis to: the symmetric, asymmetric, standardized and invariant Fourier coefficients. They reported that few principal components (Fig. 2) of Fourier coefficients were able to give information about variation in grain kernel shapes (e.g., the first two (symmetric), five (standard), eight (invariant), and seventeen (asymmetric) coefficients). The model of symmetric Fourier coefficients had a minimum number of principal components leading to a reduction in dimensionality [15].

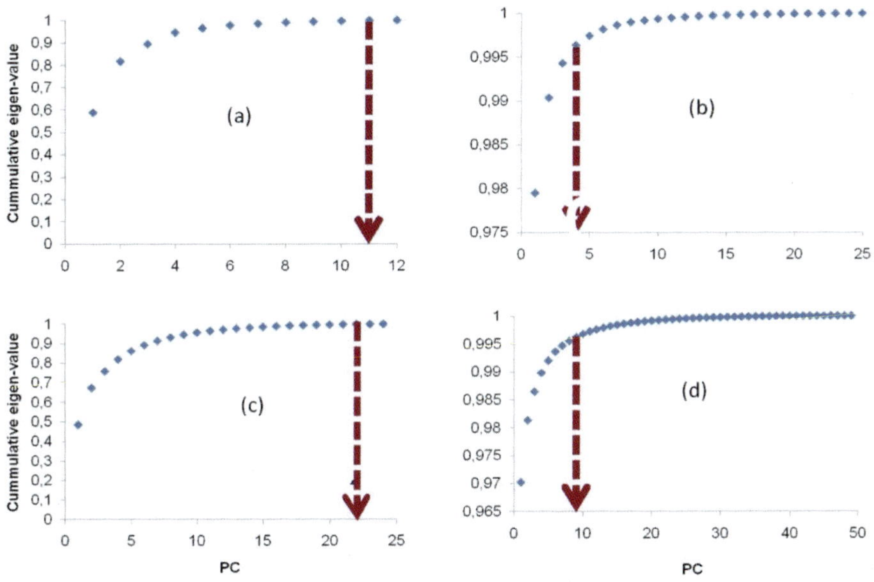

Fig. 2. The plot of principal component (PC) versus Fourier coefficients is displayed here: (a) Invariant; (b) Symmetric; (c) Asymmetric; (d) Standard. [14]. Used with permission of Elsevier.

Least squares classifier

The classification of training sets was determined by applying a least squares method that calculates the minimum difference between morphological and/or extracted color features. The least squares classification procedure is also used to estimate the prediction accuracy. First, the mean and standard deviation values of the color and morphological coefficients of four grain kernels were calculated for the training datasets. Then the deviation from the means was estimated and used to classify the features values that show the values within the variation in training sets. The feature parameters were determined for each type of grain. The values of the root square difference of the morpho-color coefficients of the multiple training sets were calculated [28]. Then, the grain was classified based on the least sums of the squares differences. Fig. 3 shows a schematic illustration of the least square classifier procedure (e.g., classification by only on EFD parameters (dotted line 3), or by calculating S_{FX} (solid line 2) or I_{FX} (dashed line 1)). The following morphological and color features alone or in combination can be used for classification:

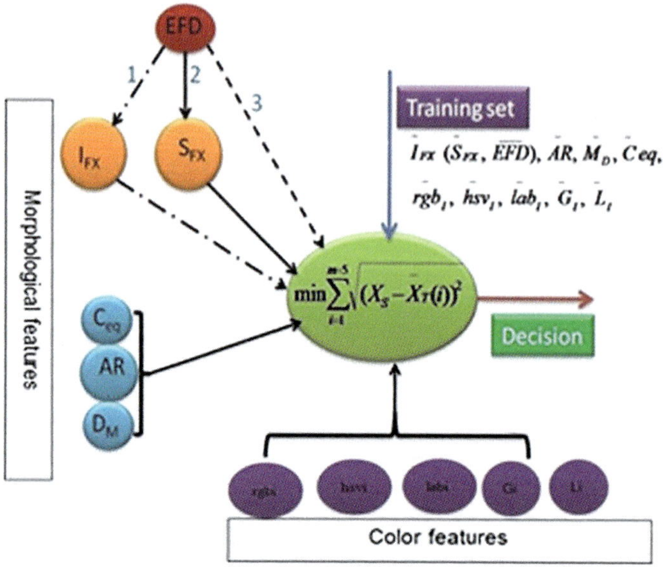

Fig. 3. Least square classification procedure displayed here. A set of morphological and color features are illustrated. The mean values of morphological and color parameters are represented with bars. The possible combinations of shape features are given by broken arrows [28]. Used with permission of Elsevier.

The following morphological and color features alone or in combination can be used for classification:

$$\textit{Symmetric Fourier index:} \quad S_{FX} = C_{eq} \times \sum_{nn=1}^{13} abs(a_{nn}^{**}) + abs(d_{nn}^{**}) \qquad (19)$$

$$\textit{Invariariant Fourier index:} \quad I_{FX} = sqrt(a_{nn}^{**2} + d_{nn}^{**2}) \qquad (20)$$

$$\textit{Roundness:} \quad C_{eq} = \frac{4\pi A}{P^2} \qquad (21)$$

$$\textit{Aspect ratio:} \quad AR \qquad (22)$$

$$\textit{Major diameter:} \quad D_M \qquad (23)$$

The accuracy of classification with the least-square classifier was estimated as a ratio of the number correctly classified grain kernels over the total number of the testing sample. The following equation defines the classification accuracy η_c:

$$\eta_c = \frac{N_{CC}}{N_{TS}} 100 , \quad (24)$$

where N_{CC} – is a number of correctly classified grains, N_{TS} is a total number of testing sets.

The EFD and geometrical feature model did not show any significant difference in separation efficiency of cereal grains: EFD (79.2%, 72.8%, 78.14%, 86.6% and 98.9%) I_X model (79.6%, 67.04%, 77.8%, 89.5% and 98.9%) for Barley, CWAD, Oats, CWRS and Rye, respectively. Therefore, Mebatsion *et al.*, [28] proposed a one-parameter S_{FX} model with other additional geometrical features. By addition of geometrical features (i.e. roundness, aspect ratio and maximum diameter) to the symmetric Fourier index, the model was improved and the classification accuracy for rye, barley and CWAD of 102%, 22.5%, 14.5%, were achieved [28]. A small improvement in the separation accuracy for oats was obtained, whereas classification accuracy for CWRS was not improved much. A visual comparison based on classification accuracy of grain classification models based on EFD coefficients and combined features is illustrated in Fig. 4. The combined model achieved a mean separation accuracy of 98.2% comparable to the result achieved by Choudhary *et al.* [8]. It is should be noted that Mebatsion *et al.*, [28] obtained these results using only four morphological features, whereas in previous studies the same accuracy results were obtained by [8] using a total of 335 morphological, textural, color and wavelet features. To the authors knowledge, the EFDs and geometrical feature model is the most efficient classification model published to date. It is obvious that the larger the number of classification features, the more computationally expensive the classification is and the classifier is prone to over-training resulting in lower prediction power [26].

Color model

The following color parameters rgb_I, hsv_I, and lab_I from RGB, HSV and Lab color systems were used in classification. They were calculated as a square root of the sum of squares of the RGB, HSV and L*a*b* color values, respectively [28]. The classification accuracy of the color model defined by 4 color parameters was found to be 100% for rye, 99.97% for CWAD and 99.63% for oats [28]. Although good accuracy results were achieved for rye, CWAD and oat, a low prediction accuracy for CWRS kernels (48.76%) was obtained. The model was unable to differentiate between two grains: CWRS and barley. Nevertheless, the S_{FX} model showed the precise classification of the two wheat types (CWAD and CWRS). The lowest accuracy was obtained for the color model of barley and CWRS and therefore, addition of extra parameters could improve the model. Therefore, the combined models based on set of morphological and colour features were built to improve the model's classification efficiency.

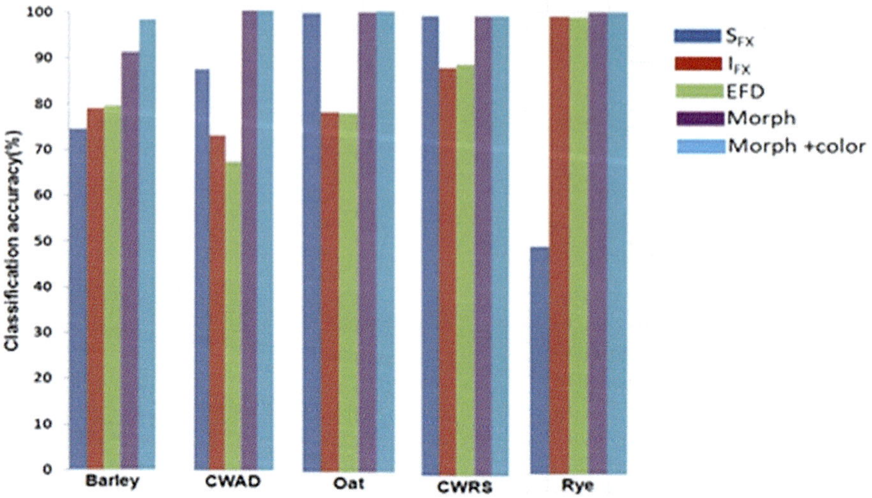

Fig. 4. Classification accuracy for different Fourier coefficients based and combined models: S_{FX} - symmetric Fourier coefficient model, I_X - the invariant Fourier coefficient model, EFDs - the symmetric coefficient model, Morph - morphological model included S_{FX}, AR, D_M and C_{eq}, Morph + color - the model of combined morphological and color features [28]. Used with permission of Elsevier.

Combined morphological and color model

The results from the models based on EFD and geometrical features showed that in order to get better classification accuracy addition of extra features is needed. Therefore, using the combined model (morphological + color) resulted in near-perfect classification for CWRS and oat, and perfect classification for CWAD and rye [28]. The barley grains were classified with accuracy of 91.4%. Overall, the very high classification accuracy (99%) was achieved by combined model. These results were much better than in previously published models and the models developed by Mebatsion *et al.*, [28] employing a smaller number of kernel features. The classification models using one feature input showed inconsistencies regarding the grain types. The separation accuracy of grains was improved by adding additional geometrical features to color models. As a result, the highest accuracy for selected cereal grains was obtained by the combined model including morphological and color features.

DISCUSSION AND CONCLUSIONS

Challenges, Advances and Future Prospects

It is clear from the aforementioned results that a machine vision system coupled with advanced classification algorithms is capable of consistent and accurate classification of the individual grain kernels in a non-contact rapid way. MVS technology has great potential in grain handling and quality monitoring. But, the main challenge is to present individual grain kernels to the machine vision systems in a non-touching manner. It is imperative that kernels do not touch one another even on vibrating beds while moving as a single layer. To date, several research studies have been focused in classification of grains using their morphological parameters [10, 19]. In the previous studies, features were obtained by extracting them from the images of the non-touching kernels that were manually presented to the imaging systems. This was done to avoid clustering of touching kernels so that the applied algorithm of features extraction could easily be applied. Although introducing vibrating beds can spread the kernels into one layer, there is still a high chance of the presence of touching grains [29]. Crowe *at al.*, [29] claimed that major the touching grains scenarios occurred for two or multiple touching kernels and they are observed only for less than 4% from all kernels. Zhang *et al.*, [1] suggested focusing on development of the algorithm that should separate two and three kernel touching scenarios because other possible multiple kernel touching scenarios can be easily avoided by using mechanical systems (e.g. vibrating beds).

To date, there are many different algorithms for separating touching kernels known and widely reported in literature. These methods are based on morphological attributes [30], a concavities based algorithm [31], an ellipse-fitting algorithm [1], watershed segmentation [32]), and a combined concavity algorithm and watershed [2]. Most of these techniques have certain limitations and disadvantages: (i) the use of morphological feature was not able to handle the segmentation of longer chains of touching kernels; (ii) the concavity algorithm is unsuccessful when the touching kernels are more than three; (iii) the ellipse-fitting based algorithm is time consuming and can deal with only two touching kernels [1]; (iv) the watershed algorithm is not able to segment elongated grains and can cause over segmentation [32]); and (v) the concavities algorithm combined with watershed method became too lengthy and requires more computation time. Moreover, broken kernels and kernels that have rough boundaries create another complexity because pseudo corner points may be observed during the segmentation process.

Recent advances in machine vision show that cereal grains can be classified by shape using EFDs with smaller number of morphological and color features. The EFDs shape determination method is more objective and precise and it also does not require *a priori* knowledge about variation in shape, and human error. The shapes of most cereal grains differ with their growing locations and highly dependent on environmental factors [26]. Thus, the shape-based identification approach has a disadvantage because shapes of plant organs also are highly dependent on environmental factors. The difference in shape variability might have resulted from agricultural practice as well. A huge challenge in implementing MVSs for grain assessment - is how to separate touching grain kernels in acquired images and to develop an algorithm for automated analysis of grain samples. Mebatsion and Paliwal [13] reported that classification of selected types of grains based only on Fourier descriptors was low (83.6%). Amongst the four Fourier series groups: standard, symmetric, asymmetric, and invariant, the symmetric model had a minimum number of principal components, leading to a reduction in dimensionality.

The accuracy of the classification was improved by adding three geometrical features resulting in accuracy of 98.2%. Moreover, the addition of the color features led in better separation efficiency. This combined model for most grain types showed the near-perfect classification (>99.9%). Further improvements in the model are still needed to achieve perfect classification for barley kernels. It should be noted that developed algorithm could achieve near-perfect classification accuracy of five different cereal grains using only a few classification features. The developed algorithm is much more efficient compared to previous separation algorithms when hundreds of features were required to get the same accuracy and efficiency.

Recent advances in separating touching kernels pave a new way for realistic grain classification [12, 14, 15, and 26]. Now, the future challenge is combining kernel separation and classification models into a realistic machine vision grading classification model that will revolutionize the industry.

REFERENCES

[1] Zhang G, Jayas DS, and White NDG (2005) Separation of touching grain kernels in an image by ellipse fitting algorithm. *Biosystems Engr* **92**: 135-142.

[2] Zhong Q, Zhou P, Yoa Q, and Mao K (2009) A novel segmentation algorithm for clustered slender-particles. *Comp Elect Agricult* **69**: 118-127.

[3] Jayas DS, Ghosh PK, Paliwal J, and Karunakaran C (2008) Quality evaluation of wheat. **In**: *Computer Vision Technology for Food Quality Evaluation.* D Sun (Ed.) Elsevier, Netherlands.

[4] Majumdar S and Jayas DS (2000). Classification of cereal grains using machine vision I: morphology models. *Trans. ASAE* **43**: 1669-1675.

[5] Paliwal J, Visen NS, Jayas DS, and White NDG (2003) Comparison of neural network and non-parametric classifier for grain kernel identification. *Biosystems Engr* **85**: 405-413.

[6] Majumdar S and Jayas DS (2000) Classification of cereal grains using machine vision II: color models. *Trans. ASAE* **43**: 1677-1680.

[7] Majumdar S and Jayas DS (2000) Classification of cereal grains using machine vision III: texture models. *Trans. ASAE* **43**: 1681-1687.

[8] Choudhary R, Paliwal J, and Jayas DS (2008) Classification of cereal grains using wavelet, morphological, colour, and textural features of non-touching kernel images. *Biosystems Engr* **99**: 330-337.

[9] Neuman M, Sapirstein HD, Shwedyk E, and Bushuk W (1987) Discrimination of wheat class and variety by digital image analysis of whole grain samples. *J Cereal Sci* **6**: 125-132.

[10] Kuhl FP and Giardina CR (1982) Elliptic Fourier features of a closed contour. *Comp Graph Image Proc* **18**: 236-258.

[11] Ohsawa R, Tsutsumi T, Uehara H, Namai H, and Ninomiya S (1998) Quantitative evaluation of common buckwheat (*Fagopyrum esculentum Moench*) kernel shape by elliptic Fourier descriptor. *Euphytica* **101**: 175-183.

[12] Mebatsion HK and Paliwal J (2011) A Fourier analysis based algorithm to separate touching kernels in digital images. *Biosystems Engr* **108**: 66-74.

[13] Mebatsion HK and Paliwal J (2012) A novel algorithm utilizing elliptical Fourier descriptors to automatically separate touching objects in digital images. *Comput Industry* **63**: 723-730.

[14] Mebatsion HK, Paliwal J, and Jayas DS (2012) Evaluation of variations in the shape of grain types using principal components analysis of the elliptic Fourier descriptors. *Comput Elect Agricult* **80**: 63-70.

[15] Mebatsion HK, Paliwal J, and Jayas DS (2012). A novel invariant elliptic Fourier coefficient based classification of cereal grains. *Biosystems Engr* **111**: 422-428.

[16] Ninomiya S, Ohsawa R, and Yoshida M (1995).Evaluation of Buckwheat and Tartary Buckwheat kernel shape by elliptic Fourier method. *Current Adv Buckwheat Res* 53: 389-396.

[17] Hirose T, Yoshida M, Nemoto H, Kitabayashi H, Minami M, Matano T, Sherchand KK, and Ujihara A (1995) Diversity of grain character of Tartary buckwheat in Nepal. **In**: Matano T and Ujihara A. (Eds.) *Current Advances in Buckwheat Research*. Shinshu Univ. Press, Asahi Matsumoto, Japan.

[18] Utku H and Koksel H (1998) Use of statistical filters in the classification of wheat by image analysis. *J Food Engr* **36**: 385-396.

[19] Paliwal J, Shashidhar NS, and Jayas DS (1999).Grain kernel identification using kernel signature. *Trans ASAE* **42**: 1921-1924.

[20] Lou X, Jayas DS, and Symons SJ (1999).Identification of damaged kernels in wheat using a colour machine vision system. *J Cereal Sci* **30**: 49-59.

[21] Gonzalez RC and Woods RE (1992) *Digital Image Processing*. Addison-Wesley Publishing Company, Inc.

[22] Hiraoka Y and Kuramoto N (2004) Identification of *Rhus succedanea* L. cultivars using elliptic Fourier descriptors based on fruit shape. *Silvae Genetica* **53**: 221-226.

[23] Neto JC, Meyer GE, Jones D, and Samal AK (2006) Plant species identification using Elliptic Fourier leaf shape analysis. *Comput Elect Agricult* **50**: 121-134.

[24] Iwata H, Nesumi H, Ninomiya S, Takkano Y, and Ukai Y (2002) The evaluation of genotype x environment interactions of citrus leaf morphology using image analysis and elliptic Fourier descriptors. *Breeding Sci* **52**: 243-251.

[25] Costa C, Manesatti P, Pagila G, Pallottino F, Aguzzi J, Rimatori V, Russo G, Recupero S, and Recupero GR (2009). Quantitative evaluation of Toroco sweet orange fruit shape using optoelectronic elliptic Fourier based analysis. *Postharvest Biol Tech* **54**: 38-45.

[26] Paliwal J, Visen NS, and Jayas DS (2001) Evaluation of neural network architectures for cereal grain classification using morphological features. Journal of Agricultural Engineering Research **79**: 361-370.

[27] Yoshioka Y, Iwata, H, Ohsawa R, and Ninomiya S (2004). Analysis of petal shape variation of Primula sieboldii by elliptic Fourier descriptors and PCA. *Ann Bot* **94**: 657-664.

[28] Mebatsion HK, Paliwal J, and Jayas DS (2013) Automatic classification of non-touching grains in digital images using limited morphological and color features. *Comput Elect* **90**: 99-105.

[29] Crowe TG, Luo XY, Jayas DS, and Bulley NR (1997) Color line-scan imaging of cereal grain kernels. Applied Engr Agricult **13**: 689-694.

[30] Shatadal P, Jayas DS, and Bulley NR (1995) Digital image analysis for software separation and classification of touching grains. *Trans ASAE* **38**: 635-643.

[31] Visen NS, Shashidhar NS, Paliwal J, and Jayas DS (2001) Identification and segmentation of occluding groups of grain kernels in grain sample images. *J Agricult Engr Resources* **79**: 159-166.

[32] Wang W and Paliwal J (2006) Separation and identification of touching kernels and dockage components in digital images. *Canadian Biosystems Engr* **48**: 1-7.

Sexual Differences in Distress Sound in the West Indian Sweet Potato Weevil *Euscepes postfasciatus*

HARUKI TATSUTA[1] and NORIKUNI KUMANO[2]

Abstract

The sweet potato weevil, *Euscepes postfasciatus* is one of a number of serious pests of sweet potato. The species is distributed throughout the Ryukyu Islands of Japan. Although a sterile insect technique (SIT) has been used to eradicate this species since 1994, it is still far from accomplishing the principal goal of eradication. One of the reasons for this delay is that the assumed mating systems including courtship behavior for this weevil is not well understood. In addition, there is also the problem of mass rearing as well. It was discovered that adult mature *E. postfasciatus* produce sounds when they are disturbed or under mating. Although it is thought that these sounds might play an important role during the weevil's communication, especially for recognizing enemies and for mating success. Thus, fundamental studies on the characteristics of these sounds would be desirable for an understanding of how mating recognition and avoidance of enemies via sound is carried out in this species. Several features were extracted from the distress sound readings of the weevil to determine differences between populations or sexes. Mass-reared lineages of varying rearing periods from six local populations in Ryukyu Islands were used as materials. It was determined that the characteristics of distress sounds were differentiated among different reared lineages. For time-domain features, females tended to produce longer period of echeme than males. In particular, although the slope of the regression between body size and the mean duration of the echeme varied significantly among lineages in males, the regression slopes were almost isotropic throughout localities in females. Such patterns of spatial and sexual variation may suggest the presence of a strong genetic basis that is responsible for the development of the stridulatory apparatus or sound emission system existing in *E. postfasciatus*.

INTRODUCTION

The West Indian sweet potato weevil, *Euscepes postfasciatus* (Fig. 1A) is known as one of the harmful invasive pests and also as an important pest of the sweet potato, *Ipomoea batatas* [1, 2]. The evidence of invasion of this species to Japan was detected at Katsuren peninsula in Okinawa in 1947 [3]. Until 1954 since the first discovery, the weevil had expanded over the entire Ryukyu Island chain [4]. In 1994, Okinawa Prefecture government started an eradication project using a sterile insect release technique (SIT) for targeting *E. postfasciatus* and another species, *Cylas formicarius* [5]. Among several behavioral traits, migration ability of target pest species is important for spreading sterile individuals over a target field; however, *E. postfasciatus* does not have flight ability, which makes the SIT project difficult [6]. Interestingly, it has been revealed that both male and female adults of this weevil produce a sound by scraping the abdomen with elytra when they are disturbed (so called "distress sound") or under mating ("calling song") [7, 8]. The distress sound is produced during an alive or death-feigning situation.

[1] Department of Ecology and Environmental Sciences, Faculty of Agriculture, University of the Ryukyus, Sembaru, Okinawa 9030213, Japan, htatsuta@agr.u-ryukyu.ac.jp.
[2] Okinawa Prefectural Plant Protection Center, Naha, Okinawa 9020072, Japan, norikumano@gmail.com.

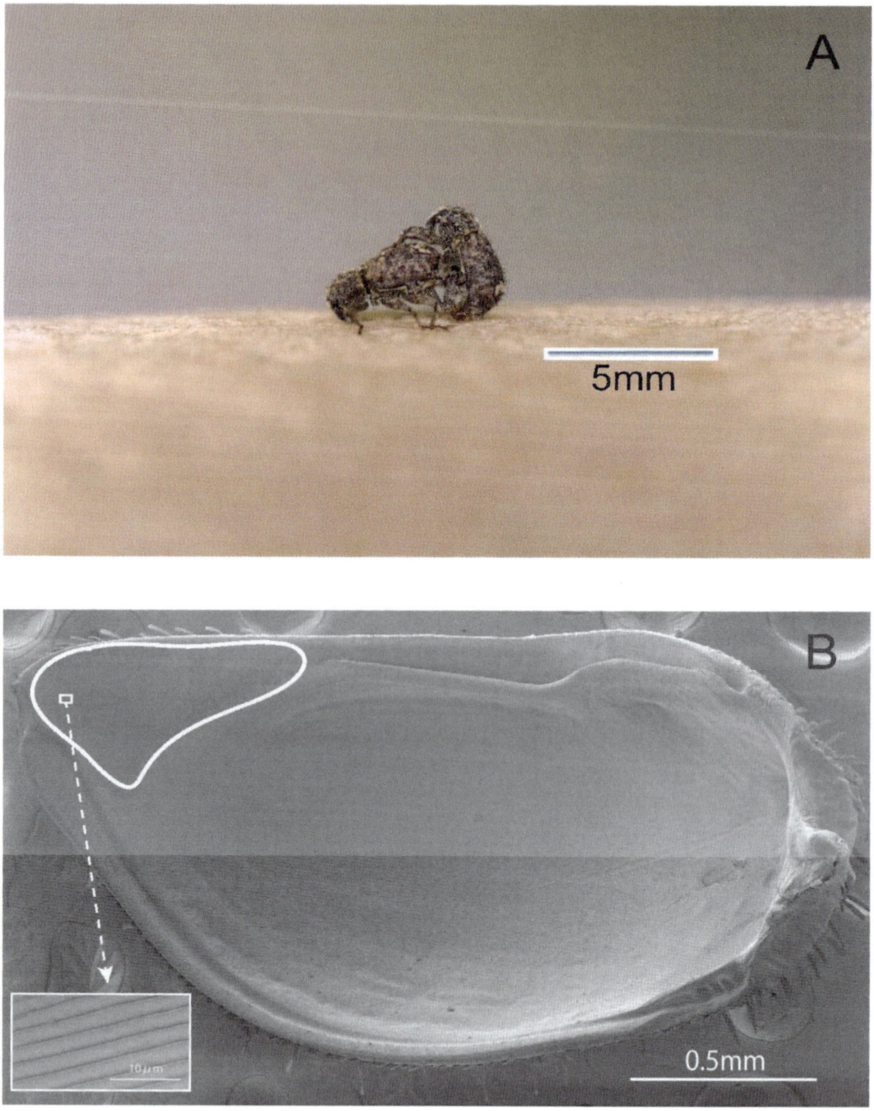

Fig. 1. (A) Photo of *Euscepes postfasciatus* (left: female, right: male) and (B) location of pars stridens (surrounded by white line) on the posterior tip of the ventral surface of elytron.

Although certain roles of the distressed signals are unknown, the signals might have a compound function such as surprising enemies or aggressive interactions [9]. A stridulatory apparatus exists on the back of posterior tip of the elytron (Fig. 1B) [7], and a possible plectrum is located on a posterior segment of the abdomen (Tatsuta, personal communication). Extensive comparisons in song structures of the closely related species (*Acalles* spp.) revealed that there are substantial species-specific characteristics in time and frequency domain of the song structures [10]. In Orthopteroid insects such as grasshoppers and crickets, for example, the species-specific songs play an important role for discriminating the exact partner and a secured system of auditory afferents strengthens the accuracy of discrimination [11, 12]. In planthopper and whitefly, they use plant-mediated vibrational signals for successful mating [13, 14].

From a phylogeographic point of view, it is assumed that the founder invaded into the Ryukyu Islands, possibly around Katsuren peninsula in Okinawajima Island, and then the descendants spread over the Ryukyus within several years after the invasion [7]. Although the expansion route and the extent of invasion event have not been yet clarified, it would be conceivable that some ways of behavioral processes vary among local populations as a result of genetic admixture and differentiation during expansion and succeeding adaptation to a local environment. Such local processes would induce subsequent variation in various levels of traits from genes to external phenotypes. In the present study, the primary aim is to examine whether variation in sound features such as peak frequency and duration of echemes exists in *E. postfasciatus* among local populations in Ryukyu archipelago. The second objective was to assess the relationship between body size that might affect the overall structure of the stridulatory apparatus and duration of signals as detected in other insect species [15]. A series of decomposing structures of sound or vibration signals was required for comparison purposes. These characteristics require techniques of the measurement of signals as well as the statistical analysis of shape comparisons. Thus, one can consider the present work to be posed in the broader framework of morphometric approaches.

MATERIALS AND METHODS

Specimens and Recording Signals

To establish experimental lineages of local populations, parental specimens of *E. postfasciatus* were collected from different localities of the Southwestern Islands in 2008 and 2009, except for one of the Yomitan populations and the Yaese population (Fig. 2). These were then brought back to the laboratory of Okinawa Prefectural Plant Protection Center and reared for three to six generations. One of Yomitan populations was collected in 1994 for use for the SIT program. Specimens that had been reared for approximately 80 generations since being captured were used (Yomitan* in Fig. 2). For Yaese population, specimens that had been reared for approximately 29 generations since 2004 were used. The rearing temperature was kept at 25 ± 1°C and relative humidity was 50-90%. In each generation, a total of 500 parental individuals (♂:♀=1:1) were put into the plastic box (Lustroware®, Iwasaki Industry Co., Ltd,160 x 223 x 82 mm) with sufficient amount of sweet potatoes for rearing (300 – 380g per box).To obtain virgin weevils, the sweet potato roots were dissected approximately 7 weeks after the inoculation of weevils. We then extracted newly-emerged weevils from pupal chambers.

Based on the observation of exoskeleton using scanning electron microscope (SEM), the stridulatory apparatus exists on the ventral side of the elytron in both sexes (Fig. 1B) [7]. To record distress sound in laboratory, an individual of *E.*

postfasciatus was set on a petri dish and settled on an adhesive tape so that the ventral side of abdomen faces to upside after the sexing. We then prickled the body gently using a pair of tweezers and recorded the distress sound. We continued recording until at least three echemes per individual were obtained (Fig. 3A). Each echeme consists of a series of pulses (Fig. 3B), each of which is probably produced by the friction of the plectrum back and forth on the pars stridens [14].

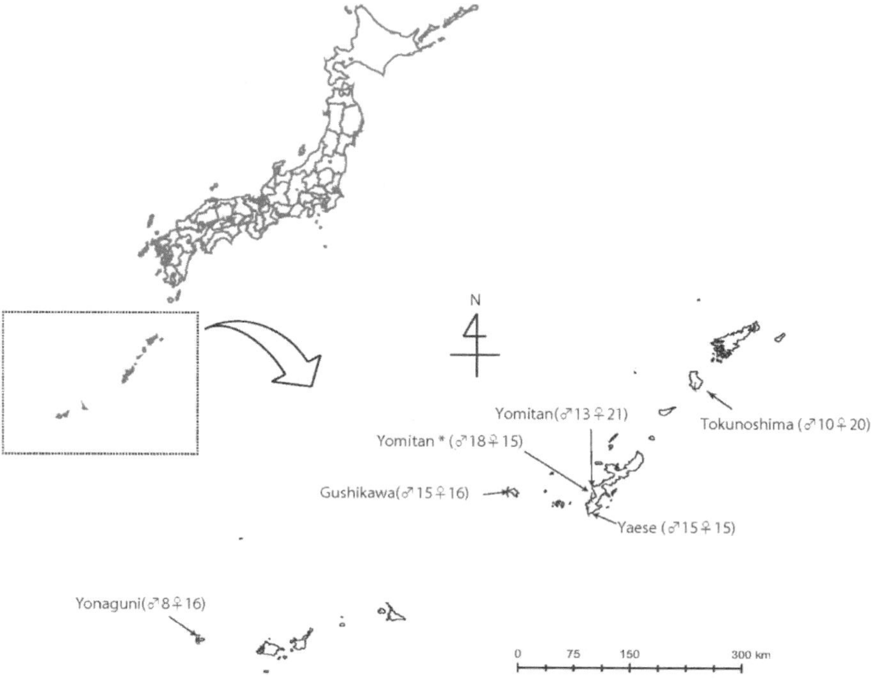

Fig. 2. Map of sampling localities. The number of specimens (males and females) in each locality is shown. The asterisk indicates reared line in laboratory for a certain period (see text for details).

The recording device was 1/4 inch free-field microphone (frequency range: 4Hz – 100 kHz, Type 4939, Brüel & Kjær, Denmark) attached on a wide range measuring amplifier (Type 2610, Brüel & Kjær, Denmark). The analog signals of sounds were converted into digital values using a DAQ Card (6062E, National Instruments, USA) connected with BNC adapter (BNC-2110, National Instruments, USA) and then stored on a hard disk in personal computer (Versa Pro, NEC, Japan). The software Avisoft Recorder (Version 4.0) was used for monitoring and recording converted signals [15]. All recordings were made in December 2009 and approximately at the same distance from weevil (ca. 5mm between a tip of microphone and specimen) in order to avoid distance interference on the frequency measurements. Room temperature was maintained about 25°C and the recording room kept quiet as much as possible throughout the recording.

Sound Analysis

Recorded stridulatory signals were analyzed using Avisoft SAS Lab Pro for Windows (Version 4.53) [17]. To compare frequency-domain features, we performed a Fast Fourier Transform (FFT) to obtain mean amplitude spectra (FFT length 512, Hann window, 50% overlap for temporal resolution). Three echemes were chosen randomly for each individual and estimated peak frequency for each echeme. Average peak frequencies were then compared among lineages. Because the homogeneity of variances among lineages and between sexes was not rejected by Levene test (lineage: $F = 0.317$, $P = 0.730$, sex: $F = 0.427$, $P = 0.516$), differences among mean values of peak frequencies were compared using an ANOVA model [18].

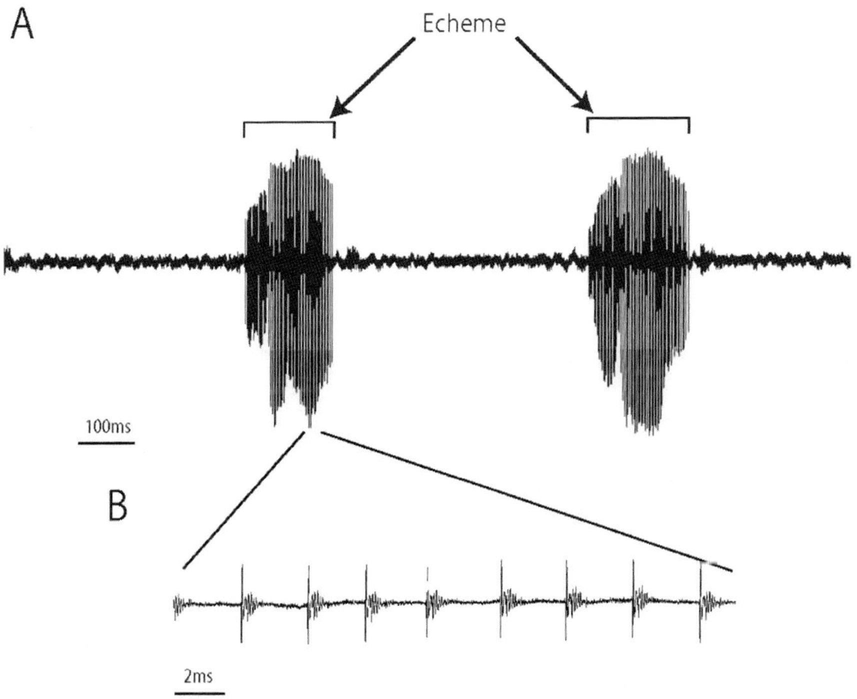

Fig. 3. (A) Echemes and (B) pulses of *E. postfasciatus* are shown here.

The explanatory variables in the model were lineage and sex. To compare time-domain features among lineages, the duration was measured of each echeme that was chosen in the frequency-domain analysis (see Fig. 3A). The duration of the three measured echemes in each individual was then averaged for the subsequent analysis. To assess the relationship between body size and the mean duration of echemes, photographs were taken and the width of pronotum measured on a computer screen as an index of body size in each individual using ImageJ software [19]. The homogeneity of variances was not rejected for body size and mean duration of echemes among lineages and between sexes (Levene test: F = 1.191, P = 0.316 for lineage, F = 0.001, P = 0.971 for sex). Body size was compared between sexes based on an ANOVA model. Relationship between body size and the mean duration of echeme was assessed by ordinary (type I) regression analysis for each sex [18]. Whether the estimated regression slopes varied between sexes was also assessed using an ANCOVA model, in which the mean duration of echeme was modeled as the dependent variable with sex as the factor and body size as the covariate [18]. All statistical analyses were implemented using R statistical package (Version 3.0.2) [20].

RESULTS

For the comparison of peak frequencies among three lineages of local populations (Tokunoshima, Gushikawa, and Yonaguni), a significant difference was detected (F = 3.35, P = 0.039), while no significant effect of sex was found (F = 1.76, P = 0.189). Average peak frequency was higher in Yonaguni (10384.3 Hz) than the other two (Gushikawa: 7852.5 Hz, Tokunoshima: 7842.1 Hz) (Fig. 4). The number of comparisons was very small, but the difference in the average of peak frequency does not seem to be correlated with geographic distance between localities.

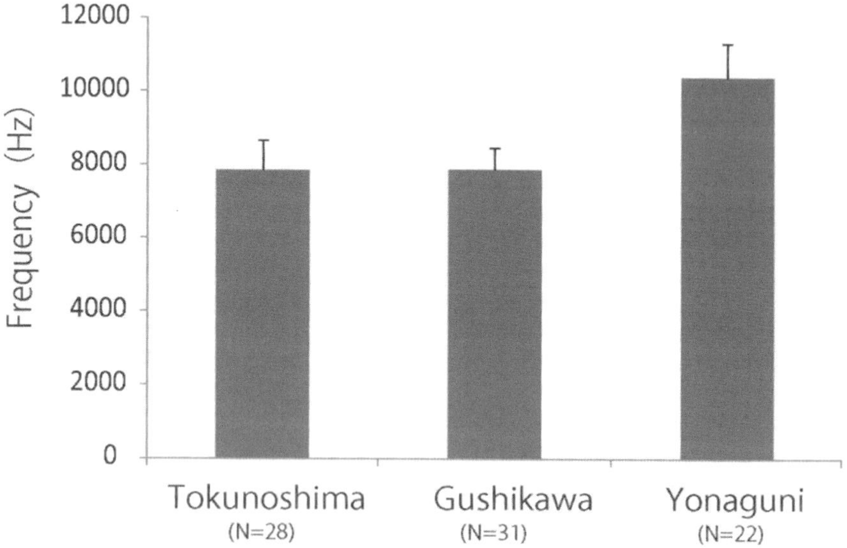

Fig. 4. Means and standard errors of peak frequencies of echemes in the three lineages of the local populations are displayed here.

Comparisons of body size between sexes revealed that females are significantly larger than males (F = 10.75, P = 0.0011). Correlation between body size and the duration of echeme was significant in both sexes (male: R^2 = 0.032, F = 4.86, P = 0.030, female: R^2 = 0.037, F = 5.93, P = 0.016). The duration of echeme tended to be longer as body size increased in both sexes (male: $b = 7.35 \times 10^{-5}$, $t = 2.20$, $P = 0.030$, female: $b = 8.96 \times 10^{-5}$, $t = 2.43$, $P = 0.016$). The slope of the regression between body size and the mean duration of the echeme was quite similar for both males and females (ANCOVA: $F = 0.006$, $P = 0.936$). Furthermore, the mean duration of the echeme was significantly longer in females than in males (ANCOVA: $F = 71.20$, $P \ll 0.0001$, Fig. 5). Thus, the regression slopes between body size and the mean duration of echeme among lineages was compared in males and females separately.

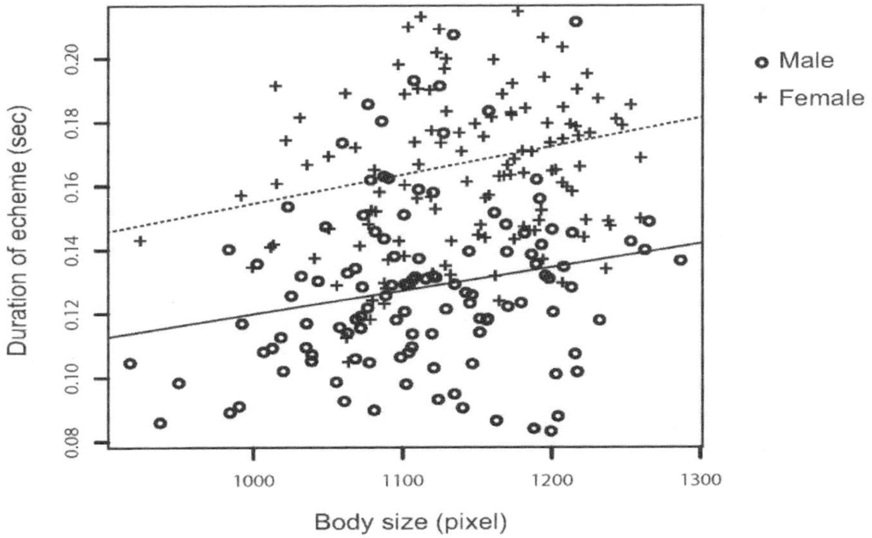

Fig. 5. Plots of body size (in pixel) and duration of echeme (sec.) with regression lines fitted to each sex. Solid and dotted line indicate male and female, respectively.

The spatial patterns of the mean duration of echeme were almost concordant between sexes (Fig. 6). Nevertheless, it was found that the regression slopes were significantly heterogeneous among lineages for the males, but they were almost isotropic throughout the localities for females (ANCOVA: $F = 2.58$, $P = 0.034$ in males, $F = 0.93$, $P = 0.465$ in females).

DISCUSSION AND CONCLUSIONS

In other species such as dung beetles, substantial variation in the structures of distress signals exists among localities and species [15]. Both males and females sometimes possess the same structure of stridulatory organs, but the opportunities for usage of such organs may not be necessarily identical in both males and females [21, 22]. For instance, calling sounds are fundamentally emitted by males, which promote sexual advertisement for increasing their fitness. In contrast, the acoustic signals are used for aggression and defense in a given context [23]. It is known that *E. postfasciatus* emits distress sounds during death-feigning. Although the possibility that such sounds are useful for menacing enemies of other species cannot be ruled out, it seems likely that their signals function as territorial defense or recognition of mating partners within species.

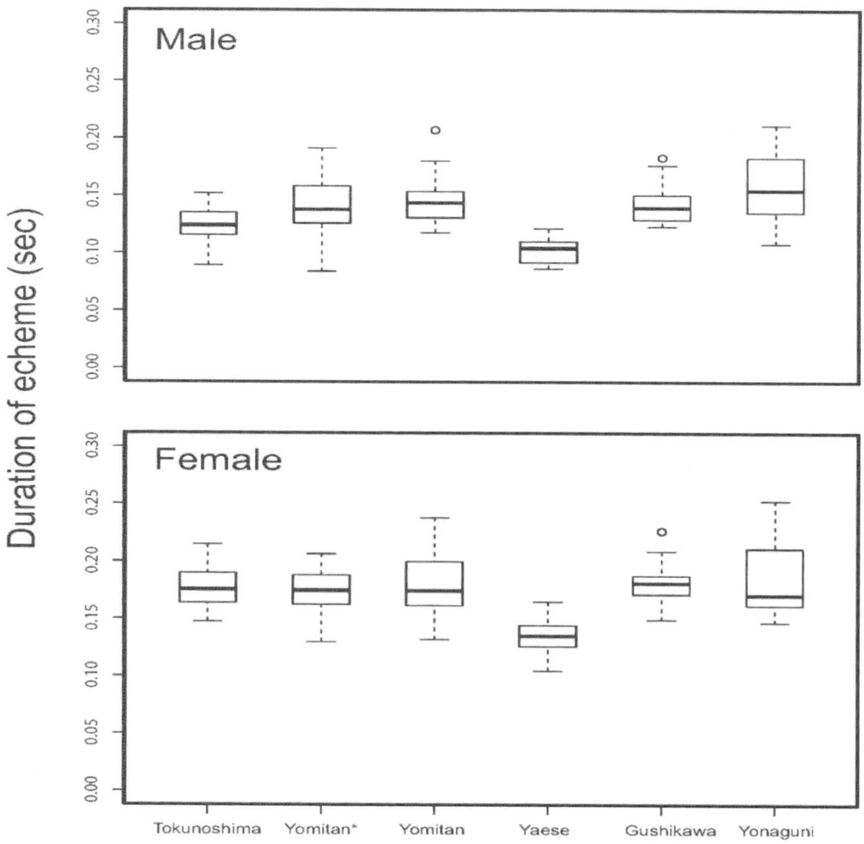

Fig. 6. Box-and-whisker plots of averaged duration of echeme in each sex.

In the present study, individuals with larger body size tended to produce a relatively longer echeme. Females possess generally larger bodies than males as shown in the present study and also produced a longer echeme than males. This indicates that a strong correlation exists between the physical structures of the pars stridens such as surface area of stridulatory files or neuro-ethological pathways for the movement of plectrums and body size. Interestingly, because characteristics in frequency-domain were similar between sexes, both males and females have probably the same fundamental features of the stridulatory organs. Although the genetic context of the stridulatory organs is completely unknown in this species, the evidence of substantial intraspecific variation in spectral features between sexes and in the duration of echeme suggests the existence of genetic basis that can be a motive force of diversification in the sound communication systems. In sexually-mediated signals such as calling songs are sometimes under the pressure of sexual selection, sexual differences in sound structures detected in the present study might have been accelerated rapidly by selective pressures in parallel with bottleneck events after invasion event in *E. postfasciatus*. While applied entomologists have been attracted to studies on rearing and management strategy for eradication programs using the SIT for *E. postfasciatus*, the ethological and genetic context of communicative signals in this species are not well understood and warrant further investigation in the future.

ACKNOWLEDGEMENTS

We cordially thank Prof. Pete Lestrel for his thoughtful suggestions and instructive advice on improving greatly the early draft. We are also grateful to T. Takanashi (Forest and Forestry Product Institute) for his advice on the analyses of sounds, K. Yasuda (Okinawa Plant Protection Station) for providing devices for recording sounds, and T. Sasaki (University of the Ryukyus) for providing a space for recording sound. Thanks are also due to D. Haraguchi and T. Matsuyama (Okinawa Plant Protection Station) for providing the weevil specimens. This study was supported in part by Grant-in-Aid for Scientific Research (KAKENHI), No. 25650134 to HT and 26450067 to NK and Sumitomo Foundation Research Fund for Environmental Studies, No. 103209 to HT.

REFERENCES

[1] Sherman M and Tamashiro M (1954) The sweet potato weevils in Hawaii; their biology and control. *Hawaii Agric Experimental Station Tech Bull* **23**: 1-36.

[2] Raman KV and Alleyne EH (1991) Biology and management of the West Indian sweet potato weevil, *Euscepes postfasciatus*. **In**: Jansson, RK and Raman KV (Eds.) *Sweet Potato Pest Management: A Global Perspective* Westview Press, Boulder, Colorado.

[3] Asato S (1950) West Indian sweet potato weevil as new sweet potato pest. *Kunigami Noho* **2**: 5-11. [In Japanese]

[4] Takara T (1954) Dissemination and control of vine borer of sweet potato and West Indian sweet potato weevil. *Plant Protection* **8**: 18-20. [In Japanese]

[5] Moriya S (1995) The possibility of eradicating two sweet potato pests, *Cylas formicarius* and *Euscepes postfasciatus*, in Japan. *FFTC Exten Bull* **403**: 1-7.

[6] Kumano N, Kawamura F, Haraguchi D, and Kohama T. (2009) Irradiation does not affect field dispersal ability in the West Indian sweet potato weevil, *Euscepes postfasciatus*. *Entomol Exp Appl* **130**: 63-72.

[7] Yasuda K and Miyara A (1992) Sound production of the West Indian sweet potato weevil, *Euscepes postfasciatus* (Fairmaire) (Coleoptera: Curculionidae) *Proc Assoc Pl Prot Kyushu* **38**: 89-91. [In Japanese]

[8] Yasuda K and Tokuzato M (1999) Sound production during mating and disturbance in the West Indian sweet potato weevil, *Euscepes postfasciatus* (Fairmaire) (Coleoptera: Curculionidae). *Appl Entomol Zool* **34**: 443-447.

[9] Hirschberger P (2001) Stridulation in *Aphodius* dung beetles: behavioral context and intraspecific variability of song patterns in *Aphodius ater* (Scarabaeidae) *J Insect Behav* **14**: 69-88.

[10] Riede K and Stüben PE (2000) Die musikalischen Acallen. **In**: (Stüben PE (Ed.) *Beobachtungen zur Stridulation bei den Cryptorhynchinae von den Kanaren* (Coleoptera: Curculionoidea). 1. Edition, SNUDEBILLER 1 (CD-Rom): CURCULIO-Institut: D-Mönchengladbach. [In German with English summary]

[11] Tunstall DN and Pollack GS (2005) Temporal and directional processing by an identified interneuron, ON1, compared in cricket species that sing with different tempos. *J Comp Physiol A* **191**: 363-372.

[12] Hedwig B (2014) Insect Hearing and Acoustic Communication. Springer.

[13] Ichikawa T (2011) Vibrational communication in planthoppers (Hemiptera). **In**: Miyatake Y (Ed.) *Insect Communication by Sounds and Vibrations* The Hokuryukan Co., Ltd, Tokyo. [In Japanese]

[14] Kanmiya K (2011) Mating behavior and vibratory signals in Aleyrodidae (Hemiptera). **In**: Miyatake Y (Ed.) *Insect Communication by Sounds and Vibrations* The Hokuryukan Co., Ltd, Tokyo. [In Japanese]

[15] Carisio L, Palestrini C, and Rolando A (2004) Stridulation variability and morphology: an examination in dung beetles of the genus *Trypocopris* (Coleoptera, Geotrupidae). *Population Ecol* **46**: 27-37.

[16] Specht R (2010) *Avisoft Recorder*. Version 4.0, Germany.

[17] Specht R (2010) *Avisoft SAS Lab Pro*. Version 4.53, Germany.

[18] Sokal RR and Rohlf FJ (2012) *Biometry. 4th Edition*. W. H. Freeman and Company, New York.

[19] Rasband WS (1997) *ImageJ*, U.S. National Institutes of Health, Bethesda, Maryland, USA, http://imagej.nih.gov/ij/.

[20] R Developmental Core Team (2013) *R: a language and environment for statistical computing*. R Foundation for Statistical Computing, Vienna, http://www.T-project.org.

[21] Greenfield MD (2002) *Signalers and Receivers: Mechanisms and Evolution of Arthropod Communication.* Oxford University Press, Oxford.

[22] Grant PBC, Roets F, and Samways MJ (2014) Sound characterization and structure of the stridulatory organ in *Gonogenia tabida* (Coleoptera: Carabidae). *Bioacoustics* **23**: 209-217.

[23] Wessel A (2006) Stridulation in the Coleoptera – an overview. **In**: Drosopoulos S, Claridge MF (Eds.) *Insect Sounds and Communication. Physiology, Behaviour, Ecology and Evolution.* Taylor and Francis, Boca Raton, Florida.

Morphometric Analysis of Intraspecific Variation of the Elytral Shape Between the Wild- and Mass-reared Sweet Potato Weevil *Cylas formicarius*

NORIKUNI KUMANO[1, 2, 3], KAORI TSURUI[1, 2, 3], TETSUYA TOYOSATO[1, 2] and HARUKI TATSUTA[3]

Abstract

In an eradication program using the sterile insect technique (SIT), a large number of individuals are mass-reared, sterilized and then released into the field. Since the ratio of sterile to normal males in the field is used to monitor the progress of the program, an extensive mass-trapping survey is conducted in the whole target area. Although a definite marker for the released individuals is indispensable in this monitoring procedure, such a definite marker has not been developed for the sweet potato weevil *Cylas formicarius*. Consequently, the monitoring of the eradication program in this weevil has been hampered by the presence of less accuracy due to using an uncertain marker. Therefore, one needs to develop another marking system. Since the mass-rearing of the beneficial insect is conducted under highly artificial conditions compared to the field, such a condition prior to the emergence in the rearing facility, might lead to differences in morphological characters in the mass-reared individuals. In this study, the shape of the elytral contour between wild and mass-reared weevils was compared to assess the potential of an intraspecific differentiation pattern in *C. formicarius*. This comparison was conducted in spring and autumn for two successive years (four comparisons in total). The morphometric analysis using elliptic Fourier descriptors (EFDs) demonstrated that the elytral shape of the mass-reared weevil differed significantly to that of wild one in each comparison and the proportion of individuals that were correctly classified into original lineages was high (> 80%). Therefore, the EFDs are considered useful for identifying the source of the trapped weevils in an eradication program using the mass-reared individuals for the sweet potato weevil. Although the fine-scale intraspecific morphological variation is generally difficult to recognize, the present study suggests that the EFDs have potential to highlight these differences between closely related lineages within the same species.

INTRODUCTION

The external morphology is one of the most important key factors as a taxonomic trait in taxonomy [e.g., 24]. Many entomologists have been fascinated with the intra- and inter-specific morphological and/or structural diversities, such as the genitalia, or wing-venation [2, 5, 8, 17, 22, 40, and 41]. In Coleoptera, there are many variations in the elytral coloration patterns, which have been used as the criteria for species/populations distinctions [10, 14, 16, 33, 34, and 44]. Likewise, because body size, shape or allometry (e.g., horn-length-body size relationships) are also useful for representing various taxonomic states [e.g., 6, 19, 28, and 27], there seems to be the possibility of also utilizing these for applied purposes.

The sweet potato weevils, *Cylas formicarius* (Summers) (Fig. 1A) is the major pest of the sweet potato, *Ipomoea batatas* (L.) Lam. in the South Pacific, the Caribbean, parts of Central and South America, and the southwestern islands of

[1] Okinawa Prefectural Plant Protection Center, Naha, Okinawa 902-0072, Japan, nrkumano@gmail.com.
[2] Ryukyu Sankei Co. Ltd., Naha, Okinawa 902-0072, Japan.
[3] Department of ecology and environmental science, Graduate school of agriculture, University of the Ryukyus, Sembaru, Nishihara, Okinawa 903-0213, Japan.

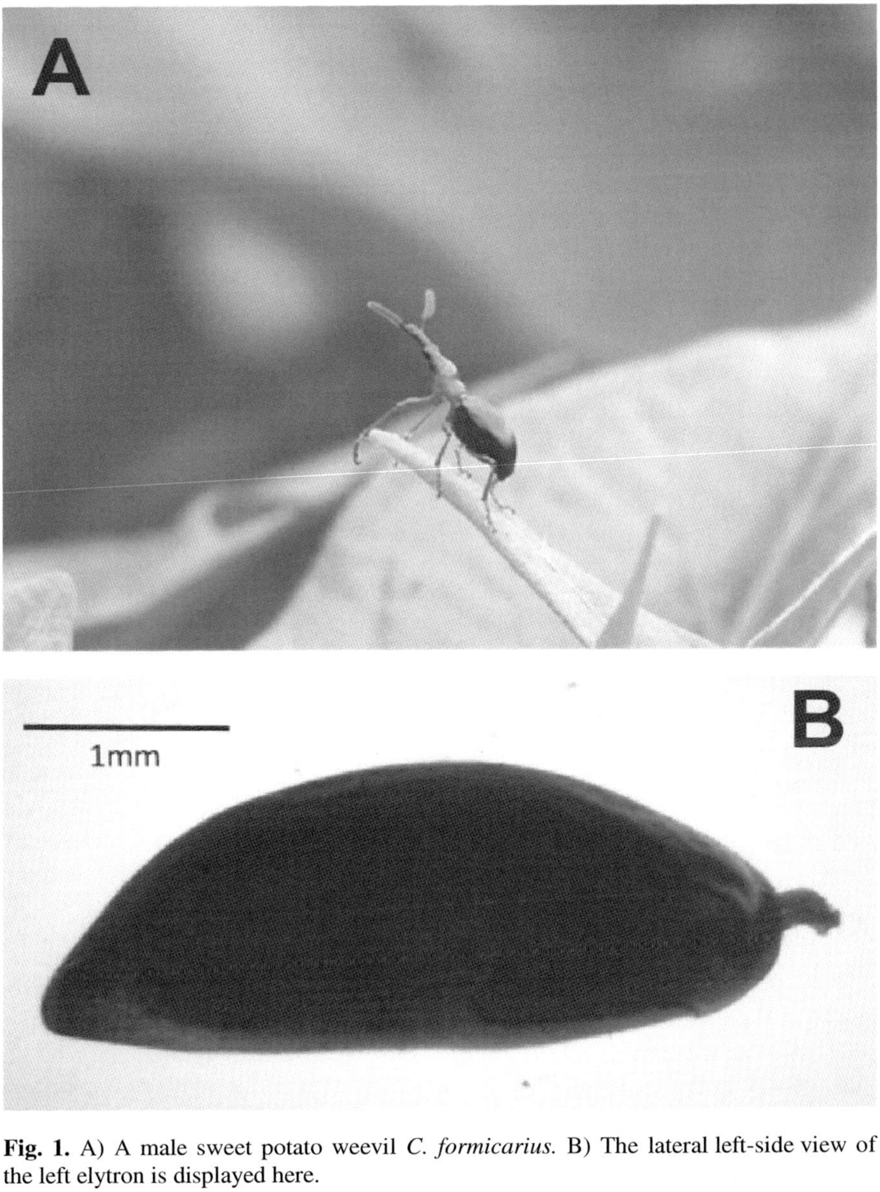

Fig. 1. A) A male sweet potato weevil *C. formicarius.* B) The lateral left-side view of the left elytron is displayed here.

Japan [4, 12, 29, 36, 42, 45, and 46]. Since its invasion of Japan in 1903 [30], *C. formicarius* has expanded its distribution throughout the Southwestern Islands of Japan [29]. If the weevils infest sweet potato roots, ipomeamarone, a furanoterpenoid toxin, is produced in blemished potatoes, which is harmful for humans and other animals [1, 38, and 43]. Therefore, even low weevil densities may cause devastating crop losses. The spread of plants infested with weevils into uninfested areas is strictly prohibited under the Japanese Plant Protection Law (Plant

Protection Act, Act No. 151 of May 4, 1950). To resolve these problems, the sterile insect technique (SIT) is currently being implemented in Okinawa and Kagoshima Prefectures to eradicate this sweet potato weevil, *C. formicarius* [18, 20].

Many insect species used as biological control agents (e.g., sterile insect technique [SIT] or as natural enemies), as pollinators, beekeeping and/or sericulture are reared under the artificial condition [e.g., 3]. The mass-rearing conditions of these insects are decidedly different from that in the natural environment [3]. For example, the diet is highly nutritious. Furthermore, the population density becomes extraordinarily high and the rearing temperature is kept stable in the mass-rearing facility because of the economic reasons [3, 23]. Therefore, the mass-rearing condition is very dissimilar to that found in nature. It is well known that not only the genetic components but also the environmental components, such as the nutrition or temperature before emergence, affect the development of morphological characteristics in coleopteran species [e.g., 7, 13]. Although elytral contour shape (or outline) in Coleoptera seem to bear a close resemblance, the artificial conditions before the emergence in the rearing facility might induce some of the differences in the elytral contour shape. Moreover, because it is necessary to distinguish wild and sterilized weevils with an easy, convenient, and cheap marking approach, one needs to be able to evaluate its effect in the SIT program. If the rearing environment before the emergence induces some variation in the elytral shape of the mass-rearing individuals, then the shape would then be available as phenotypic marker to discriminate between wild and reared weevils.

In the 2012 eradication program using the SIT for *C. formicarius*, about 205,000 weevils were released every week into the target areas. Of these released weevils, about 20,000 trapped weevils were evaluated every week at the Okinawa Prefectural Plant Protection Center (OPPPC) to monitor the progress of this program [32]. Although many eradication projects use a fluorescent powder dye as marker for sterile insects including *C. formicarius*, this monitoring technique is not fully appropriate for this weevil. For example, the disappearance or contamination of the dye has frequently occurred during the period of the eradication program. Thus, the weevil monitoring has been less accurate due to the uncertainty in the marking technique, which has hampered progress of the eradication program [18, 39].

C. formicarius shows three elytral color polymorphisms: bluish elytra (BE), greenish elytra (GE), and brownish piceous elytra (PE) [16, 34, and 43]. The distribution pattern of these three color morphs is geographically different in Japan [15]. For example, the PE weevils are only distributed in the southern Ryukyus at very low frequency while the GE color morph is dominant in the current target area, the northern Ryukyus, using the SIT eradication program [15]. Therefore, elytral color morph has been used as a visual marker together with a fluorescent powder dye to monitor the progress of the eradication program in Kume Is. for *C. formicarius* in Okinawa Prefecture since 2010 [32, 39]. This visual marker has facilitated the progress of this eradication program and the eradication of *C. formicarius* from this island was accomplished in 2012. Consequently, this is the first success in the world of the SIT program in Coleoptera. This eradication program demonstrates that the visual discrimination system using elytral color morph is successful for the sweet potato weevil *C. formicarius*.

However, the elytral color morph raises a new operational problem. Although the mass-reared PE strain has been established as a consequence of the selection for coloration at the OPPPC, the color morph is not genetically fixed. Therefore, a decision in regard to the trapped weevils that have an intermediate color morph (e.g., greenish piceous elytra) had to be deferred. The emergence frequency of the weevils having the intermediate color morph is one in 35,454 weevils (T. Matsuyama, unpubl. data). However, as noted earlier, the number of released weevils is huge [32, Table 1]. Although the weevils having an intermediate color are judged as wild to avoid misclassification in the early stage of the eradication program, this overestimation of the intermediate elytral color morph causes a delay in the progress of the final part of the eradication program. Therefore, what is needed is an additional criterion to correctly distinguish those sterilized weevils deemed uncertain from the trapped individuals.

Table 1. Overview of Samples Used in the Present Study, Derivation of the Weevil, Year, Month Sampling Site, Type of Trap Survey, and the Number of Released and Trapped Weevils

Weevil sampling					Number of		
Derivation	**Year**	**Month**	**Locality**	**Type of survey**	**Analyzed weevils**	**released PE weevils per month**	**Trapped weevils per month**
Wild	2011	May	Uruma Is.	Occurrence forecast	60	---	675
		October	Uruma Is.	Occurrence forecast	60	---	3,675
	2012	May	Uruma Is.	Occurrence forecast	60	---	238
		October	Uruma Is.	Occurrence forecast	60	---	203
	2011	May	Miyako Is.	Occurrence forecast	60	---	63
		October	Miyako Is.	Occurrence forecast	60	---	4,570
	2012	May	Miyako Is.	Occurrence forecast	60		392
		October	Miyako Is.	Occurrence forecast	60	---	9,465
Mass-reared (PE)	2011	May	Kume Is.	Sterile insect release	60	---	63,971
		October	Kume Is.	Sterile insect release	60	---	19,843
	2012	May	Kume Is.	Sterile insect release	60	---	92,923
		October	Kume Is.	Sterile insect release	60	---	74,821

The conditions before the emergence of this weevil are completely different between the mass-reared production facility and the field. For example, in the field the wild weevils use the stem or tuberous root of *Ipomoea* (e.g., sweet potato *I. batatas* and blue morning glory *I. indica*) and some weevils experience the winter before the field emergence. On the other hand, the mass-reared weevils are reared on the fresh sweet potato tuberous roots under the constant temperature with a high population density (25 ± 1; 2,000 adult weevils with ca. 1,800g diet per a container (29*36*12cm)) at the mass-rearing facility of the OPPPC [31]. However, some immature individuals exist in the stem or tuberous root of *Ipomoea* (e.g., sweet potato *I. batatas* and blue morning glory *I. indica*) even in the winter because this weevil has multiple generations per year in Okinawa (i.e. multivoltine). Thus, larval and pupal stages of this weevil are observed even in the winter season [36]. Therefore, the morphological differences between mass-reared and wild weevils may be highly variable especially after the winter season because the temperature in the laboratory is likely to vary substantially from that in the wild. These differences in environmental conditions might induce the morphological variation seen in the elytral shape of *C. formicarius*. In order to assess that possibility, the elytral shape between the wild and trapped mass-reared weevils were compared using geometric morphometric techniques. The feasibility of the discrimination system using the elytral shape as a diagnostic in the eradication program using the SIT for *C. formicarius* is then discussed. This chapter represents a continuation of research to elucidate various morphological characteristics of associated with the sweet potato weevil as well as other beetles [25, 26].

MATERIALS AND METHODS

The adult elytral shape of the mass-reared PE sweet potato weevil *C. formicarius* was compared to the wild one for the two target months, May and October, in 2011 and 2012. Since wild weevils were collected from two different localities (Uruma and Miyakojima Cities), two comparisons were performed independently (PE vs. Uruma and PE vs. Miyakojima). All the individuals, including mass-reared PE weevils, were collected by trap survey in the field using the synthesized sex pheromone for *C. formicarius* because the flight ability might to affect the elytral shape of this weevil. Since the sex pheromone attracts only male weevils, these comparisons were restricted to males.

PE Weevils

As stated earlier, the brownish piceous elytra (PE) is one of three major elytral color polymorphisms in *C. formicarius* [15, 16, 34, 39, and 43] and the weevils with PE are distributed in southern Ryukyus with very low density in Japan [15]. Since this color morph is available as a visual marker to discriminate between the mass-reared sterile males and the wild ones after the trap survey, the PE strain has been used in an eradication program for this weevil in Okinawa Prefecture since 2010 [39]. The OPPPC has continued rearing a number of PE weevil for the SIT program at the facility (25 ± 1°C and a photoperiod of L14:D10, light between 04:00 and 18:00 hours) after collecting adult founders from Yonaguni Town, Okinawa Island, Japan, in November 2007. These mass-reared PE weevils were released every week after sterilization on Kume Island (Kumejima town, 59.1km^2) from 2010 to 2013 because this island is the target area of the SIT program. In the eradication program, 36,725 to 165,000 mass-reared sterile PE weevils were released per month in the target area in the survey period (see Table 1). The derivation of the mass-reared PE weevils in the present study was the recaptured one in Kume Island (see below).

Sampling of Weevils

In the trap survey, a natural rubber ring (white, 9 mm in diameter) to which 0.1 mg of the synthetic sex pheromone (*Z*)-3-dodecen-l-ol (*E*)-2-butenoate was applied (synthesized by Shin-etsu Chemical Co., Tokyo, Japan) was used [15]. To attract the male weevils, a small piece of sterilizer plate (Panaplate®; vinyl chloride resin containing 16 percentage of DDVP) was used as an insecticide. Since the sampling manner and the handling of weevils after the trap survey differ slightly between the two sources of weevil (mass-raised and wild), the details of the handling procedures are described below.

Wild weevils

The OPPPC has conducted the extensive trap survey twice in every month for the occurrence forecast of *C. fo*rmicarius at several sites in the Ryukyus. The wild males trapped in Uruma (86.1 km^2) and Miyakojima (204.6 km^2) cities were used in the present study (Fig. 2). In this trap survey, ten and five traps are arranged by the OPPPC in the sweet potato cultivation area of Uruma and Miyakojima cities, respectively. The trapped wild weevils were collected, counted and stored at the stock room (25 ± 1°C) of the OPPPC after drying (60°C, 48 hours).

Mass-reared weevils

PE weevils were used as the specimens for the mass-reared weevil, which are also used as a visual marker strain in the eradication program for this weevil in Okinawa prefecture. The OPPPC arranged about 170 traps throughout the target area in Kume Island to monitor the progress of the eradication program from 2010 to 2013, and captured a large number of released males by the monitoring the pheromone traps. The trapped male weevils were brought back to the OPPPC and the elytral coloration checked to judge the origin of the male weevil (whether mass-reared sterile male or wild one) in the laboratory [39]. These trapped weevils were also stored at the OPPPC as in the same manner of the wild weevils (see above). The detail sampling data is shown in Table 1. Morphometric analysis was restricted in the trapped mass-reared PE males as judged by the visual survey based on the elytral color morph.

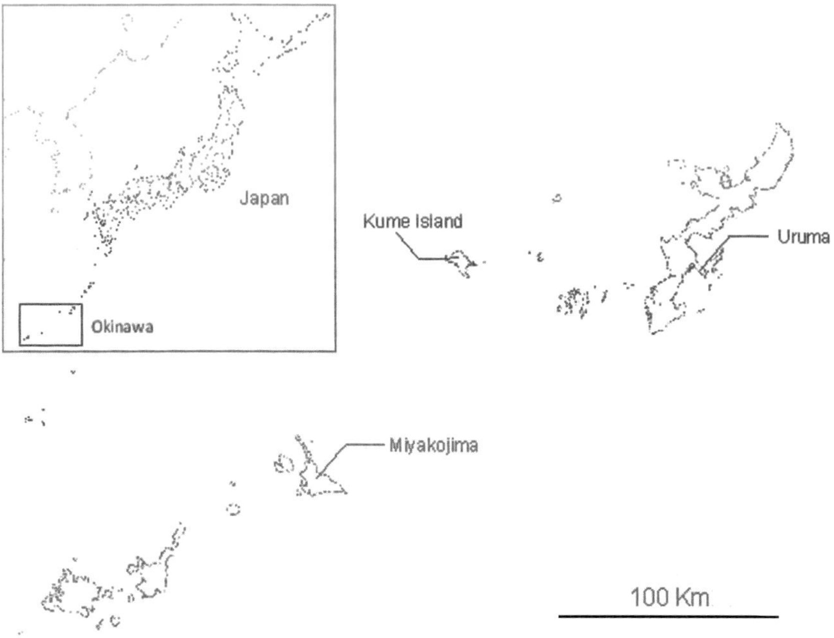

Fig. 2. Map of sampling sites for the sweet potato weevil *C. formicarius*. Wild weevils were collected in Uruma and Miyakojima cities and mass-reared brownish piceous elytral PE) weevils were collected in Kumejima town.

Imaging and Comparison of the Elytral Shape

A sample of 720 male weevils was randomly chosen as specimens from the stock of the OPPPC (Table 1). Since the contour shape was used in the image analysis, the males with injured elytra were excluded. To image the elytral shape of the weevils, the right-side elytron was peeled using tweezers and put on clear mounting tape (Scotch removable clear mounting squares, Scotch®, USA) and placed on glass slides. In order to take digital images of the lateral view of elytron, the middle line of the right-side elytron was bonded to the mounting tape (Fig.1b). Fifteen elytra were arranged on a mounting tape (17 by 17 mm). The subsequent digital image was saved in bitmap format for each sample using a digital microscope eyepiece/USB camera (AM4023X Dino-Eye, AnMo Electronics Corporation, Taiwan) with a stereo-microscope (Leica MZ6, Germany) and DinoXcope software (Duno-Lite™, AnMo Electronics Corporation, Taiwan) using a computer with the Windows 7 operating system (O/S). To obtain high contrast digital images, the transmitted light function of the stereo-microscope was used. Each digital image was stored at 1280 x 1024 pixels with a color depth of 24-bit per pixel. Before image analysis, the GNU Image Manipulation Program (GIMP) software package (v. 2.8.2) on Mac O/S X (ver. 10.9.4) was used to remove noise in the images as appropriate. For example, the basal portion of elytron or small size dust particles around the elytron in the images were removed by hand with either pen or brush tool under screen magnification.

Elliptic Fourier Analysis

Elytral shape diversity in the sweet potato weevil, *C. formicarius* was evaluated using elliptic Fourier analysis [21]. Outlines were drawn from digital images and these represented a two dimensional projection of the lateral shape of the elytra. A series of the calculations of standardized Elliptic Fourier approximation were carried out using the SHAPE software [11]. The EFDs are useful for the delineation of any kind of biological form and have been applied to the evaluation of various biological shapes [11, 25, and 26]. This method is based on the separate Fourier decompositions of the incremental changes of the x- and y-coordinates, as a function of the cumulative length along the outline. The Chain Corder module (ChainCorder.exe) of the SHAPE software extracts the contours of the objects from digital images and describes the relevant information as a chain code. The Chc2Nef module (Chc2Nef.exe) provides the normalized EFDs from the chain-coded contour. These normalized coefficients are calculated with the discrete Fourier transformation following the procedures of Kuhl and Giardina [21] and are invariant with respect to size, rotation and starting point. In this study, the lateral contour shape of the elytron is described for *C. formicarius* using the first twenty harmonics. This procedure generates 77 nontrivial normalized coefficients, which are then available for subsequent analyses.

Multivariate Analysis

The data analysis was conducted with both the SHAPE software [11] on a Windows 7 O/S computer and the R statistical analysis package version 3.0.2[35] on Mac O/S X machine as appropriate. In order to summarize the information on the variation of the normalized EDF coefficients, a principal components analysis (PCA) using the PrinComp module (PrinComp.exe) of the SHAPE software [37] was carried out. The Principal components (PCs) were analyzed using a multivariate analysis of variance (MANOVA). A canonical discriminant analysis [2] was then conducted in addition to the MANOVA comparison. The accuracy of the canonical discriminant analysis was evaluated by the leave-one-out cross validation procedure using the R software. In these analyses, the data of each season for the two years was not pooled because the elytral morphology of mass-reared weevils might be seasonally variable due to the quality of sweet potato roots diet even when the weevils are reared under a stable environment in the mass-rearing facility. On the other hand, since the PCs were analyzed to explain the variation of the elytral shape between the mass-reared males and wild ones in each season, the data was pooled for each season. In this analysis, the generalized linear mixed models (GLMM) with a Gaussian error distribution, was used. The origin of the weevils was considered as an independent fixed effect, and the sampling year was treated as a random effect as appropriate (using the glmer function in the lme4 package). The elytral shape between mass-reared males and wild ones (PE vs. Uruma and PE vs. Miyakojima) was then compared.

RESULTS

Comparison of the Elytral Shape between Mass-reared and Wild Weevils

A sample of 720 male weevils that were collected in the field (480 wild weevils and 240 mass-reared PE weevils) were analyzed. The results of the PCA based on the EFDs are shown in Table 2.

Table 2. A Summary of the Results of the Principal Component Analysis for Each Population is Displayed Here

Principal component	Uruma			Miyako		
	Eigenvalue	Proportion (%)	Cumulative (%)	Eigenvalue	Proportion (%)	Cumulative (%)
PC1	1.02E-04	44.4	44.4	8.97E-05	39.6	39.6
PC2	3.73E-05	16.2	60.6	4.64E-05	20.5	60.2
PC3	3.07E-05	13.4	73.9	3.15E-05	13.9	74.1
PC4	1.24E-05	5.4	79.3	1.18E-05	5.2	79.3
PC5	9.19E-06	4.0	83.3	9.90E-06	4.4	83.7
PC6	7.37E-06	3.2	86.5	7.65E-06	3.4	87.1
PC7	4.80E-06	2.1	88.6	4.47E-06	2.0	89.1
PC8	4.36E-06	1.9	90.5	4.08E-06	1.8	90.9

The contributions of the first, second and third principal components were 44.4%, 16.2% and 13.4% for the PE from Uruma and 39.6%, 20.5% and 13.9% for the PE from Miyakojima, respectively. The first five principal components cumulatively explained 83.3% and 83.7% of the total variance, respectively. The estimated elytral shapes obtained by redrawing the contour with the principal components are shown in Figs. 3a and 3b. The elytral shape of the sweet potato weevil, *C. formicarius* differed significantly between the mass-reared males and wild ones for each comparison using the MANOVA (Table 3). The identification rate was more than 80 percent (Table 3). PC1 was related to the length to height ratio and the PC2 and 3 reflect the front and rear upper roundness of the elytron. The mean values of PC1 and PC2 for the mass-reared weevil were consistently smaller than that of wild weevils. These results show that the elytral contour shape of mass-reared weevil displays more flatness compared to that of wild weevils. Although the variance of the ratio of the length to height, which is summarized by the PC1, was remarkable, the variance of the front upper roundness of the elytron which is summarized by the PC2, was not large.

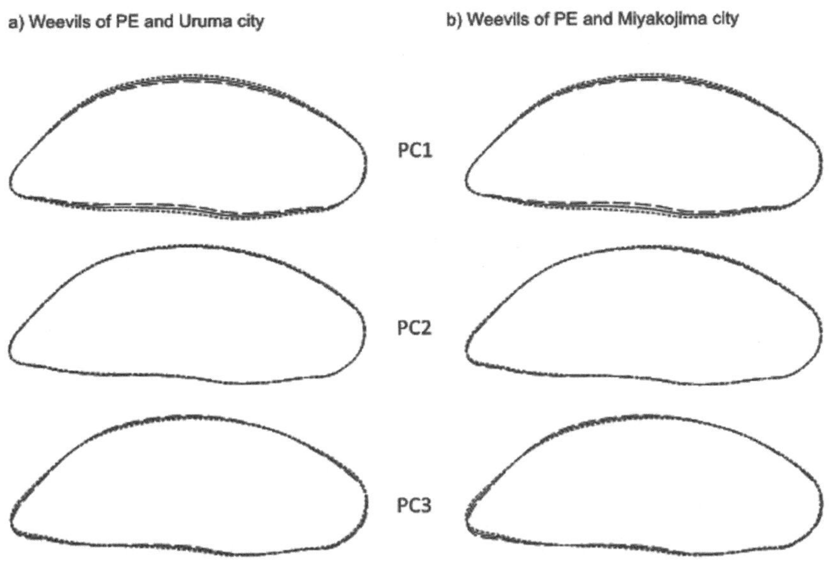

Fig. 3. The contour shape of the elytra in mass-reared PE and wild weevils drawn using the elliptic Fourier descriptors estimated under three principal component scores (PC). The contour shape of elytra was drawn using the elliptic Fourier descriptors for the datasets of the (a) mass-reared PE and wild Uruma weevils and (b) mass-reared PE and wild Miyakojima weevils, respectively are shown here. Lines show +2SD (dotted line), mean (solid line) and −2SD (dashed line).

Table 3. Summary of the Multivariate Analysis of Variance (MANOVA) Tests for the Comparison of Elytral Shape between Wild and Mass-reared PE Weevils (Uruma or Miyakojima) for each season is illustrated here

Comparison between mass-reared PE and wild weevils					
Compared wild weevil			MANOVA		Results of leave-one out cross-validation (%)
Locality	Year	Month	Wilk's Λ	P	
PE vs. Uruma weevils					
Uruma	2011	May	0.0528	P<0.001	94.17
		October	0.0764	P<0.001	83.33
	2012	May	0.0909	P<0.001	81.67
		October	0.0692	P<0.001	84.75
PE vs. Miyakojima weevils					
Miyako	2011	May	0.0452	P<0.001	95.00
		October	0.0520	P<0.001	90.00
	2012	May	0.0401	P<0.001	95.00
		October	0.0528	P<0.001	91.67

The difference of partial elytral morphology between the populations was constant throughout the season. The length to height ratio described in PC1 differed significantly between wild and mass-reared populations except for one of three comparisons (Table 4). On the other hand, the roundness of the elytron that was described in PC2 and PC3 differed significantly between wild and mass-reared populations in each comparison (Table 4).

Table 4. The Results of the Generalized Linear Mixed Model (GLMM) for the Analysis of the Elytral Shape using the Principal Component Score are shown here

Comparison	Month	PC1		PC2		PC3	
		χ^2	P	χ^2	P	χ^2	P
PE vs. Uruma	May	10.418	<0.002	34.166	<0.001	79.113	<0.001
	October	0.2817	ns	84.021	<0.001	60.619	<0.001
PE vs. Miyako	May	36.106	<0.001	46.421	<0.001	19.791	<0.001
	October	83.227	<0.001	67.767	<0.001	7.7023	<0.006

DISCUSSION AND CONCLUSIONS

In this study, the elytral contour shape using by the EFDs of the trapped males in May and October was compared and one is able to detect the morphological differences between the mass-reared and wild male weevil *C. formicarius*. The statistical difference that was demonstrated by the MANOVA and the classification performance using the discrimination analysis was significant for each comparison during the two years of observation. Although the length to height ratio described with PC1 was not able to show the differentiation of elytral shape in one of four cases. However, the roundness of the elytron that was summarized with PC2 and PC3 was significantly different from that of the wild one. These findings suggest that the elytral contour shape analysis using the EFDs in spring and autumn can be potentially useful for identifying the source of the trapped weevils in the eradication program for *C. formicarius*.

The difficulty of this analysis in the eradication program for this weevil is the investment of time before the imaging process. Since the eradication program needs to classify a large number of trapped insects after the extensive trap survey is regularly carried out, one would like to avoid the time-consuming tasks as much as possible. For example, about 20,000 *C. formicarius* males are checked per week at the OPPPC in the eradication program [32 and Table 1]. Although one is not able to conduct a complete census for the trapped weevils using the contour shape analysis, preliminary screening based on the elytral coloration of *C. formicarius,* it is possible to reduce the number of the weevils to be re-analyzed before applying the morphometric analysis. Since the emergence frequency of the weevil with the intermediate elytral color is extremely low in *C. formicarius*, the number of specimens is very limited. The morphometric analysis using the elytral shape would be expected to be a promising tool to increase discrimination accuracy, which enables the promotion of a smoothly run eradication program for *C. formicarius*.

Although this study investigated 720 male weevils (480 wild weevils and 240 mass-reared PE weevils), one was not able to detect weevils having intermediate color morph among these specimens. The survey to date for the emergence of the intermediate elytral color morph in the PE strain demonstrated that the frequency of the weevils with greenish piceous elytra is extremely low (1 of 35,454 male weevils). Therefore, the number of specimens analyzed in the present study would not be enough to detect the target weevil. Furthermore, the seasonal variation of the elytral color morph has not been studied in this weevil. In a future study, one need to compare the elytral shape between the weevils displaying an intermediate color morph and those that have a clearly defined color morph for the brownish piceous elytron using larger-scale surveys throughout the season.

ACKNOWLEDGEMENTS

We thank the members of OPPPC and Ryukyu Sankei Co. Ltd. for their assistance. This study was supported in part by Grant-in-Aid for young scientist from University of the Ryukyus and KAKENHI No. 26450067 to NK.

REFERENCES

[1] Akazawa T, Uritani I, and Kubota H (1960) Isolation of ipomeamarone and two coumarin derivatives from sweet potato roots injured by the weevil, *Cylas formicarius elegantulus*. *Arc Biochem Biophys* **88**: 150-156.

[2] Blackith R and Reyment R (1971) *Multivariate Morphometrics*. Academic Press. New York. USA.

[3] Calkins CO and Parker AG (2005) Sterile insect quality. **In**: VA Dyck, J Hendrichs and AS Robinson (Eds.) *Sterile insect technique: principles and practice in area-wide integrated pest management*. Springer, Dordrecht, The Netherlands.

[4] Chalfant KL, Jansson RK, Seal DR, and Schalk JM (1990) Ecology and management of sweet potato insects. *Ann Rev Entomol* **35**: 157-180.

[5] Eberhard WG (2009) Static allometry and animal genitalia. *Evolution* **63**: 48-66.

[6] Emlen D (1996) Artificial selection on horn length-body size allometry in the horned beetle Onthophagus acuminatus (Coleoptera: *scarabaeidae*). *Evolution* **50**: 1219-1230.

[7] Gotoh H, Cornette R, Koshikawa S, Okada Y, Lavine LC, Emlen DJ, and Miura T. (2011) Juvenile hormone regulates extreme mandible growth in male stag beetles. *PLoS ONE* **6**: e21139.

[8] Hamilton K. 1972. The insect wing, Part III. Venation of the orders. *J Kansas Entomol Soc* **45**: 145-162.

[9] Heath RR, Coffelt JA, Sonnett RE, Proshold, FI, Dueben B, and Tumlinson JH (1986) Identification of sex pheromone produced by female sweet potato weevil, *Cylas formicarius elegantulus* (Summers). *J Chem Ecol* **12**: 1489-1503.

[10] Hodek I, Honek A, and van Emden HF (2012) *Ecology and Behaviour of the Ladybird Beetles (Coccinellidae).* John Wiley and Sons, Oxford, UK.

[11] Iwata H and Ukai Y (2002) SHAPE: a computer program package for quantitative evaluation of biological shapes based on elliptic Fourier descriptors. *J Heredity*, **93**: 384-385.

[12] Jansson RK and Raman KV (1991) Sweet potato pest management: a global overview. **In**: RK Jansson, KV Raman (Eds.), *Sweet potato pest management - a global perspective.* Westview Press, Boulder, Colorado, USA.

[13] Karino K, Seki N, and Chiba M (2004) Larval nutritional environment determines adult size in Japanese horned beetles *Allomyrina dichotoma. Ecol Res* **19**: 663-668.

[14] Kawakami Y, Yamazaki K, Ohashi K. 2013. Geographical variations of elytral color polymorphism in *Cheilomenes sexmaculata* (Fabricius) (Coleoptera: Coccinellidae). *Entomol Sci* **16**: 235-242.

[15]Kawamura K, Ohno S, Haraguchi D, Kawashima S, and Kohama T (2009) Geographic variation of elytral color in the sweet potato weevil, *Cylas formicarius* (Fabricius) (Coleoptera: Brentidae), in Japan. *Appl Entomol Zool* **44**: 505-513.

[16] Kissinger DG (1968) *Curculionidae Subfamily Apioninae of North and Central America: With Reviews of the World Genera of Apioninae and World Subgenera of Apion Herbst (Coleoptera).* Taxonomic Publications, Massachusetts, USA.

[17] Knell RJ, Pomfret JC, and Tomkins JL (2004) The limits of elaboration: curved allometries reveal the constraints on mandible size in stag beetles. *Proc Biol Sci R Soc* **271**: 523-528.

[18] Kohama T and Kuba H. (2008) The eradication program of the sweet potato weevil, *Cylas formicarius* (Fabicius) (Coleoptera: Brentidae), using both of sex-pheromone and sterile insect technique in Kume Island. **In**: Y Itô (Ed.) *Sterile insect technique: eradication technique for invasive species.* Kaiyusha, Tokyo, Japan.

[19] Konuma J, Sota T, and Chiba S (2013) Quantitative genetic analysis of subspecific differences in body shape in the snail-feeding carabid beetle *Damaster blaptoides*. *Heredity* **110**: 86-93.

[20] Kuba H, Kohama T, and Haraguchi D (2003) Eradication projects of exotic sweet potato weevils using SIT in Okinawa. **In**: M Oka, M Matsui, T Shiomi, Y Ogawa and K Tsuchiya (Eds.) *Proceedings of the NIAES-FFTC Joint International Seminar on Biological Invasions: Environmental Impacts and the Development of a Database for the Asian-Pacific Region*. National Institute for Agro-Environmental Sciences and Food and Fertilizer Technology Center for the Asian and Pacific Region, Tsukuba, Japan.

[21] Kuhl FP and Giardina CR (1982) Elliptic Fourier features of a closed contour. *Comput Graph Image Proc* **18**: 236-258.

[22] Kukalová-Peck J and Lawrence JF (1993) Evolution of the hind wing in Coleoptera. *Canada Entomol* **125**: 181-258.

[23] Kuriwada T, Kumano N, Shiromoto K, and Haraguchi D (2009) High population density and egg cannibalism reduces the efficiency of mass-rearing in *Euscepes postfasciatus* (Coleoptera: Curculionidae). *Florida Entomol* **92**: 221-228.

[24] Leschen RAB and Beutel RG (2014) *Coleoptera, Beetles: Morphology and Systematics (Phytophaga)*. Walter De Gruyter Inc, Berlin, Germany.

[25] Tatsuta H, Iwata H, and Goka K (2011) **In**: Lestrel PE (Ed.) *Biological Shape Analysis: proceedings of the 1st International Symposium. Proc 1st Int Symp.* World Scientific Pub., Singapore.

[26] Kumano N, Kuriwada T, Shiromoto K and Tatsuta H (2013) **In**: Lestrel PE (Ed.) *Biological Shape Analysis: Proceedings of the 2nd International Symposium. Proc 2nd Int Symp.* World Scientific Pub., Singapore.

[27] McCullough EL, Weingarden PR, and Emlen DJ (2012) Costs of elaborate weapons in a rhinoceros beetle: How difficult is it to fly with a big horn? *Behav Ecol* **23**: 1042-1048.

[28] Moczek AP (1998) Horn polyphenism in the beetle *Onthophagus taurus*: larval diet quality and plasticity in parental investment determine adult body size and male horn morphology. *Behav Ecol* **9**: 636-641.

[29] Moriya S. 1997. Is it possible to eradicate the two weevil pests of sweet potato [Ipomoea batatas], *Cylas formicarius* and *Euscepes postfasciatus* from Japan? *Bull Okinawa Agric Exp Station* **18**: 19-27.

[30] Nawa U (1903) A note of the sweet potato weevil. *Konchu-Sekai*, **7**: 327-330.

[31] Okinawa Prefectural Government, Department of Agriculture, Forestry and Fisheries. 2000. Ann Rep on Eradication Program Destructive Pest Okinawa 1998. Okinawa Prefectural Government, Department of Agriculture, Forestry and Fisheries, Okinawa, JPN.

[32] Okinawa Prefectural Government, Department of Agriculture, Forestry and Fisheries. 2013. Ann Rep on Eradication Program Destructive Pest Okinawa 2012. Okinawa: Okinawa Prefectural Government, Department of Agriculture, Forestry and Fisheries, Okinawa, JPN.

[33] Osawa N and Nishida T (1992) Seasonal variation in elytral color polymorphism in *Harmonia axyridis* (the ladybird beetle): the role of non-random mating. *Heredity* **40**: 395-407.

[34] Pierce WD (1918) Weevils which affect Irish Potato, Sweet Potato and Yam. *J Agric Res* **12**: 601-611.

[35] R Development Core Team (2008) R: a language and environment for statistical computing. *R Foundation for Statistical Computing*. Vienna, Austria, http://www.r-project.org.

[36] Raman KV and Alleyne EH (1991) Biology and management of the West Indian sweet potato weevil, *Euscepes postfasciatus*. **In**: RK Jansson and KV Raman (Eds.) *Sweet Potato Pest Management*. Westview Press, Boulder, Colorado, USA.

[37] Rohlf J and Archie J (1984) A comparison of Fourier methods for the description of wing shape in mosquitoes (Diptera: Culicidae). *Syst Zool* **33**: 302-317.

[38] Sato K, Uratani I, and Saito T (1981) Characterization of the terpene-inducing factor isolated from the larvae of the sweet potato weevil, *Cylas formicarius* Fabricius (Coreoptera: Brenthidae). *Appl Entomol Zool* **16**: 103-112.

[39] Shiromoto K, Kumano N, Kuriwada T, and Haraguchi D (2011) Is elytral color polymorphism in sweet potato weevil (Coleoptera: Brentidae) a visible marker for sterile insect technique?: comparison of male mating behavior. *J Econ Entomol* **104**: 420-424.

[40] Soto IM, Carreira VP, Fanara JJ, and Hasson E (2007) Evolution of male genitalia: environmental and genetic factors affect genital morphology in two Drosophila sibling species and their hybrids, *BMC Evol Biol* **7**: 77.

[41] Soto IM, Carreira VP, Soto EM, Márquez F, Lipko P, and Hasson E (2013) Rapid divergent evolution of male genitalia among populations of *Drosophila buzzatii*. *Evol Biol* **40**: 395-407.

[42] Sutherland JA (1986) A review of the biology and control of the sweet potato weevil *Cylas formicarius* (Fab.). *Trop Pest Manage* **32**: 304-315.

[43] Uritani I, Saito T, Honda H, and Kim WK (1975) Induction of furano-terpenoids in sweet potato roots by the larval components of the sweet potato weevils. *Agric Biol Chem* **37**: 1857-1862.

[44] Wolfe GW (1991) The origin and dispersal of the pest species of Cylas with a key to the pest species groups of the world. **In**: RK Jansson and KV Raman (Eds.) *Sweet Potato Pest Management: A Global Perspective*. Westview Press, Boulder, Colorado, USA.

[45] Yasuda K (1993) Seasonal prevalence of the sweet potato weevil *Cylas formicarius* and the West Indian sweet potato weevil *Euscepes postfasciatus* in Okinawa, Japan. *Proc Assoc Pl Prot Kyushu* **39**: 88-90.

[46] Yasuda K and Kohama T (1990) Distribution of the sweet potato weevil, *Cylas formicarius* (Fabricius) and the West Indian sweet potato weevil, *Euscepes postfasciatus* (Fairmire) in Okinawa prefecture. *Proc Assoc Pl Prot Kyushu* **36**: 123-125.

Genome-wide Association Study of Biological Shape Based on Elliptic Fourier Analysis: A Case Study in Rice Grain Shape Variation

HIROYOSHI IWATA[1], KAWORU EBANA[2], YUSAKU UGA[3] and TAKESHI HAYASHI[4]

Abstract

Cultivars of a crop species sometimes show large infraspecific morphological variability in the organs or parts of a plant of interest to humans. The Genome-wide association study (GWAS) based on elliptic Fourier analysis is suited for the genetic dissection of infraspecific shape variation. In this chapter, a series of analyses are described for GWAS of biological shape based on elliptic Fourier analysis, using rice grain shape data as an example. Genotype data of 3,183 SNPs was used and grain shape data of 77 elliptic Fourier descriptors collected for 179 accessions from a world rice collection. Scores of the first three principal components of elliptic Fourier descriptors were used as grain shape phenotypes. In the rice accession, there was a subpopulation structure, among which the grain shape was differentiated. In GWAS, a Bayesian regression model was employed to estimate subpopulation structure effects and effects of all genome-wide SNPs simultaneously. In the first principal component, which accounted for the length-to-width ratio of the grain, two strong associations were detected on chromosome 3 and 5 at the locations that were close to known grain shape genes, *GS3* and *qSW5*. In the second and third principal components, each of which explained minor (less than 2%) shape variation, relatively strong associations were found at several SNPs. These SNPs may be closely linked to genes controlling grain shape variation. The rice dataset and R scripts used in the analysis described in this chapter are available at http://lbm.ab.a.u-tokyo.ac.jp/~iwata/gwashape/.

INTRODUCTION

In a crop species, large infraspecific morphological variability is sometimes found in organs or parts of a plant of human interest [1]. For example, a vast range of fruit shape, size, and color was found in peppers, in contrast with the poor variation of the fruits of its wild relatives [1]. Cultivated tomatoes vary tremendously for fruit size and shape but have little genetic variation elsewhere in their genome, whilst its wild relatives contain tremendous genetic variation in their genomes but show little variation in fruit size and shape [2]. Large morphological variation in organs or parts of a plant of crop species can enhance the versatility of the crop and the wide adaptation to human preference. For example, a long, slender grain of rice is generally preferred by consumers in Southern China, the USA, and South and Southeast Asian countries, whereas consumers in Japan, Korea, and Northern China prefer a rice grain that is short and round [3]. Genetic dissection of infraspecific morphological variability has been conducted in some model crop species [2, 3].

[1] Department of Agricultural and Environmental Biology, Graduate School of Agricultural and Life Sciences, University of Tokyo, Bunkyo, Tokyo, Japan, iwata@ut-biomet.org.
[2] Genetic Resources Center, National Institute of Agrobiological Sciences, Tsukuba, Ibaraki, Japan.
[3] Agronomics Research Center, National Institute of Agrobiological Sciences, Tsukuba, Ibaraki, Japan.
[4] Agroinformatics Division, National Agricultural Research Center, National Agriculture and Food Research Organization, Tsukuba, Ibaraki, Japan.

Genes or loci controlling shape variation, however, are still unknown in most crop species, despite the importance of shape variation in plant breeding.

The genome-wide association study (GWAS) is a promising method for detecting genes responsible for phenotypic variation observed in a population of unrelated individuals. GWAS is a well-established approach for the genetic dissection of human disease. In crop species, GWAS is coming into use in the genomic dissection of complex agronomical traits. Current advancement in DNA sequencing technologies allows one to use a large number of genome-wide SNPs not only in model organisms but also in a wide range of plant species. Two especially novel genotyping methods called restriction-site associated DNA sequencing (RAD-Seq) and Genotyping-By-Sequencing (GBS), both of which use the next generation sequencing (NGS) technologies, dramatically increase throughput and decrease cost for genotyping genome-wide SNP markers [4]. Because cost and time required for genotyping genome-wide SNPs are getting lower, GWAS is becoming one of the most feasible methods for genetic dissection of complex traits. GWAS has several advantages over bi-parental QTL mapping. The most important advantages are (1) GWAS does not require line-crossing experiments and can use samples from previously well-studied populations, and (2) GWAS provides higher resolution, often to the gene level [5]. Thus, GWAS is ideally suited for the genetic dissection of phenotypic variation observed in accessions of a germplasm collection [6].

By using elliptic Fourier analysis [7], GWAS of biological shape has been conducted in rice grain shape [8] and grape leaf shape [9] in order to investigate the loci and genes that were responsible for infraspecific shape variation. Because elliptic Fourier analysis has the potential to evaluate various types of biological shapes [10, 11], the method is expected to be applicable to the genetic dissection of infraspecific shape variation in a range of plant and animal species. The method, however, has not been widely used in the genetic research community. One of the reasons for this is that the method requires various types of analyses including image analysis, Fourier analysis, genome data analysis, and statistical genetic analysis. In order to make the method much easier to use, the series of analyses required for the method needs be explained in some detail.

In this chapter, the method for GWAS of biological shape, using a rice dataset that has been analyzed in [12] is described. The software SHAPE [13] and R [14] enables one to perform a series of analyses required for the GWAS of biological shape based on elliptic Fourier analysis. The rice dataset and R scripts used in the analysis described in this chapter are freely available from the senior author at http://lbm.ab.a.u-tokyo.ac.jp/~iwata/gwashape/.

MATERIALS AND METHODS

The Analyzed Dataset

In this chapter, one of three datasets that had been utilized for accessing the potential of genomic prediction of biological shape [12] is used. The dataset included 179 of 332 representative accessions [15] in the rice germ plasm collection at the National Institute of Agrobiological Science Genebank, Tsukuba, Japan. The 179 accessions had been genotyped with 3,753 SNPs and imputed (i.e., replacement with substitutions) for missing and untyped SNPs as described in [12]. The imputation step was repeated 100 times, and the mean imputation was obtained by averaging imputed genotypes over the 100 replications. Two homozygous non-missing genotypes were scored as 0 or 2 while missing genotypes were imputed with continuous scores ranging from 0 to 2. Excluding markers with minor allele frequency (MAF) less than 0.05, one had 3,138 SNPs in the following analysis.

As quantitative measurements of brown rice grain shape of the 179 accessions, the elliptic Fourier descriptor data that had been analyzed in [8] and [12] was used. In Iwata *et al.* [8], digital images were taken of six brown rice grains for each accession using a digital camera (EOS10D; Canon Inc., Japan) at ca. 0.002 mm/pixel resolution. By using the software SHAPE [13], the contours of rice grains from the digital images were extracted, and were then characterized with elliptic Fourier descriptors as described below. This software, identified as SHAPE, is freely available to all interested users at http://lbm.ab.a.u-tokyo.ac.jp/~iwata/shape/.

Quantitative Delineation of Rice Grain Shape with Elliptic Fourier Descriptors

Rice grain contours were extracted using digital image analysis of the rice grain images. Each contour was represented as a sequence of x and y coordinates of ordered points that were measured contour-clockwise from an arbitrary starting point. Assuming that the contour between the $(i-1)$th and the ith points is linearly interpolated, the length of the contour from the starting point to the pth point is:

$$t_p = \sum_{i=1}^{p} \Delta t_i \quad (1)$$

where Δt_i is the distance between the $(i-1)$th and the ith points. The perimeter of the contour can be represented as $T = t_P$, where P is the number of ordered points. Notice that the Pth point is equivalent to the starting point. The x and y coordinates of the pth point are:

$$x_p = \sum_{i=1}^{p} \Delta x_i \quad (2)$$

and

$$y_p = \sum_{i=1}^{p} \Delta y_i \quad (3)$$

where Δx_i and Δy_i are the distance along the x and y axes between the $(i-1)$th and the ith points. Thus, the coordinates on the contour can be approximated with Fourier series as:

$$x_p \approx x_{Np} = A_0 + \sum_{n=1}^{N} \left(a_n \cos \frac{2n\pi t_p}{T} + b_n \sin \frac{2n\pi t_p}{T} \right) \quad (4)$$

and

$$y_p \approx y_{Np} = C_0 + \sum_{n=1}^{N} \left(c_n \cos \frac{2n\pi t_p}{T} + d_n \sin \frac{2n\pi t_p}{T} \right) \tag{5}$$

where the elliptic Fourier coefficients of the nth harmonic, a_n, b_n, c_n, and d_n, are given as:

$$a_n = \frac{T}{2n^2\pi^2} \sum_{p=1}^{P} \frac{\Delta x_p}{\Delta t_p} \left(\cos \frac{2n\pi t_p}{T} - \cos \frac{2n\pi t_{p-1}}{T} \right), \tag{6}$$

$$b_n = \frac{T}{2n^2\pi^2} \sum_{p=1}^{P} \frac{\Delta x_p}{\Delta t_p} \left(\sin \frac{2n\pi t_p}{T} - \sin \frac{2n\pi t_{p-1}}{T} \right), \tag{7}$$

$$c_n = \frac{T}{2n^2\pi^2} \sum_{p=1}^{P} \frac{\Delta y_p}{\Delta t_p} \left(\cos \frac{2n\pi t_p}{T} - \cos \frac{2n\pi t_{p-1}}{T} \right), \tag{8}$$

and

$$d_n = \frac{T}{2n^2\pi^2} \sum_{p=1}^{P} \frac{\Delta y_p}{\Delta t_p} \left(\sin \frac{2n\pi t_p}{T} - \sin \frac{2n\pi t_{p-1}}{T} \right) \tag{9}$$

In the present study, the contour coordinates were approximated by the Fourier series using the first 20 harmonics (i.e., $N = 20$).

Because the coefficients of elliptic Fourier descriptors can vary depending on the size, rotation of a contour and the starting position of the contour trace, the coefficients were standardized to avoid variation related to these factors according to the size and direction of the long axis of the first harmonic ellipse. Refer to [7] for this procedural step. After this standardization, three Fourier coefficients became constant ($a_1 = 1, b_1 = 0, c_1 = 0$), and the remaining coefficients could be used as "descriptors" of shape. Hereafter the vector notation was used:

$$\mathbf{f} = (\mathbf{a}_{-1}^{\mathrm{T}}, \mathbf{b}_{-1}^{\mathrm{T}}, \mathbf{c}_{-1}^{\mathrm{T}}, \mathbf{d}^{\mathrm{T}})^{\mathrm{T}} = (a_2, \ldots, a_N, b_2, \ldots, b_N, c_2, \ldots, c_N, d_1, \ldots, d_N)^{\mathrm{T}} \tag{10}$$

for denoting the standardized elliptic Fourier descriptors, where the subscript "−1" represents the exclusion of the first harmonic coefficients, a_1, b_1, and c_1. When $N = 20$ as set in the present study, the dimensionality of the vector $\mathbf{f}$ was 77. The average of the standardized elliptic Fourier descriptors over six grains was then calculated for each accession.

Because of the high dimensionality of elliptic Fourier descriptors, it is often difficult to relate the variation of each coefficient to the morphological variation in the entire contour of an object. In the present study, principal component analysis was performed to summarize the information included in the 77 coefficients and to reduce the data to a lower number (dimensions) of principal components. The principal component analysis was conducted based on the among-accession covariance matrix of $\mathbf{f}$. Denoting the average of standardized elliptic Fourier descriptors of *i*th accession as $\mathbf{f}_i$, the principal component scores of elliptic Fourier descriptors of *i*th accession can be calculated as:

$$\mathbf{v}_i = \mathbf{\Phi}\left(\mathbf{f}_i - \bar{\mathbf{f}}\right) \tag{11}$$

where $\mathbf{v}_i$ is a vector of length $(4N-3)$ whose *j*th element, v_{ij} is the *j*th principal component score of the *i*th accession, $\mathbf{\Phi}$ is a $(4N-3) \times (4N-3)$ matrix whose *k*th row vector is the *j*th eigenvector of the variance and covariance matrix of the standardized elliptic Fourier descriptors, and $\bar{\mathbf{f}}$ is the grand mean of $\mathbf{f}_i$, i.e., $\sum_{i=1}^{I} \mathbf{f}_i$. Here, it is assumed that the number of accessions is larger than $4N-3$ and the among-accession covariance matrix is of full rank. When the first $j$principal components explain most of the variation of the standardized elliptic Fourier descriptors, the contour shape variations is well-approximated by the $j$components and thus the contour shape of the ith grain is well-described by the first $j$elements of principal component score vector $\mathbf{v}_i$.

Using Eq. 11, one can derive the equation for calculating elliptic Fourier descriptors that correspond to a particular principal component score vector $\mathbf{v}^*$. Because eigenvectors are mutually orthogonal to each other and are normalized to have unit length, $\mathbf{\Phi\Phi}^{\mathrm{T}} = \mathbf{I}$. From this equation, one gets $\mathbf{\Phi}^{\mathrm{T}} = \mathbf{\Phi}^{-1}$. Multiplying by $\mathbf{\Phi}^{\mathrm{T}}$on the left of the both sides of Eq. 11 and rearranging the equation, one gets:

$$\mathbf{f}^* = \bar{\mathbf{f}} + \mathbf{\Phi}^{\mathrm{T}}\mathbf{v}^* \tag{12}$$

Using Eq. 12, one can calculate $\mathbf{f}^*$corresponding to a particular point $\mathbf{v}^*$in the $(4N - 3)$ dimensional principal component space, and reconstruct a contour via the inverse Fourier transformation with the coefficients $\mathbf{f}^*$. This mechanism can be used for the visualization of shape variation accounted for by principal components. To visualize shape variation accounted for by the *j*th principal component, the *j*th element of $\mathbf{v}^*$, v_j^*,is set to plus or minus *s* times of the square root of *j*th eigenvalue, and the remaining elements to zero. For example, to visualize shape variation explained by the first principal component, one can set:

$$\mathbf{v}^* = \left(\pm s\sqrt{\lambda_1}, 0, \ldots, 0\right)^{\mathrm{T}} \tag{13}$$

where λ_1 is the eigenvalue of the first principal component. In the present study, $s = 2$ was set. Based on the coefficients $\mathbf{f}^*$corresponding to the $\mathbf{v}^*$, the *x* and *y* coordinates of the contour can be calculated as:

$$x(t) = \sum_{n=1}^{N} \left(a_n^* \cos \frac{2n\pi t}{T} + b_n^* \sin \frac{2n\pi t}{T} \right) \quad (14)$$

and

$$y(t) = \sum_{n=1}^{N} \left(c_n^* \cos \frac{2n\pi t}{T} + d_n^* \sin \frac{2n\pi t}{T} \right) \quad (15)$$

where $a_1^* = 1, b_1^* = 0, c_1^* = 0$ and $(a_2^*, \ldots, a_N^*, b_2^*, \ldots, b_N^*, c_2^*, \ldots, c_N^*, d_1^*, \ldots, d_N^*)^{\mathrm{T}} = \mathbf{f}^*$. Setting *T* to 1.0 and *t* to 0, 0.001, …, 0.999, one can calculate the coordinates of 1,000 points on a contour, and then draw the shape of the contour based on the calculated coordinates.

Population Structure Analysis

The population structure of the 179 accessions was estimated via the *k*-means clustering and the principal component analysis of marker genotype data. In the *k*-means clustering, the number of clusters was set as 5 based on the relationship between the number of clusters and the within-cluster sum of squares and also according to the population structure analysis conducted by Zhao *et al.* [16]. In the principal component analysis, the scores of 3,138 markers were treated as multivariate data, and performed the analysis based on the variance-covariance matrix of the marker scores.

Genome-wide Association Study

One can consider the *j*th principal component score of elliptic Fourier descriptors of *i*th accession, v_{ij}, as the phenotypic value of the *i*th accession, y_i, for the shape variation accounted for by the component, and performed GWAS for each of the first three principal components. In the GWAS, the phenotypic value of *i*th accession was described by the following model:

$$y_i = \mu + \sum_{k=2}^{K} q_{ik}\,\alpha_k + \sum_{l=1}^{L} x_{il}\,\beta_l + e_i \quad (16)$$

where μ is the intercept, q_{ik} is the indicator variable representing whether accession *i* belongs to the cluster *k* (*k* = 2, …, *K*), α_k is the effect associated with cluster k,x_{il} denotes the genotype score of accession *I* at marker *l*(*l* = 1, 2, …, *L*), β_l is the effect of marker *l*, and e_i is the residual error assumed to follow $N(0, \sigma_e^2)$. The above model was considered as a Bayesian regression model and the parameters of the model were estimated using the R-package BGLR [17]. In Bayesian regression, the intercept μ and each of the cluster effects, α_k, were assigned a flat prior, respectively. Each of marker effects, β_l, was assigned a scaled-*t* density. As described in [17], this prior density setting makes the model one called a BayesA [18]. The residual variance σ_e^2 was assigned a scaled-inverse chi-square density. In this study, the default hyper-parameter setting in BGLR was simply used, although careful tuning of the values may increase the power and estimation accuracy in GWAS. Markov chain Monte Carlo (MCMC) cycles were repeated 1×10^5 times, and the first 1×10^4 cycles (burn-in) were not used for estimating the parameter values.

RESULTS AND DISCUSSION

The first three principal components of elliptic Fourier descriptors explained more than 97% of among-accession grain shape variation (Fig. 1). The contribution of the first principal component is larger than 94%, indicating that most of the grain shape variation can be summarized by scores of this component.

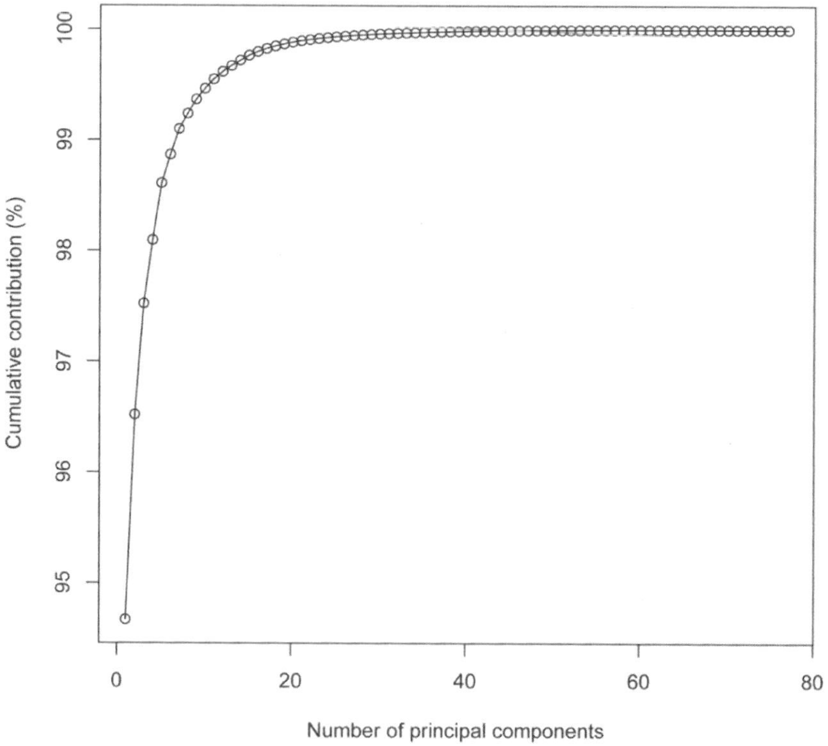

Fig. 1. Cumulative contribution (%) of principal components of elliptic Fourier descriptors. The principal component analysis was conducted based on among-accession covariance matrix of the standardized elliptic Fourier descriptors.

From the visualization of the shape variation explained by the first three principal components (Fig. 2), one found that the first, second, and third components were associated with length-to-width ratio of the grain, convex or concave shapes of the grain, and the relative width of the grain proximal and distal to the embryo, respectively. In this study, the principal component analysis was performed based on the among-accession covariance matrix. On the other hand, Iwata *et al.* [8] performed the principal component analysis based on the among-grain covariance matrix. The morphological meaning of the first three components was same between the two analyses, and the contribution of the first three components was larger in this study than in [8]. This result indicates that grain shape variation explained by the first three components well represented the genetic variation of grain shape existing in the rice accessions.

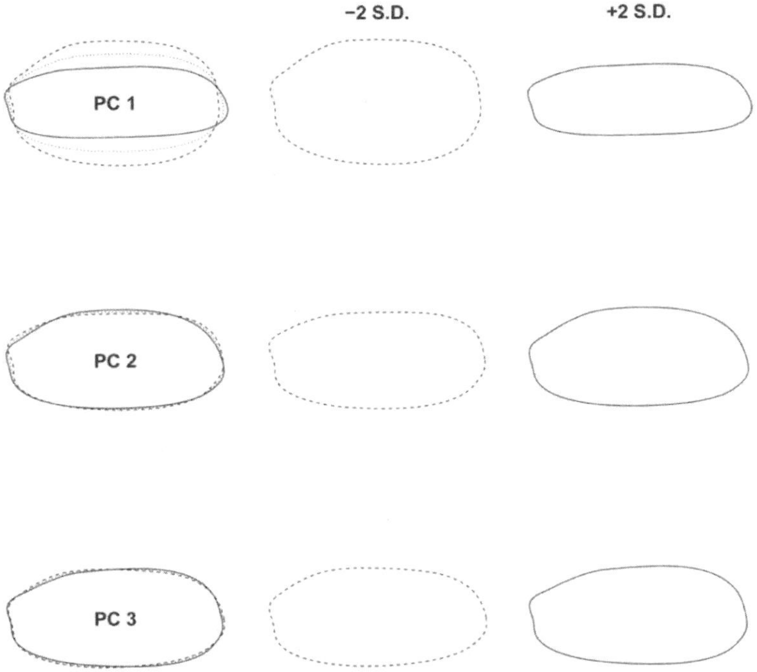

Fig. 2. Shape variation explained by the first three principal components of elliptic Fourier descriptors. Outlines of the brown rice grain are shown with the embryo on the upper left-hand side of the kernel. Dotted, dashed, and solid lines show the cases in which the score takes mean, − 2 SD (standard deviation), and + 2 SD, respectively.

Because population structure can lead to spurious association in GWAS [19], the variation in marker genotype data was analyzed using *k*-means clustering and principal component analysis of marker genotype scores. The analyses suggest the existence of subpopulation structure in the population of 179 rice accessions (Fig. 3). Among the clusters, there are differences in the scores of the first three principal components of elliptic Fourier descriptors (Fig. 4). The accessions in cluster 1 have small scores in the first principal component, indicating the accessions have rice grains with small length-to-width ratio. There are among-cluster differences in a range of the score also in the second and third principal component. These results indicate the genetic differentiation of rice grain shape among subpopulations, and suggest the need for controlling the influence of the subpopulation structure in GWAS.

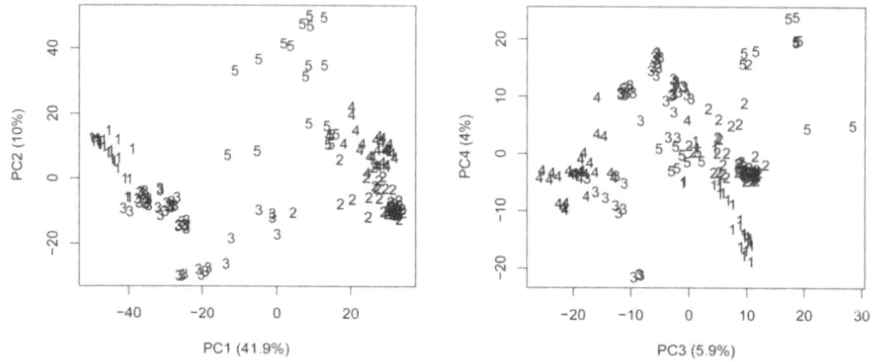

Fig. 3. Principal component analysis summarizing variation in marker genotype data is shown here. The numbers represent the cluster IDs assigned for 179 accessions in the *k*-means clustering.

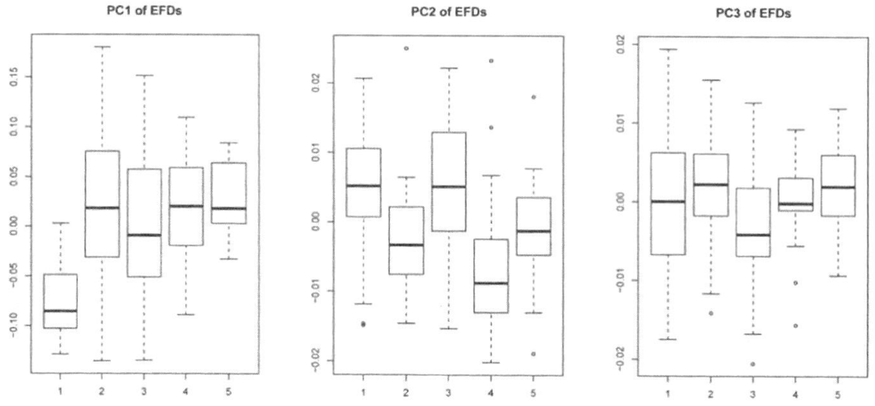

Fig. 4. Box plots showing differences in the scores of the first three principal components of elliptic Fourier descriptors among five clusters of 179 accessions estimated in the *k*-means clustering.

In GWAS, the subpopulation structure was taken into account and estimated effects of subpopulations and effects of all genome-wide SNP markers simultaneously based on the model of Eq. 13 with the Bayesian regression method. In the first principal component of elliptic Fourier descriptors, the strongest association signal was found on chromosome 5 (Fig. 5). The SNP that showed the largest squared-marker effect, β_l^2, was located at 5,434,540 bp on the chromosome 5 sequence of the IRGSP Build 4.0 psuedomolecules (http://rgp.dna.affrc.go.jp/E/IRGSP/Build4/build4.html).

Around this SNP, four other SNPs, which were located at 4,769,725 bp, 4,835,191 bp, 5,422,630 bp and 6,352,064 bp, showed the large squared-marker effects. In this chromosome region, the *qSW5* gene [20] is located from 5,347,177 bp to 5,348,756 bp. A strong association signal was also found on chromosome 3 (Fig. 5).

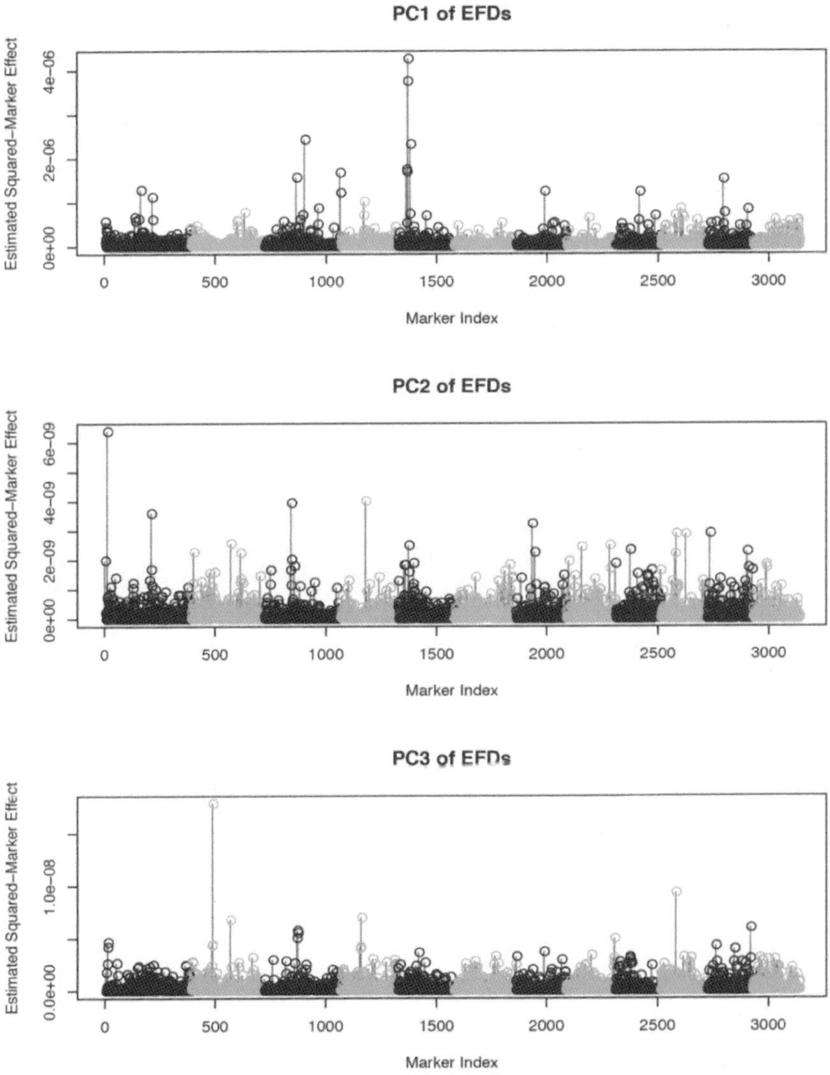

Fig. 5. Square of estimated effects, β_l^2, of genome-wide SNP markers. Black and gray circles represent SNPs on odd- and even- number chromosomes, respectively.

On this chromosome, SNP at 18,912,783 bp showed the largest squared-marker effect. Around this SNP, SNP at 14,327,750 bp also showed the third largest effects among SNPs on this chromosome. In this chromosome region, the *GS3* gene [21, 22] is located from 17,365,305 bp and 17,371,032 bp. In the first principal component, relatively strong associations were also found at 19,672,373 bp and 26,014,200 bp on chromosome 1; at 36,717,641 bp and 37,231,462 bp on chromosome 3; at 18,990,981 bp on chromosome 7; at 12,544,004 bp on chromosome 9; and at 9,866,549 bp on chromosome 11. These SNPs may be located closely to genes controlling the length-to-width ratio of rice grains. Although the contribution of the second and third principal components was less than 2% of total shape variation, relatively strong associations were found in both principal components (Fig. 5).

This chapter will not go into a detailed discussion on the result of GWAS, because the primary objective of this work is describe how to perform a GWAS of biological shape based on elliptic Fourier analysis. The series of analyses described in this chapter are generally applicable to any biological shape that can be well delineated by elliptic Fourier descriptors. These analyses will be useful for the genetic dissection of infraspecific shape variability observed in a range of plant and animal species, and will contribute to the discovery of genes responsible for the variability.

ACKNOWLEDGEMENT

This work was supported by Grants-in-Aid for Scientific Research (A), MEXT, No. 25252002.

REFERENCES

[1] Hawkes JG (1983) *The diversity of crop plants.* Harvard University Press, Cambridge.

[2] Tanksley SD (2004) The genetic, developmental, and molecular bases of fruit size and shape variation in tomato. *Plant Cell* **16** (Suppl.): 181-189.

[3] Huang R, Jiang L, Zheng J, Wang T, Wang H, Huang Y, and Hong Z (2013) Genetic bases of rice grain shape: so many genes, so little known. *Trend Plant Sci* **18**: 1360-1385.

[4] Davey JW, Hohenlohe PA, Etter PD, Boone JQ, and Catchen JM (2011) Genome-wide marker discovery and genotyping using next-generation sequencing. *Nat Rev Genet* **12**: 499-510.

[5] Brach B, Morris GP, and Borevitz JO (2011) Genome-wide association studies in plants: the missing heritability is in the field. *Genome Biol* **12**: 232.

[6] McCouch SR, McNally KL, Wang W, and Hamilton RS (2012) Genomics of gene banks: A case study in rice. *Amer J Bot* **99**: 407-423.

[7] Kuhl FP and Giardina CR (1982) Elliptic Fourier analysis of a closed contour. *Comput Graph Image Process* **18**: 259-278.

[8] Iwata H, Ebana K, Uga Y, Hayashi T, and Jannink JL (2010) Genome-wide association study of grain shape variation among *Oryza saliva* L. germplasms based on elliptic Fourier analysis. *Mol Breeding* **25**: 203-215.

[9] Chitwood DH, Ranjan A Martinez CC, Headland LR, Thiem T, Kumar R, Covington MF Hatcher T, Naylor DT, Zimmerman S, Downs N, Raymundo N, Buckler ES, Maloof JN, Aradhya M, Prins B, Li L, Myles S, and Sinha NR (2014) A modern ampelography: a genetic basis for leaf shape and venation patterning in grape. *Plant Physiol* **164**: 259-272.

[10] Iwata H (2011) Quantitative genetic analyses of crop organ shape based on principal components of elliptic Fourier descriptors. **In**: PE Lestrel (Ed.) *Biological shape analysis: Proc 1st Int Sym Biol Shape Anal* (ISBSA) World Scientific Pub, Singapore.

[11] Lestrel PE (2013) *Biological shape analysis: Proc 2nd Int Sym Biol Shape Anal* (ISBSA) World Scientific Pub, Singapore.

[12] Iwata H, Ebana K, Uga Y, and Hayashi T (in press) Genomic prediction of biological shape: elliptic Fourier analysis and kernel partial least squares (PLS) regression applied to grain shape prediction in rice (*Oryza sativa* L.). PLoS ONE.

[13] Iwata H and Ukai Y (2002) SHAPE: a computer program package for quantitative evaluation of biological shapes based on elliptic Fourier descriptors. *J Hered* **93**: 384-385.

[14] R Core Team (2014). R: A language and environment for statistical computing. R Foundation for Statistical Computing, Vienna, Austria. ISBN 3-900051-07-0, URL http://www.R-project.org/.

[15] Kojima Y, Ebana K, Fukuoka S, Nagamine T, and Kawase M (2005) Development of an RFLP-based rice diversity research set of germplasm. *Breed Sci* **55**: 431-440.

[16] Zhao K, Tung CW, Eizenga GC, Wright MH, Ali ML, Price AH, Norton GJ, Islam MR, Reynolds A, Mezey J, McClung AM, Bustamante CD, and McCouch SR (2011) Genome-wide association mapping reveals a rich genetic architecture of complex traits in *Oryza sativa*. *Nat Commun* **2**: 467.

[17] Pérez P de los Campos G (2014) Genome-wide regression and prediction with BGLR statistical package. *Genetics* **198**: 483-495.

[18] Meuwissen THE, Hayes BJ, and Goddard ME (2001) Prediction of total genetic value using genome-wide dense marker maps. *Genetics* **157**: 1819-1829.

[19] Lander ES and Schork NJ (1994) Genetic dissection of complex traits. *Science* **265**: 2037-2049.

[20] Shomura A, Izawa T, Ebana K, Ebitani T, Kanegae H, Konishi S, and Yano M (2008) Deletion in a gene associated with grain size increased yields during rice domestication. *Nat Genet* **40**: 1023-1028.

[21] Fan C, Xing Y, Mao H, Lu T, Han B, Xu C, Li X, and Zhang Q (2006) *GS3*, a major QTL for grain length and weight and minor QTL for grain width and thickness in rice, encodes a putative transmembrane protein. *Theor Appl Genet* **112**: 1164-1171.

[22] Mao H, Sun S, Yao J, Wang C, Yu S, Xu C, Li X, and Zhang Q (2010) Linking differential domain functions of the *GS3* protein to natural variation of grain size in rice. *Proc Natl Acad Sci USA* **107**: 19579-19584.

Integration of Shape Mapping and Genome-wide Association Studies to Characterize Human Body Shape Genes

HAN HAO[1], LIDAN SUN[1], ZHONG WANG[2] and RONGLING WU[1]

Abstract

As a complex trait underlying human health, body shape is believed to be under genetic control. While the traditional description of body shape based on length-width ratios and angles is criticized due to its simplicity, more sophisticated geometric morphometrics has been recently integrated with genetic mapping, leading to the birth of shape mapping, which can precisely capture subtle variation in body shape and its genetic underpinnings. In this chapter, it is argued that the incorporation of shape mapping into genome-wide association studies (GWAS) is an essential approach for elucidating the complete genetic architecture of global and local variation in body shape. Shape mapping allows GWAS researchers to test and estimate how specific genes pleiotropically affect co-variation of body shape with complex diseases and how such co-variation can be used to predict human health risk. Through this review, it is hoped that it will stimulate an in-depth exploration of body shape gene identification and highlight new directions for future work in this increasingly expanding genetic research area.

INTRODUCTION

Despite enormous difficulty in unveiling the genetic secrets of quantitative traits, the development and use of genome-wide association studies (GWAS) arising from single nucleotide polymorphism (SNP) genotyping techniques have revolutionized our understanding of genetic control over these polygenic traits. By systematically characterizing an entire set of individual genes throughout the genome [1, 2], GWAS is equipped with a capacity to quantify how genes contribute to phenotypic variation singly or through their interactions with other genes and environmental factors. A typical GWAS includes scanning thousands or tens of thousands of individuals, simultaneously for several hundreds of thousands to more than a million SNPs using DNA microarrays (chips), associating each of these SNP with a phenotypic trait and obtaining a significance test for these associations through multiple testing corrections. More recently, thousands of significant associations have been found for a variety of complex traits or diseases [2].

It has been increasingly recognized that the success of GWAS does not only depend on the genome-wide genotyping of SNPs throughout the entire genome, but also on how a phenotypic trait is measured and described precisely and comprehensively. Traditional ways of measuring single values for a trait phenotype at a static time point or a fixed space are insufficient to capture biological attributes of the trait. Functional GWAS or *f*GWAS derived from functional mapping, a statistical model for mapping quantitative trait loci (QTLs) that control developmental trajectories of a complex trait [3, 4, 5, and 6], has been developed to accommodate to the dynamic feature of phenotypic data over time [7]. The central

[1] Center for Statistical Genetics, The Pennsylvania State University, Hershey, PA 17033, USA, rwu@phs.psu.edu.
[2] The Baker Institute for Animal Health, College of Veterinary Medicine, Cornell University, Ithaca, NY 14853, USA.

idea of *f*GWAS is to integrate biological principles of trait formation into GWAS, allowing hypotheses about the interplay between genes and development to be tested. Thus, in addition to increasing statistical power, *f*GWAS can shed more light on the biological mechanisms involved in cellular and biological processes.

While functional mapping was derived to study the temporal pattern of genetic variation over time, a question naturally arises about the spatial pattern; that is, how genes expressed differentially in an organ from one position to the next can be studied. The spatial change of allometric growth for an organ determines its shape. Thus, the identification of specific QTLs for shape variation has become one of the most fundamental issues in developmental and evolutionary genetics [8]. Many morphological QTL mapping studies are based on descriptive measures of a shape through length-width ratios and angles [9], which cannot quantitatively characterize numerous subtle structures that collectively contribute to shape variation [10]. Langlade and colleagues [10] pioneered the application of a geometric morphometric (GM) model to describe leaf shape and map its underlying QTLs. Wu and colleagues have for the first time developed the fundamental theory of mapping morphological shape by integrating shape analysis (including shape recognition and shape registration) into a framework of genetic mapping [4, 11]. For example, Fu *et al.* [12] incorporated a binary model to map shape QTLs through modeling black and white shape data. Implemented with a Procrustes analysis for aligning shape with different poses, scales and rotations, a radius-centroid-contour analysis approach was developed to map the QTLs underlying morphological shape [13]. Bo and co-workers [14] integrated an elliptic Fourier-based outline analysis approach into QTL mapping by fitting the shape's outline. Hao and colleagues [15] capitalized on a high-dimensional functional ANOVA to test the genetic variation of any part in a biological shape. All these approaches, called shape mapping, show their ability to characterize the control patterns of specific QTLs on the origin, properties and functions of shape variation.

In this chapter, it is proposed that a confluence of two currently disjoined areas, shape mapping and GWAS, be joined to precisely study the genetic architecture of human body shape. Body shape is a complex trait with sophisticated detail and function, which has been thought to be associated with human health [16, 17, and 18]. To describe body shape, body mass index (BMI), defined as an individual's body mass divided by the square of his or her height, and the waist-to-hip ratio (WHR), as measures of body fat distribution, are generally used. Although these two parameters have served as a measure of body shape for genetic mapping [19, 20, 21, and 22], they are limited in elucidating the genetic control mechanisms of body shape. First, BMI can only measure the overall adiposity of a body, failing to characterize body composition. Second, WHR cannot precisely quantify the spatial distribution of shape changes across the body and, more importantly, it cannot provide an index of why fat accumulates in different places. Thus, it is argued that shape mapping, when it is implemented into a GWAS setting, can refresh our view of the genetic control of body shape and improve our understanding of the biological mechanisms involved in the pattern of where our body stores fat.

METHODOLOGICAL REVIEW

Integrating Shape Mapping into GWAS

GWAS design

A GWAS can be designed by sampling two groups of genotyped subjects (cases and controls) who are matched for demographical factors and comparing and testing the differences in allelic frequencies between the two groups. Alternatively, a GWAS can be established from a random sample of genotyped subjects from a population in which a phenotypic trait is measured for each subject. For the GWAS aimed to identify shape genes, a population-based design composed of many randomly sampled subjects was used. Unlike traditional single-valued phenotypic traits, such as body height, BMI or WHR, the phenotypic trait considered here is body shape shown as digitalized images (Fig. 1). For each subject, three photos from the front, back and side are taken to comprehensively describe body shape. Shape analysis can then be implemented to quantify the form of body shape.

Shape analysis

D'Arcy Thompson in 1917 was the first who attempted to quantitatively describe the mechanism of shape change [23]. However, until the early 1980's, shape analysis has been revolutionized by geometric morphometrics (GM) [24, 25, and 26]. The central idea of GM is to study shape variation by determining, analyzing, and connecting meaningful anatomical landmarks of the organism [27]. A landmark is defined here as an anatomical point that can locate other structures of a shape from a particular orientation [24]. To describe a human body shape, a set of key homogeneously distributed landmarks are chosen as shown in Fig. 2. These landmarks are presented in the Cartesian coordinate system from which the geometric information about the relative positions of landmarks is extracted. Through filtering location, scale, and rotational effects out from the photo, shape can be aligned, described, and compared [24]. There have been many approaches developed to align shape, including the commonly used Procrustes analysis [26].

Apart from the landmark-based approach, outline analysis has also been used to quantify human body shape [28]. Different from the former, the latter uses a mathematical function to fit those ordered points recorded along the contour of the body which are placed in the Cartesian coordinate system [29]. By testing how the mathematical parameters specifying the function vary from shape to shape, the differences between shapes can be characterized. Outline analysis has been based on many approaches, like e.g., "Eigenshape" analysis, standard Fourier descriptors, and Elliptic Fourier analysis.

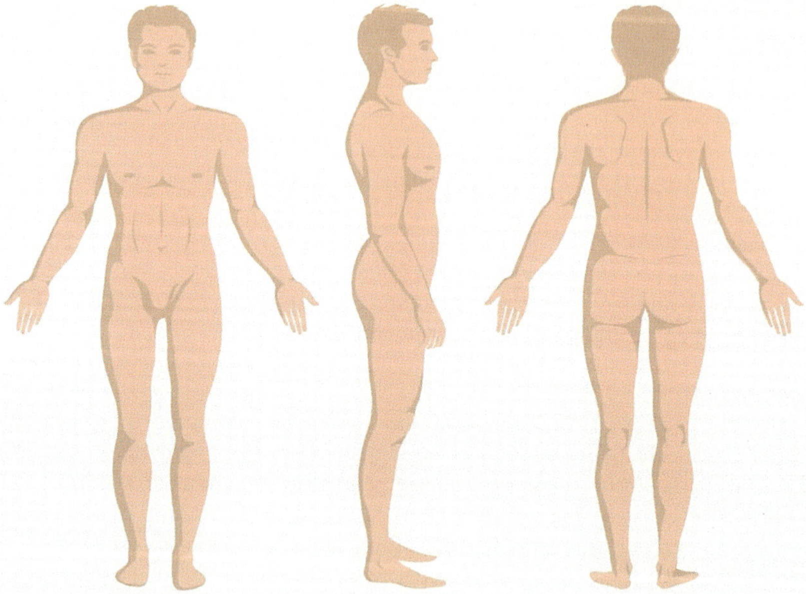

Fig. 1. Photos of body shape from the front, side and back for a typical sample in a GWAS study. Adapted from http://www.123rf.com/photo_24619903_man-body-anatomy-front-back-and-side-standing-human-pose.html.

Association mapping

Both landmark-based and outline analysis approaches obtain a high-dimensional set of data that describe a body shape. For the landmark-based approach, a multivariate statistical analysis, such as principal component analysis (PCA), can be used to decompose the original high-dimensional space into a space of reduced dimension. By mapping the high-dimensional data to the subspace that accounts for the distribution of the original pattern, PCA removes redundant information. Each component of PCA represents an eigenvector of the covariance matrix among landmark variables. The first component contributes to maximum variation among subjects, with other components contributing in other ways the sample variation. In general, the first component determines the global variation of body shape, whereas the other components are responsible for the more local variations in body shape. Other dimension reduction approaches include wavelet-based transform techniques that compress the data based on a wavelet decomposition approach [30, 31].

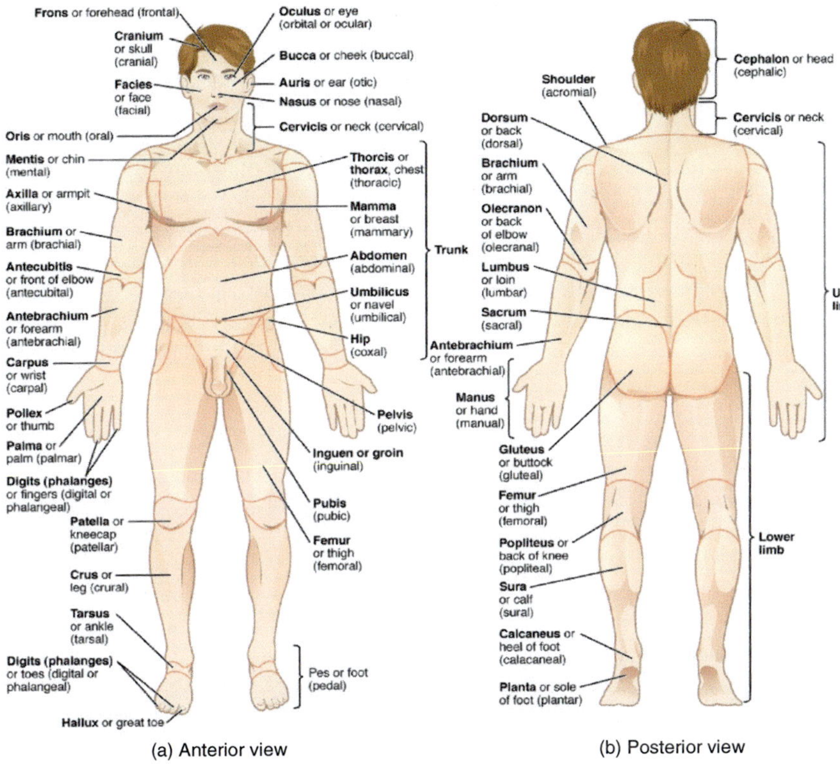

Fig. 2. Key anatomical points used in locating body landmarks on the front and back of the body. Adapted from http://en.wikipedia.org/wiki/Anatomical_terminology.

By treating component values from a PCA or compressed values from wavelet techniques as a phenotypic trait, shape mapping is formulated to map shape QTLs by regression analysis or maximum likelihood analysis [11]. Shape mapping can estimate genotypic values for each significant QTL, implemented with a procedure to test how QTL genotype-specific values differ. Through such a test, one can characterize how the QTL determines overall and detailed structures of body shape. QTL genotypic values are then transformed back to the original scale from which the pattern of how QTL affects shape variation can be visualized as configurations of landmarks.

Unlike the landmark-based approach, outline analysis integrates mathematical functions that describe body shape (such as some form of Fourier analysis) into the GWAS model [14]. The spatial effect of QTLs on shape variation can be quantified by estimating and testing mathematical parameters of shape. Analogously, one can transform Fourier coefficients that describe QTL effects back into the physical space of body shape visualized as outlines. Bo and co-workers [14] provide a procedure for examining the control patterns of specific QTLs on the origin, properties, and functions of biological shape using outline analysis.

Extensions

In this chapter two basic shape analysis approaches have been outlined and their incorporation with GWAS to detect shape QTLs discussed. Given a possible relationship with coronary heart disease and adverse metabolic risk causing diabetes, hypertension and dyslipidemia [19, 17, and 18], body shape may share the common genetic basis with these human health risks. Shape mapping can be extended to jointly model longitudinal Cartesian coordinate data of shape and risk variables within a GWAS setting. By adopting a joint modeling framework proposed by [32, 33], genetic parameters associated with shape variation and risk variation can be estimated. The extended shape mapping model will not only test how the same genes pleiotropically determine body shape and disease, but also provide a basis on which to build a predictive model for assessing health risk based on genetic information and body shape parameters.

DISCUSSION AND CONCLUSIONS

Human diseases related with body shape, such as cardiovascular disease, diabetes, hypertension and dyslipidemia, are generally highly inherited. Thus, the integration of shape mapping into a family-based GWAS design can increase the power of shape gene identification. This design requires genotyping both subjects and their parents, allowing the more complete elucidation of genetic architecture by estimating an additional genetic effect – genetic imprinting [34, 35]. If the body shape of each member from a family is measured by GM (Fig. 3), this design will provide more insight into the genetic control mechanisms of body shape variation, such as the mechanism of how shape genes trigger their transgenerational and sex-specific differences for body shape. Such information is crucial for building a precise genotype-phenotype map for body shape and its relationship with health risks.

The current 2D-based shape mapping model may not adequately describe comprehensive variation of body shape. Shape is thought to govern the extent and pattern of energy, matter and signal transduction through the surface and inner structure of the biological object. For this reason, an understanding of biological curvature and texture has received considerable attention in modern biology. A 3D model for the shape and texture of human body should be integrated with shape mapping to better characterize the biochemical mechanisms of genetic effects.

Vision technologies have been developed to estimate the 3D shape of an object from 2D image data without information about its texture, its pose and the illumination environment [36]. These technologies include a 3D morphable model (3DMM) that represents the 3D shapes and textures as a linear combination of shapes and textures principal components. A stochastic Newton optimization algorithm that fits the 3DMM to a single image can estimate the 3D shape, the texture and the imaging conditions. By integrating 3DMM into shape mapping, a series of hypothesis tests about the genetic control of topological features (such as steepness and ridgeness) and texture of body shape can be formulated.

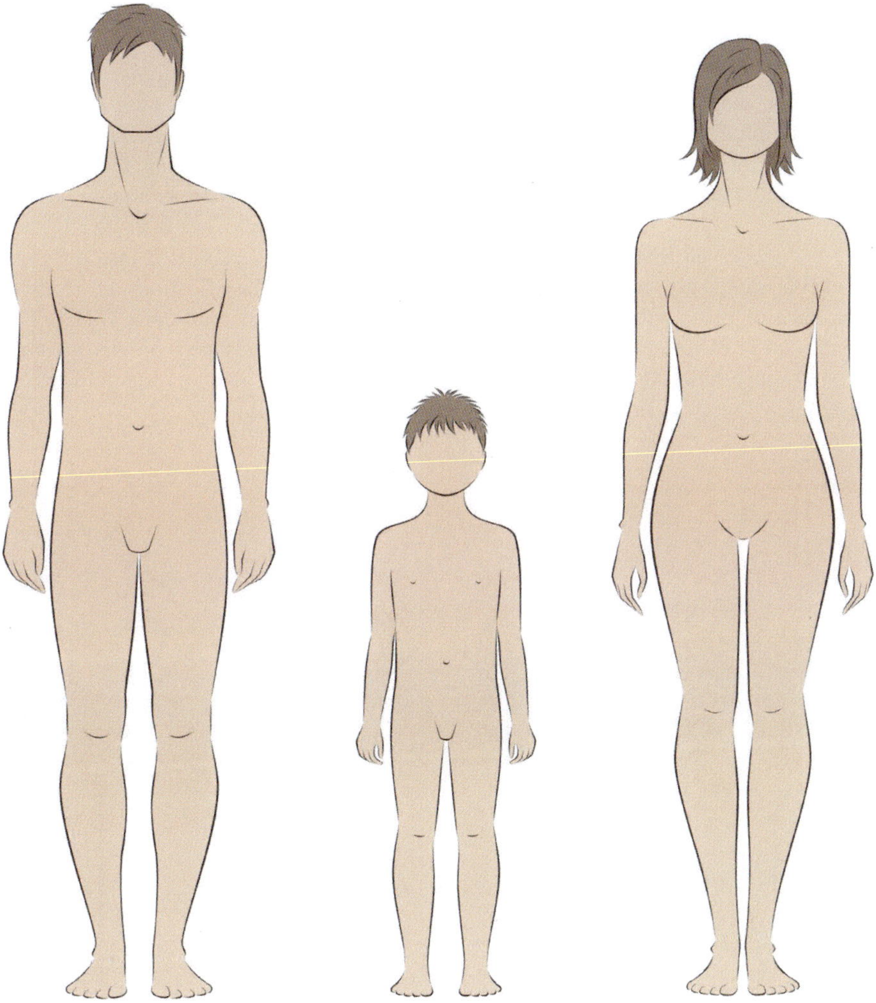

Fig. 3. Photos of body shape for three members, both parents and their children, from a family in a family-based GWAS design. Adapted from http://depositphotos.com/6411414/stock-illustration-male-female-and-child-body.html.

Prospects

The integration of shape mapping and GWAS can address many integrative biological and genetic questions as well as challenges in body shape variation faced by the health sciences community. A diversity of body shape naturally exists, but the question of how it is associated with, or causes, health risks has received increasing interest only recently. Many studies have provided and continue to provide the evidence that the pattern of where one is hormonally predisposed to store fat to produce a particular body shape is also closely related with health risks. Unfortunately, the genetic dissection of body shape by powerful GWAS studies is only based on two simple parameters, body mass index and waist-to-hip ratio, limiting our detailed understanding of body shape variation and its relationships with health risks. More recently, a computational model, shape mapping, has been derived on the basis of geometric morphometrics and has been shown to possess great potential to reveal the developmental mechanisms that underlie different shapes. Its integration with GWAS should yield immediate inferences for precisely and comprehensively identifying genes and their genetic interactions affecting body shape.

It is observed that female body shape may be roughly classified into different types: banana, apple, pear and hourglass (Fig. 4). If these classifications are realistic then each of these body types may be controlled by a set of particular QTLs. The identification of these QTLs from shape mapping can not only help predict the type of body shape based on a subject's genetic information, but also provide guidance to design new therapies or drugs for possible obesity prevention.

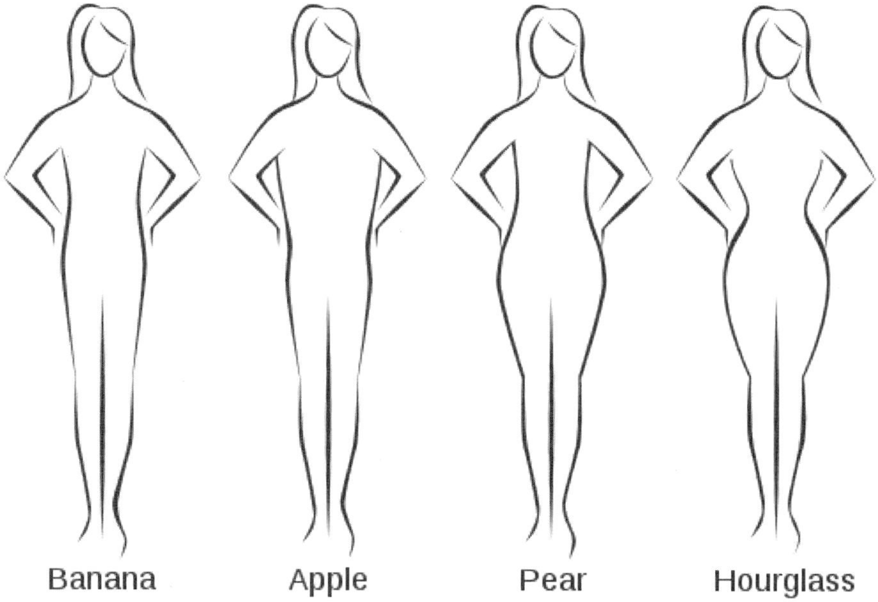

Fig. 4. Different types of female body shape. Adapted from http://en.wikipedia.org/wiki/Female_body_shape.

ACKNOWLEDGEMENTS

This research was funded by grants NIH/NHLBI-1U10HL098115, U01 HL119178 and UL1 TR000127.

REFERENCES

[1] Lettre G, Jackson AU, Gieger C, Schumacher FR, Berndt SI, Sanna S, Eyheramendy S, Voight BF, Butler JL, and Guiducci C (2008) Identification of ten loci associated with height highlights new biological pathways in human growth. *Nat Genet* **40**: 584-591.

[2] Hindorff LA and MacArthur J (European Bioinformatics Institute), Morales J (European Bioinformatics Institute), Junkins HA, Hall PN, Klemm AK, and Manolio TA. A Catalog of Published Genome-Wide Association Studies. Available at: www.genome.gov/gwastudies. Accessed, December 27, 2014.

[3] Ma CX, Casella G, and Wu R (2002) Functional mapping of quantitative trait loci underlying the character process: A theoretical framework. *Genetics* **161**: 1751-1762.

[4] Wu R and Lin M (2006) Functional mapping – how to map and study the genetic architecture of dynamic complex traits. *Nat Rev Genet* **7**: 229-237.

[5] Li Y and Wu R (2010) Functional mapping of growth and development. *Biol Rev* **85**: 207-216.

[6] He QL, Berg A, Li Y, Vallejos CE, and Wu R. Modeling genes for plant structure, development and evolution: Functional mapping meets plant ontology. *Trends Genet 2010* **26**: 39-46.

[7] Das K, Li JH, Wang Z, Fu G, Li Y, Mauger D, Li R, and Wu RL (2011) A dynamic model for genome-wide association studies. *Hum Genet* **129**: 629-639.

[8] Klingenberg CP (2010) Evolution and development of shape: integrating quantitative approaches. *Nat Rev Genet* **11**. 623-635.

[9] Rohlf FJ (1990) Fitting curves to outlines. **In**: Rohlf FJ, and Bookstein FJ (Eds.). *Proceedings of the Michigan Morphometrics Workshop*, The University of Michigan Museum of Zoology, Ann Arbor, Michigan, pp 167-177.

[10] Langlade NB, Feng X, Dransfield T, Copsey L, Hanna AI, Thébaud C, Bangham A, Hudson A, and Coen E (2005) Evolution through genetically controlled allometry space. *Proc Natl Acad Sci* USA **102**: 10221-10226.

[11] Wu RL, Ma C-X, and Casella G (2007) *Statistical Genetics of Quantitative Traits: Linkage, Maps, and QTL*. Springer-Verlag, New York.

[12] Fu GF, Berg A, Das K, Li JH, Li R, and Wu RL (2010) A statistical model for mapping morphological shape. *Theor Biol Med Model* **7**: 28.

[13] Fu GF, Bo WB, Pang XM, Wang Z, Chen L, Song YP, Zhang ZY, Li J, and Wu RL (2013) Mapping shape QTLs using a radius-centroid-contour model. Heredity **110**: 511-519.

[14] Bo WH, Wang Z, Xu F, Fu GF, Sui YH, Wu WM, Zhu XL, Yin DN, Yan Q, and Wu RL (2014) Shape mapping: Genetic mapping meets geometric morphometrics. *Brief Bioinform* **15**: 571-581.

[15] Hao H, Sun LD, Zhu XL, Jiang LB, Cheng TR, Zhang QX, and Wu R (2014) A high-dimensional ANOVA-based *f*GWAS model for mapping biological processes. *Brief Bioinform* (accepted).

[16] Wang Y, Rimm EB, Stampfer MJ, Willett WC, and Hu FB (2005) Comparison of abdominal adiposity and overall obesity in predicting risk of type 2 diabetes among men. *Am J Clin Nutr* **81**: 555-563.

[17] Canoy D (2008) Distribution of body fat and risk of coronary heart disease in men and women. *Curr Opin Cardiol* **23**: 591-598.

[18] Pischon T, Boeing H, Hoffmann K, Bergmann M, Schulze MB, Overvad K, van der Schouw YT, and Spencer E (2008) General and abdominal adiposity and risk of death in Europe. *N Engl J Med* **359**: 2105-2120.

[19] Snijder MB, *Dekker JM, Visser M, Bouter LM, Stehouwer CD, Kostense PJ, Yudkin JS, Heine RJ, Nijpels G, and Seidell JC (2003)* Associations of hip and thigh circumferences independent of waist circumference with the incidence of type 2 diabetes: the Hoorn Study. *Am J Clin Nutr* **77**: 1192-1197.

[20] Heid IM, Jackson AU, Randall JC, Winkler TW, Qi L, Steinthorsdottir V, Thorleifsson G, and Zillikens MC (2010) Meta-analysis identifies 13 new loci associated with waist-hip ratio and reveals sexual dimorphism in the genetic basis of fat distribution. *Nat Genet* **42**: 949-960.

[21] Lindgren CM, Heid IM, Randall JC, Lamina C, and Steinthorsdottir V (2009) Genome-wide association scan meta-analysis identifies three loci influencing adiposity and fat distribution. *PLoS Genet* **5**(6): e1000508.

[22] Emilsson V, Thorleifsson G, Zhang B, Leonardson AS, Zink F, Zhu J, Carlson S, and Helgason A (2008) Genetics of gene expression and its effect on disease. *Nature* **452**: 423-428.

[23] Thompson D'Arcy W (1917) *On Growth and Form*. Cambridge University Press, New York.

[24] Kendall DG (1984) Shape manifolds, procrustean metrics, and complex projective spaces. *Bull of London Math Soc* **16**: 81-121.

[25] Adams DC, Rohlf FJ, and Slice DE (2004) Geometric morphometrics: ten years of progress following the "revolution". *Ital J Zool* **71**: 5-16.

[26] Zelditch ML, Swiderski DL, and Sheets HD (2004) *Geometric Morphometrics for Biologists*. Elsevier Academic Press, Amsterdam.

[27] Bookstein FL (1991) *Morphometric Tools for Landmark Data: Geometry and Biology*. Cambridge University Press, New York.

[28] Courtiol A, Ferdy JB, Godelle B, Raymond M, and Claude J (2010) Height and body mass influence on human body outlines: a quantitative approach using an elliptic Fourier analysis. *Am J Phys Anthropol* **142**: 22-29.

[29] Kuhl FP, and Giardina CR (1982) Elliptic Fourier features of a closed contour. *Comp Graph Image Proc* **18**: 236-258.

[30] Zhao W, Li HY, Hou W, and Wu RL (2007) Wavelet-based parametric functional mapping of developmental trajectories. *Genetics* **176**: 1811-1821.

[31] Zhao W, and Wu RL (2008) Wavelet-based nonparametric functional mapping of longitudinal curves. *J Am Stat Assoc* **103**: 714-725.

[32] Lin M, and Wu RL (2006) A joint model for nonparametric functional mapping of longitudinal trajectories and time-to-events. *BMC Bioinform* **7**(1): 138.

[33] Das K, Li JH, Fu GF, Wang Z, Li R, and Wu R (2012) Dynamic semiparametric Bayesian models for genetic mapping of complex trait with irregular longitudinal data. *Stat Med* **32**: 509-523.

[34] Li X, Sui YH, Liu T, Wang Z, Li YC, Lin ZW, Hegarty J, Koltun WA, and Wu RL (2014) A model for family-based case-control studies of genetic imprinting and epistasis. *Brief Bioinform* **15**: 1069-1079.

[35] Sun LD, Zhu XL, Bo WH, Xu F, Cheng TR, Zhang QX, and Wu RL (2014) An open-pollinated design for mapping imprinting genes in natural populations. *Brief Bioinform* **15**: doi: 10.1093/bib/bbu019 (published online).

[36] Romdhani S, Ho J, Vetter T, and Kriegman DJ (2006) Face recognition using 3-D models: Pose and illumination. *Proc IEEE* **94**: 1977-1999.

Evo-devo of Spiral Shell Growth in Gastropods

KEISUKE SHIMIZU[1] and KAZUYOSHI ENDO[1]

Abstract

Various shapes of gastropod shells have evolved ever since the Cambrian. The fossil record suggests that ancestral snails had a coiled shell but during evolution some snail lineages like limpets have independently lost the coiled shell. Although a theoretical account of shell morphogenesis exists, the molecular basis of shell development remains unclear. In order to understand the development of shell shape in gastropods, focus was placed on the *decapentaplegic* (*dpp*) gene that is expressed around the shell gland. Expression patterns of *dpp* in the shell gland and the mantle tissues between coiled and non-coiled shelled gastropods were compared. In the limpets, *Patella vulgata*, *dpp* showed symmetric expression patterns throughout ontogeny. On the other hand, in the coiled shelled snail *Lymnaea stagnalis*, *dpp* showed an asymmetric expression and mirror image patterns between the dextral and sinistral lineages. Moreover when the embryos were treated with the Dpp regulator or signal inhibitor dorsomorphin at the veliger stage, juvenile shells grew showing a cone-like form rather than a normal coiled form. These results demonstrate that the shell coiling is controlled by using Dpp as a morphogen. The loss of coiling in Patellogastropoda was likely caused by loss of the asymmetric expression of *dpp* in the shell gland at the trochophore stage, a change which led to the symmetric expression of *dpp* at the veliger and adult stages.

INTRODUCTION

The Mollusca constitute one of the most diverse animal phyla in the sea, and some groups have adapted to freshwater or terrestrial ecological niches. The phylum consists of seven living classes (Solenogastres, Caudofoveata, Polyplacophora, Monoplacophora, Bivalvia, Scalophoda, Cephalopoda and Gastropoda) and at least one extinct class (Rostroconchia). Most molluscs except for members of the two classes Solenogastres and Caudofoveata have one or more hard exoskeletons called "shell". In Gastropoda, their various shell morphology has evolved ever since the late Cambrian. In order to understand the origin of the shell shape diversity, one needs to look into the ontogeny of the shell forming region. From previous experimental studies on molluscan embryonic cell lineages, it is known that 2d blastomere of the 12-cell stage is important for the formation of the shell field (e.g. [1, 2]). Thickening of the ectodermal cells that are the descendants of 2d blastmere and invagination of these cells occur in the dorsal region at the end of gastrulation [3]. It is considered that this invagination of ectodermal cells is due to cell multiplication, which results in an enlargement of the shell field [2, 3]. After the invagination of the shell field, the reversal of the invaginated region (known as evagination) occurs, and subsequently, the initial shell is formed along with the secretion of shell matrix proteins by the evaginated cells in the shell field at the trochophore stage [3]. The evaginated shell field will then form an epithelial fold, which is called the free mantle margin, and the mantle becomes responsible for the shell growth during and after the veliger stage.

[1] The University of Tokyo, Department of Earth and Planetary Science, Japan, shimizu@eps.s.u-tokyo.ac.jp.

Morphological Understanding of the Various Shell Forms in Gastropods

In the field of theoretical morphology, the coiled shells of molluscs have been a major target of biological shape analysis for many years [4, 5]. From this research developed the idea of accretionary growth in the aperture using moving reference frame models without a fixed coordinate system as used in the preceding models pioneered by Raup [6]. The models using a moving reference frame were the first to be based upon the viewpoint of "shell growth", and allowed one to theoretically understand the formational mechanisms involved in various shell shapes, some of which had been inexplicable with the inorganic Raup's model.

Okamoto's model [4] is therefore considered to better mimic the possible conditions in a living system. Okamoto's model can explain the various shell shapes using only three parameters; *Whorl Expansion* (*E*), *Curvature* (*C*), and *Torsion* (*T*) ([4]; Fig. 1). *Whorl Expansion* represents the ratio of aperture expansion per given time, and if this parameter is very large, the aperture rapidly enlarges like in limpets or abalones. *Curvature* represents the degree of shell growth asymmetry, and if *Curvature* = 0, the shell grows straight like in *Orthoceras*. *Torsion* represents the revolution rate of the maximum growth point. Rice [7] proposed a new model basing it on the idea that the animal must keep a constant 'gradient' of shell growth rate between the outer and inner edges to make a coiled shell. This model is more restricted than the Okamoto model, which allows analysis of even heteromorph ammonoides, or *any* tube shaped shells. But the Rice model is supposed to sufficiently reflect the actual biological processes that produce regularly coiled shells, and has been incorporated into many recent models of shell growth (e.g., [8, 9]). On the other hand, the molecular basis of shell coiling has been poorly understood thus far. Probably a morphogen-like gradient substance exists, but no one had identified a candidate for the "gradient" until recently [10].

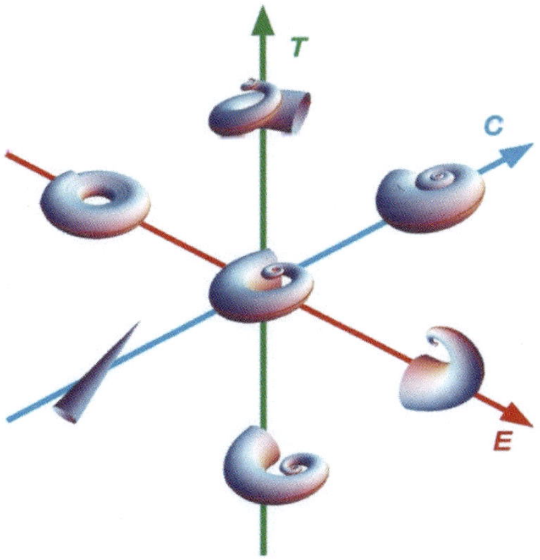

Fig. 1. Growing tube model [4]. The growing tube model can explain the various shell shapes using only three parameters; *Whorl Expansion* (*E*), *Curvature* (*C*) and *Torsion* (*T*).

Molecular Basis of Shell Development in Gastropods

Early shell development (initial shell formation)

The initial shell of gastropods is formed by the shell-secreting cells in the shell gland in the trochophore larva. Thus, in order to understand the molecular basis of shell formation, previous studies focused on the shell gland, revealing six morphogenetic genes (*hox1*, *hox4*, *post1*, *post2*, *engrailed* and *dpp*) that are specifically expressed there [11, 12, 13, and 14]. With respect to comparisons of expression patterns between coiled shelled snails and non-coiled shelled limpets, three genes (*hox1*, *engrailed* and *dpp*) have been studied so far [11, 14, 15, and 16]). The *hox1* and *engrailed* are expressed symmetrically around the shell gland in both coiled and non-coiled gastropods (Fig. 2 **A**, **B**, **D** and **E**), while the *dpp* gene shows different expression patterns between the coiled shelled gastropods and non-coiled shelled gastropods; *dpp* is expressed around the shell gland in the non-coiled shelled limpet, but it is expressed in the right half side only in the dextral pond snail *Lymanea stagnalis* (Fig. 2 **C** and **F**).

The *dpp* gene belongs to one of the signal molecule superfamilies called the TGF-β superfamily, and it has been known that these signal molecules work as morphogens and interact with other kinds of signal molecules. In vertebrates, the main functions of *bmp2/4* (bone morphogenetic proteins 2 and 4, which are the vertebrate homolog of *dpp*) have already been reported. BMP4 plays a role in the determination of the basic embryonic body axis especially the dorso-ventral axis. BMP2 and BMP4 also play roles in bone formation [17]. In molluscs, however, *dpp* is expressed around the shell gland in gastropods and hinge region in bivalves [14, 15, and 18]. Based on the results of functional analysis of *dpp* using RNA interference (RNAi), Hashimoto *et al.* [16] suggested that *dpp* regulates *grainyhead*, a marker gene of the exocrine cell, and moreover, it activates the cell proliferation at the edges of the mantle.

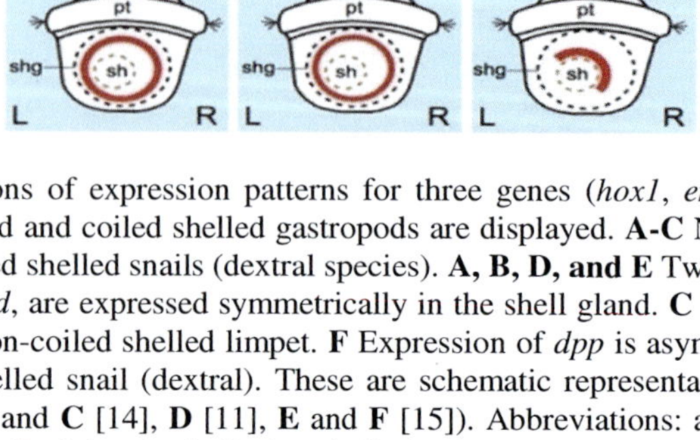

Fig. 2. Comparisons of expression patterns for three genes (*hox1*, *engrailed* and *dpp*) between non-coiled and coiled shelled gastropods are displayed. **A-C** Non-coiled shelled limpets. **D-F** Coiled shelled snails (dextral species). **A, B, D, and E** Two homeotic genes, *hox1* and *engrailed*, are expressed symmetrically in the shell gland. **C** Expression of *dpp* is symmetric in non-coiled shelled limpet. **F** Expression of *dpp* is asymmetric (right half side) in coiled shelled snail (dextral). These are schematic representations of published results (**A** [16], **B** and **C** [14], **D** [11], **E** and **F** [15]). Abbreviations: ap, apical plate; L, left; pt, prototroch; R, right; sh, shell; shg, shell gland.

Post-embryonic shell development (shell growth)

Later shell development is more poorly understood than the earlier growth stages, with most previous studies on late shell development being focused on shell matrix proteins. Thus, the molecular mechanisms for shell growth remain meager. However, understanding the molecular basis of shell growth is essential to understand shell shape evolution, because shell morphology is formed as a result of shell growth. Shimizu *et al.* [19] has reported interesting differences in gene expression patterns of *dpp* and pSMAD1/5/8 that lead to intracellular propagation of the Dpp signal between coiled shelled pond snails *L. stagnalis* and non-coiled shelled limpets *Patella vulgate* (Fig. 3). The limpets show symmetric expression patterns of *dpp* and pSMAD1/5/8 in the shell forming tissues throughout ontogeny, whereas the results of coiled shelled snail indicated asymmetric and mirror image patterns between the dextral and sinistral lineages. Moreover, Shimizu *et al.* [10] reported that the Dpp regulator signal inhibition at certain larval stages changed the shell growth pattern from coiled to non-coiled; indicating that Dpp localization in the shell gland and in the mantle edge is a key factor controlling their shell morphology.

Considering these results, Shimizu *et al.* [10] hypothesized that Dpp induces local mantle expansion, and the presence of a left-right asymmetric gradient of the Dpp protein causes the formation of a coiled shell. In fact, it has been known that Dpp correlates with the cell proliferation in the gastropod mantle [16], and the cell division is correlated positively with the concentration of Dpp gradient in fruit fly [20].

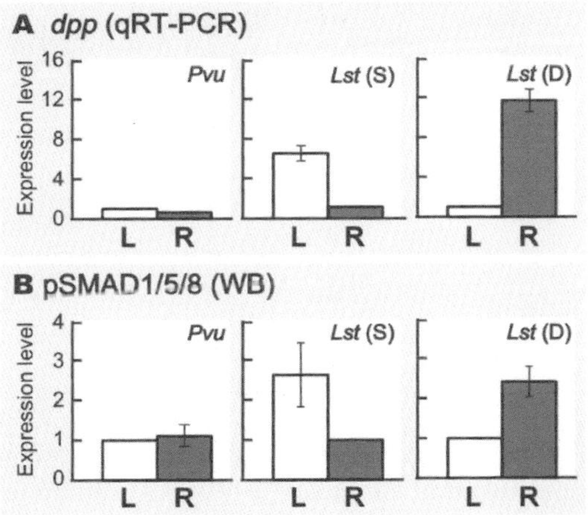

Fig. 3. Expression levels of the genes *dpp* and pSMAD1/5/8 in adult mantle edge tissues. (A) Comparisons of the *dpp* expression levels between left and right sides of the mantle tissues by qRT-PCR (data from [19]). (B) Comparisons of the pSMAD1/5/8 expression levels between left and right sides of the mantle tissues by western blotting (data from [19]. Abbreviations: L, left; *Lst* (S), *Lymnaea stagnalis* (Sinitral strain); *Lst* (D), *L. stagnalis* (Dextral strain) *Pvu*, *Patella vulgata*; R, right.

Molecular hypothesis of shell coiling evolution in gastropods

By integrating the molecular basis of shell development with the theoretical models, one can provide a molecular hypothesis for the shell coiling system. Considering the growing tube model [4], *dpp* expression patterns and results of its functional analysis of *dpp* [10, 19, and 16], at least *Curvature*, which is one of the parameters of the theoretical model, can be explained by the Dpp gradient, which regulates cell division activity in the shell gland and in the mantle edge.

In recent phylogenetic and paleontological studies of gastropods, it is reported that the last common ancestor of gastropods likely had a coiled shell, with the extinct Paragastropoda, which had a coiled shell, having diverged at the basal branch, and that the non-coiled shelled limpets including both true and false limpets evolved independently several times [21, 22] (Fig. 4). Thus, the shell morphological evolution from coiled to non-coiled might have lead by a change in the *dpp* expression patterns in the shell gland and in the mantle from asymmetric to symmetric (Fig. 4). However, it remains an important question as to how the asymmetric or symmetric expression of *dpp* is regulated in gastropods.

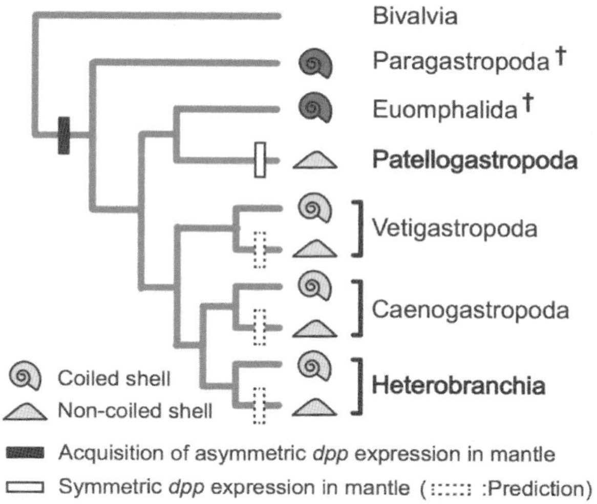

Fig. 4. Hypothesis of developmental evolution of spiral shell growth in Gastropoda. The last common ancestor of Gastropoda acquired the asymmetric *dpp* expression pathway in the mantle at one stage (black bar). Later, the Patellogastropoda lost this pathway and the non-coiled shell shape evolved in this group (gray bar). Other species with a non-coiled shell in Vetigastropoda, Caenogastropoda and Heterobranchia most likely evolved like Patellogastropoda (broken white bars). Illustration is modified from the figure in [19].

Perspectives on Understanding the Shell Shape Evolution in Mollusca

Various coiled shells can be observed not only in gastropods, but also in fossil cephalopods such as ammonoides. Unfortunately, most of the externally shelled cephalopods had become extinct by the end of Mesozoic, and coleoids, which have no external shell but have an internal shell have instead radiated since then. Thus, almost all modern cephalopods, except for *Nautilus*, have no external shell, making the mechanisms of shell formation or shell coiling difficult to understand in cephalopods. If Dpp expression is involved in the shell growth of *Nautilus* like in gastropods, it would aid in understanding how various shell shapes are formed and have evolved in extinct cephalopods with an external shell. However, in order to answer this question, one needs more investigations using the living coiled shelled cephalopod *Nautilus*.

In previous studies, the focus was only on "shell coiling" [10, 19], but that is only one of the features of molluscan shell morphology. In order to fully understand the shell morphology of molluscs, theoretical models such as the Okamoto model (Fig. 1) provide a powerful tool. As stated above, at least *Curvature* can be explained by the Dpp gradient. However, it remains to be answered how the asymmetric expression of *dpp* is regulated in early development. The parameter *Whorl Expansion* is likely explained by something like a growth hormone, and the remaining parameter *Torsion* could be explained by one of other signal molecules involved in axial patterning. However, the molecular bases of the two parameters *Whorl Expansion* and *Torsion* remain unclear. To reveal their genetic background, comparative gene expression analyses between related species with extremely different shell shapes; e.g. flat-spired shell vs. tall-spired shell would probably assist in finding the key molecule related to *Torsion*.

DISCUSSION AND CONCLUSIONS

In order to have insight into evo-devo of spiral shell growth in gastropods, previous studies of shell development were reviewed focusing on morphogenetic genes, especially on one of the signal molecule coding genes, namely *dpp*. Previous results [10, 19] suggested that the gradient of Dpp signal regulator correlates with the shell growth gradient, and that the evolution of their shell morphology might have been heralded by mutations in Dpp expression patterns in the shell forming tissues. In future studies, one would like to reveal the genetic background of shell formation and integrate those findings with the theoretical models. Thus a new integrated study interfacing three different research fields, i.e., evolutionary biology, developmental biology, and theoretical morphology should assist in revealing the underlying mechanisms of both evolutionary and developmental processes in the morphological diversification of molluscan shells.

ACKNOWLEDGEMENTS

We would like to thank Koji Noshita (University of Kyushu) for useful discussions on mathematical models. This study was supported by the JSPS Grants-in-Aid for Scientific Research 15104009.

REFERENCES

[1] Conklin EG (1897) The embryology of Crepidula. *J Morph* **13**: 1-226.

[2] Cather JN (1967) Cellular interactions in the development of the shell gland of the gastropod, Ilyanassa. *J Exp Zool* **166**: 205-223.

[3] Kniprath, E (1981) Ontogeny of the molluscan shell field. *Zoologica Scripta* **10**: 61-79.

[4] Okamoto T (1988) Analysis of heteromorphy ammonoids by differential geometry. *Palaeontology* **31**: 35-52.

[5] Ackerly SC (1989) Shell coiling in gastropods: analysis by stereographic projection. *Palaios* **4**: 374-378.

[6] Raup DM (1966) Geometric analysis of shell coiling: general problems. *J Paleontology* **44**: 1178-1190.

[7] Rice SH (1998) The bio-geometry of mollusc shells. *Paleobiology* **24**: 133-149.

[8] Hammer O and Bucher H (2005) Models for the morphogenesis of the molluscan shell. *Lethaia* **38**: 111-122.

[9] Urdy S, Goudemand N, and Bucher H. (2010) Allometries and the morphogenesis of molluscan shell: A quantitative and theoretical model. J *Exp Zool* **314B**: 280-302.

[10] Shimizu K, Sarashina I, Kagi H, and Endo K. (2011) Possible functions of Dpp in gastropod shell formation and shell coiling. *Dev Genes Evol* **221**: 59-68.

[11] Hinman VF, O'Brien EK, Richards GS, and Degnan BM (2003) Expression of anterior Hox genes during larval development of the gastropod Haliotis asinina. *Evol Dev* **5**: 508-521.

[12] Samadi L, Steiner G. (2009) Involvement of Hox genes in shell morphogenesis in the encapsulated development of a top shell gastropod (Gibbula varia L.). *Dev Genes Evol* **219**: 523-530.

[13] Moshel SM, Levine M, and Collier JR (1998) Shell differentiation and engrailed expression in the Ilyanassa embryo. *Dev Genes Evol* **208**: 135-141.

[14] Nederbragt AJ, van Loon AE, and Dictus WJ (2002) Expression of Patella vulgata orthologs of engrailed and dpp-BMP2/4 in adjacent domains during molluscan shell development suggests a conserved compartment boundary mechanism. *Dev Biol* **246**: 341-355.

[15] Iijima M, Takeuchi T, Sarashina I, and Endo K (2008) Expression patterns of engrailed and dpp in the gastropod Lymnaea stagnalis. *Dev Genes Evol* **218**: 237-51.

[16] Hashimoto N, Kurita Y, and Wada H (2012) Developmental role of *dpp* in the gastropod shell plate and co-option of the *dpp* signaling pathway in the evolution of the operculum. *Dev Biol* **366**: 367-373.

[17] De Robertis EM and Sasai Y (1996) A common plan for dorsoventral patterning in Bilateria. *Nature* 380: 37-40.

[18] Kin K, Kakoi S, and Wada H (2009) Novel role for dpp in the shaping of bivalve shells revealed in a conserved molluscan developmental program. *Dev Biol* **329**: 152-166.

[19] Shimizu K, Iijima M, Setiamarga DH, Sarashina I, Kudoh T, Asami T, Gittenberger E, and Endo K. (2013) Left-right asymmetric expression of dpp in the mantle of gastropods correlates with asymmetric shell coiling. *Evodevo* **4**: 15.

[20] Rogulja D and Irvine KD (2005) Regulation of cell proliferation by a morphogen gradient. *Cell* **123**: 449-461.

[21] Ponder WF, Lindberg DR (1997) Towards a phylogeny of gastropod molluscs: an analysis using morphological characters. *Zool J Linn Soc* **119**: 88-265.

[22] Aktipis SW, Giriibet G, Lindberg DR, and Ponder WF Gastropoda: an overview and analysis. (2008) **In**: *Phylogeny and Evolution of the Mollusca.* Ponder WF and Lindberg DR (Eds.) University of California Press, Berkeley and Los Angeles.

Morphological Variation of Pelvic Skeletal Elements of Hadrosaurid Dinosaurs Quantified Using Riemannian Analysis of Elastic Curves

ALBERT PRIETO-MÁRQUEZ[1] and SHANTANU H. JOSHI[2]

Abstract

Hadrosaurids were arguably the dominant megaherbivore dinosaurs during the last 15 million years of the Mesozoic Era. The morphological variation of a sample of ilia, pubes, and ischia corresponding to a taxonomically representative sample of hadrosaurid and basal hadrosauroid species is explored here. Patterns of inter-taxonomic variation that might unravel skeletal features useful for characterizing taxa were sought for. Each bone was decomposed into homologous structures represented by planar images and characterized by their anatomical boundaries. The morphological variation was quantified via a Riemannian Analysis of Elastic Curves (RAEC). In this technique, the boundaries of interest are represented by open continuous curves, parameterized by a square-root velocity function that maps each point of the curve to a vector in R^n. Scale-invariance is imposed by dividing this function by its norm. The space of such translation and scale invariant functions becomes an infinite-dimensional Hilbert sphere equipped with a well-defined Riemannian metric. Distances between shapes were calculated using geodesics (shortest paths) analytically specified on the sphere. Rotation and re-parameterization invariance was further imposed enabling a true elastic matching. The results showed that these skeletal structures are primarily useful in distinguishing among higher taxonomic levels (i.e., main hadrosaurid subclades Saurolophinae and Lambeosaurinae) and between Hadrosauridae and the out-group Hadrosauroidea. Only the supra-acetabular crest of the ilium and the pre-pubic process of the pubis allowed characterization at lower taxonomic levels.

INTRODUCTION

Hadrosaurids represent one of the most successful radiations of ornithischian dinosaurs [1]. Their fossil remains have been found in the Upper Cretaceous (Santonian to Upper Maastrichtian) strata of Europe, Asia, the Americas, and Antarctica [2]. The hadrosaurid record is among the richest of Mesozoic vertebrates and includes fully articulated skeletons, thousands of disarticulated bones from multindividual assemblages, growth series, soft tissue impressions, eggs, nests and neonates, and trackways [3]. Hadrosaurids are characterized by possessing hundreds of closely packed teeth (dental batteries) for slicing and/or grinding plant materials, mediolaterally expanded rostra, and hypertrophied nasal passages often associated with cranial crests [4, 5].

Because of their abundance, widespread distribution, and excellent preservation, hadrosaurids provide an excellent model for studying dinosaurian evolution and paleobiology. Such studies can be best approached in a phylogenetic context to provide an all-encompassing evolutionary understanding for the group. In recent years, various cladistic analyses have greatly improved the knowledge of hadrosaurid phylogenetic relationships [1, 6–9]. Approximately, 70% of the characters used in these phylogenetic analyses correspond to the skull. Within the

[1] School of Earth Sciences, University of Bristol, Life Sciences Building, 24 Tyndall Avenue, Bristol, BS8 1TQ, United Kingdom, a.prietomarquez@bristol.ac.uk.
[2] Ahmanson Lovelace Brain Mapping Center, UCLA School of Medicine, University of California Los Angeles, 635 Charles E. Young Dr. S., Los Angeles, California, USA.

hadrosaurid postcranium, the pelvic girdle contains most of the informative elements [10–12]. These elements, however, have hitherto provided little information to discriminate lower hadrosaurid taxa.

The morphological variation of the three bones that constitute the hadrosaurid pelvic girdle: the ilium, ischium, and pubis, is explored here. The aim was to find patterns of intra- and inter-taxonomic variation that might unravel skeletal features useful for characterizing hadrosaurid taxa and inclusion in phylogenetic inference of these dinosaurs (Fig. 1).

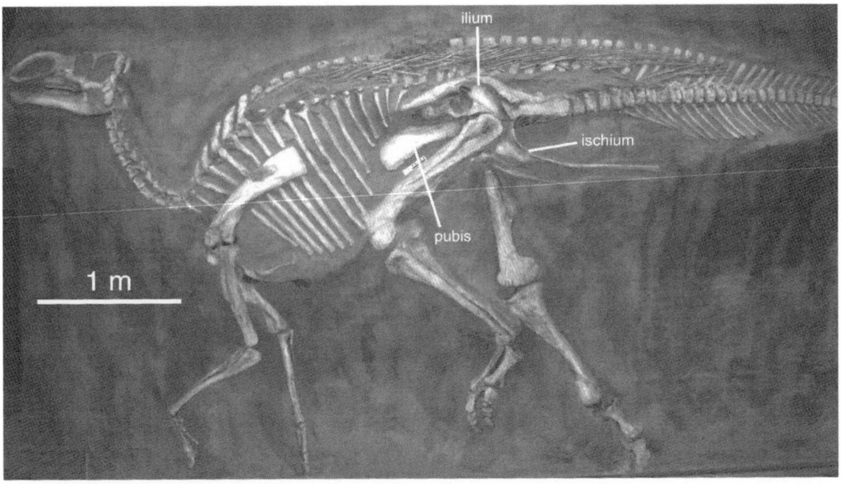

Fig. 1. Lateral view of the cast of a skeleton of the hadrosaurid dinosaur *Gryposaurus notabilis* (ROM 764) exhibited in the dinosaur galleries of the Royal Tyrrell Museum of Paleontology (Drumheller, Canada), showing the anatomical position of the ilium, pubis, and ischium. Photograph reproduced with permission of the Royal Tyrrell Museum of Paleontology.

MATERIALS AND METHODS

The Hadrosaurid Pelvic Girdle and the Sampled Material

As in other dinosaurs, the hadrosaurid pelvic girdle consists of three elements. The ilium lies dorsally and attaches medially to the sacral vertebrae. The pubis extends anteriorly and lays antero-ventral to the ilium. The ischium projects anteriorly and lays postero-lateral to the ilium.

The hadrosaurid ilium (Fig. 2) is an elongate and mediolaterally compressed bone. It consists of a central plate that extends anteriorly and posteriorly. A long pre-acetabular process projects antero-ventrally; the process is typically longer than the central iliac plate. A relatively shorter and deeper post-acetabular process extends posteriorly from the central iliac plate. The dorsal margin of the central plate of the ilium is sinuous and extends latero-ventrally into a large U or V-shaped supra-acetabular crest. The ventral margin of the central iliac plate forms the dorsal border of the acetabulum, which is bounded anteriorly and posteriorly by the pubic and ischiadic processes, respectively. In the current analyses, the boundaries of the pre-acetabular, post-acetabular, pubic, and ischiadic processes were isolated, as well as that of the supra-acetabular crest.

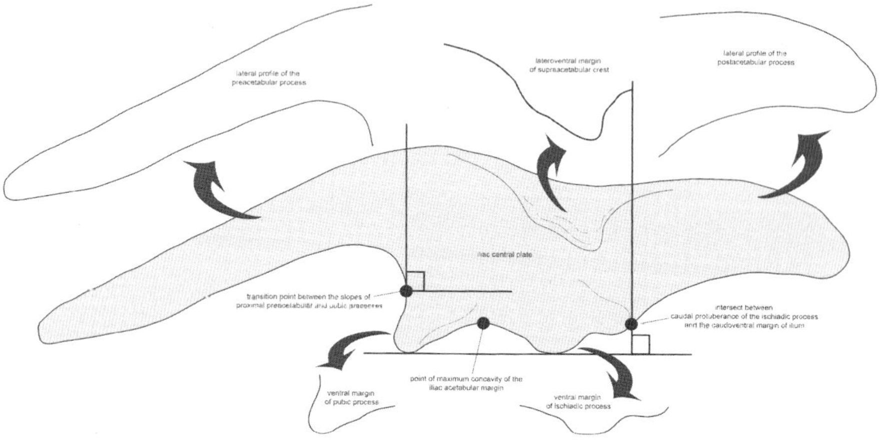

Fig. 2. Model of a typical hadrosaurid ilium (in lateral view) is displayed here showing the analyzed boundaries of the smallest homologous structures being considered and how these were delimited.

The pubis (Fig. 3) articulates with the pubic process of the ilium. Most of the pubis is formed by an extensive paddle-like plate (pre-pubic process) that extends anteriorly (as well as variably ventrally or dorsally) from the proximal region of the bone. The latter consists of the tetrahedral iliac process dorsally; the acetabular margin posteriorly, the short ischiadic and long rod-like process postero-ventrally. For the purpose of the current analyses, the pubic variation was decomposed with respect to the three structures; namely, the pre-pubic, iliac, and ischiadic processes (Fig. 3).

Finally, the ischium (Fig. 4) is mostly composed of a long shaft. Anteriorly, the ischium greatly expands dorso-ventrally, forming a bony plate. This proximal plate consists of the antero-dorsally-projected iliac process, a concave anterior margin that is the ischiadic contribution to the acetabulum (hip socket), and the short anteriorly-directed pubic process. In the current study, the morphological variation of the iliac and pubic processes, and that of the distal end of the ischiadic shaft was quantified (Fig. 4).

The sample includes 46 ilia, 21pubes, and 33 ischia. These bones correspond to a taxonomically representative sample of hadrosauroid taxa, consisting mostly of species of Hadrosauridae and a selection of out-groups to hadrosaurids. The two major subclades of hadrosaurids, Saurolophinae and Lambeosaurinae, are extensively represented in the sample. Only adults or large subadult specimens were included in the analyses. The smallest homologous structures of the ilium, pubis, and ischium were isolated in order to allow for inclusion of incompletely preserved bones in the analyses (Figs. 2–4). Had one used the entire contour of the ilium, ischium or pubis, one would have had to exclude those fragmentary or partially preserved elements from the analyses, thus considerably reducing sample size.

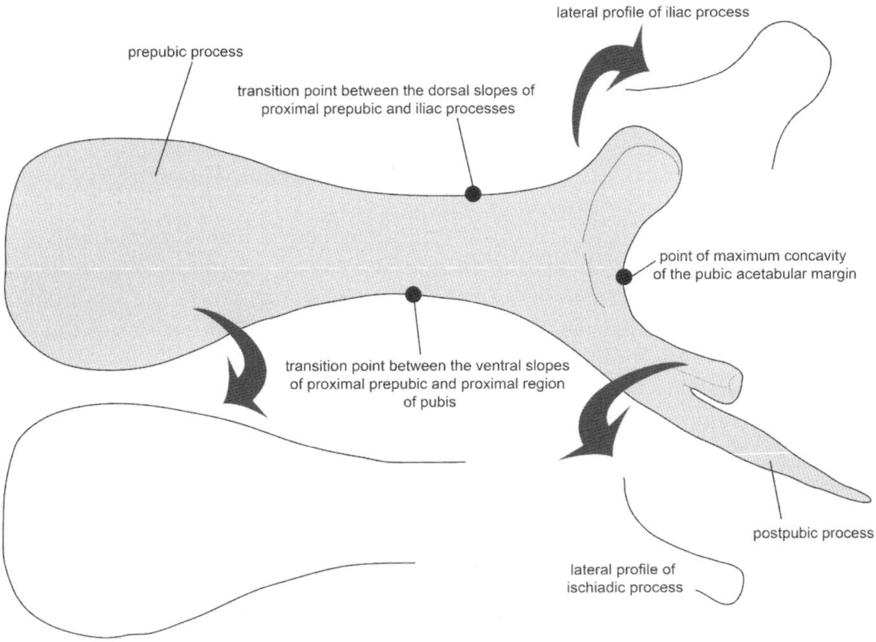

Fig. 3. Model of a typical hadrosaurid pubis (in lateral view) is shown here displaying the analyzed boundaries of the smallest homologous structures being considered and how these were delimited.

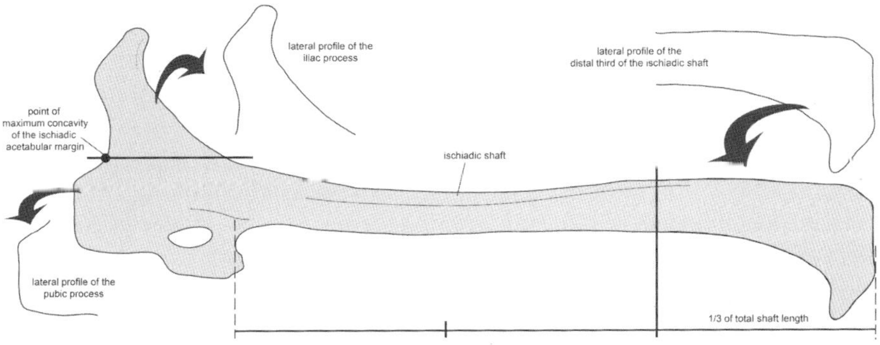

Fig. 4. Model of a typical hadrosaurid ischium (in lateral view) is displayed here showing the analyzed boundaries of the smallest homologous structures being considered and how these were delimited.

Morphometric Analyses

The morphological variation present in the samples of ilia, ischia, and pubes is quantified using Riemannian Analysis of Elastic Curves (RAEC) [13–15]. The skeletal boundaries of interest are represented by open continuous curves, parameterized by a square-root velocity function that maps each point on the curve to a vector in R^n. The square-root velocity function is invariant to translation. Scale-invariance is imposed by dividing this function by its norm. The space of such translation and scale invariant functions becomes an infinite-dimensional Hilbert sphere connected with a well-defined Riemannian metric. Distances between shapes are calculated using geodesics (shortest paths) analytically specified on the sphere. Further, rotation and re-parameterization invariance is imposed enabling true elastic matching in comparison to non-elastic matching methods [16–19].

Advantages of this approach are the representation of a multi-dimensional curve by a single function, L^2, a Riemannian metric enabling the elastic warping of shapes, efficient computing of geodesics using path-straightening flows, and an intrinsic statistical analysis on the tangent space using shape space differential geometry. Further details on this morphometric technique can be found in Joshi *et al.* [13].

Shape variability was summarized via Principal Components Analysis (PCA), a widely used dimension reduction technique that represents sample variation on a reduced number of axes [20]. It involves computing a covariance matrix of the observed samples, and then factoring it using singular value decomposition. The eigenvectors associated with this covariance matrix capture the dominant directions or modes of variation in the shape population. Based upon the eigenvalue distribution, one can calculate the variance explained by each principal component or eigenvector. Here, the percentage of the explained variance was chosen as a criterion to select the number of dominant eigenvectors that account for maximum variability in the data and discard the rest. This procedure provides one with a finite-dimensional compact Euclidean representation for the observed shape data and can also be useful for visualization of discriminative morphological patterns in the population [21], as well as hierarchical shape organization and retrieval [20, 22, and 23].

RESULTS

Ilium

In general, no clear pattern of variation was evident among hadrosauroid iliac structures (Figs. 5–8), with exception of the supra-acetabular crest (Fig. 9). Most of the variation of the pre-acetabular process was in regard to the degree of deflection and width of its proximal region. The sample of pre-acetabular processes appears scattered through the morphospace, with a mixing of saurolophine and lambeosaurine species. Notable outliers are the broad and relatively horizontal process of the basal hadrosauroid *Claosaurus agilis* (YPM 1190) and the strongly deflected and narrower processes of the saurolophines *Secernosaurus koerneri* (FMNH P13423) and *Saurolophus osborni* (AMNH 5220) (Fig. 5).

The triangular outline of the pubic process of the ilium displayed a similar scatter and mixing of saurolophine, lambeosaurine, and basal hadrosauroid taxa (Fig. 6). Notable outliers are the distinctly wide and shallow pubic process of the lambeosaurine *Parasaurolophus cyrtocristatus* (FMNH P27393) and the narrow and relatively deep process of the basal hadrosauroid *Claosaurus agilis* (YPM 1190).

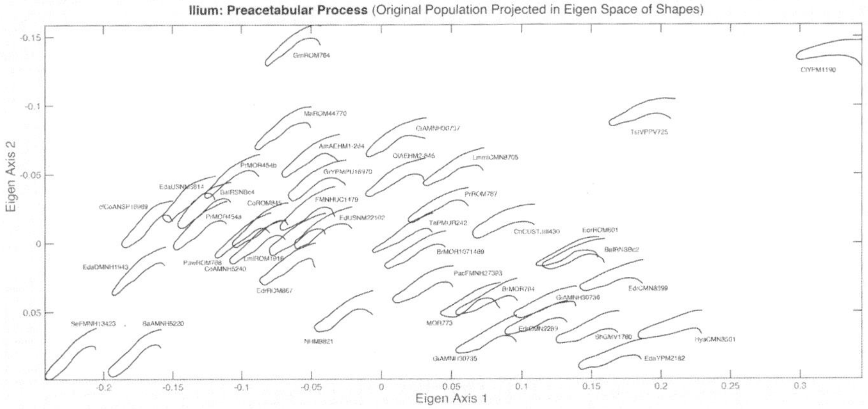

Fig. 5. Scatterplot of the PCA performed on the RAEC distance matrix for the pre acetabular process of the ilium of a sample of hadrosaurids and hadrosaurid outgroup taxa.

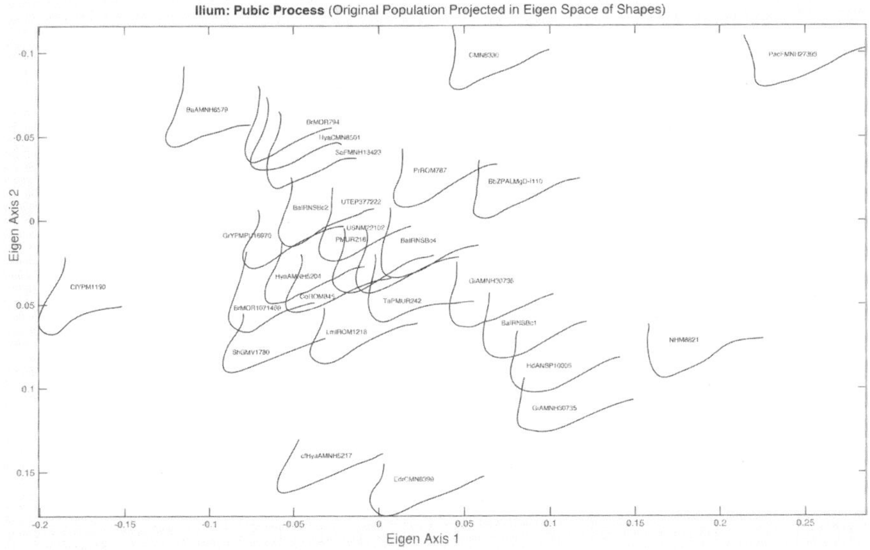

Fig. 6. Scatterplot of the PCA performed on the RAEC distance matrix for the pubic process of the ilium of a sample of hadrosaurids and hadrosaurid outgroup taxa.

The ventral profile of the ischiadic process of the ilium also displayed a widespread scatter of shapes through the morphospace encompassing the basal hadrosauroids and the two subclades of hadrosaurids (Fig. 7). Variation of this structure was in regard to the relative development of the two lobes of the ischiadic process and the relative offset between them. At one end of a continuous spectrum lies the relatively horizontal lobes of *Saurolophus osborni* (AMNH 5220), whereas on the other end lies the ischiadic process of the basal hadrosauroid *Tanius sinensis*

(PMU R242) with its dorsally offset and reduced caudal lobe. In general, basal hadrosauroids and a few saurolophines like *Brachylophosaurus canadensis* show a greater dorsal offset of the caudal lobe of the ischiadic process.

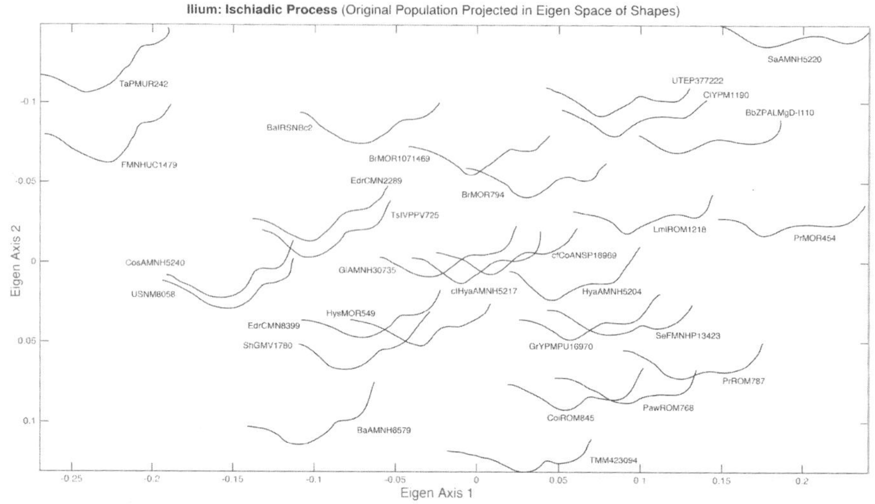

Fig. 7. Scatterplot of the PCA performed on the RAEC distance matrix for the ischiadic process of the ilium of a sample of hadrosaurids and hadrosaurid outgroup taxa.

The post-acetabular process of the ilium ranges from sub-triangular in basal hadrosauroids like *Tanius sinensis* (PMU R242) to sub-rectangular in the more derived hadrosaurids (Fig. 8). Within the latter, there was no apparent pattern of variation that allows distinction of lower taxa. A few outliers show extreme morphologies: the deep and sub-elliptical process of the lambeosaurine *Hypacrosaurus stebingeri* (MOR 549), the elongate and shallow process of the basal hadrosauroid *Claosaurus agilis* (YPM 1190), and the triangular and pointy process of the basal hadrosauroid *Bactrosaurus johnsoni* (IRSNB specimen) (Fig. 8).

Finally, the most informative structure of the ilium, the supra-acetabular crest, allowed characterization of the two major clades of hadrosaurids and basal hadrosauroids (Fig. 9). In the basal hadrosauroids, the latero-ventral profile of the supra-acetabular crest is shallow and elongate. Within the hadrosaurids, the saurolophines (except *Prosaurolophus maximus*) are characterized by a strongly asymmetrical (i.e., with much longer cranial limb) and D-shaped supra-acetabular crest. In contrast, *P. maximus* and the lambeosaurines exhibit V-shaped and slightly asymmetrical supra-acetabular crests. Notable exceptions are species of *Parasaurolophus* and *Olorotitan ararhensis* (AEHM 2/845) which show a D-shaped supra-acetabular crest that clustered with that of the saurolophines.

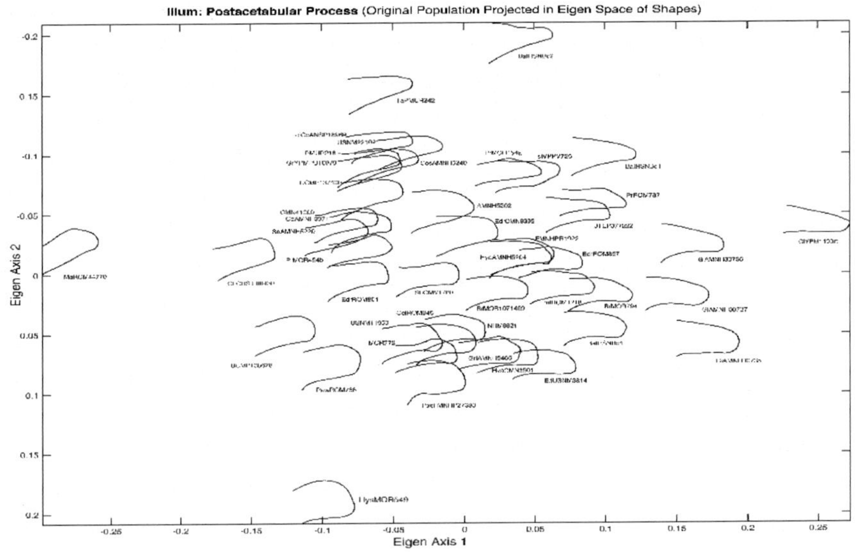

Fig. 8. Scatterplot of the PCA performed on the RAEC distance matrix for the supra-acetabular crest of the ilium of a sample of hadrosaurids and hadrosaurid outgroup taxa.

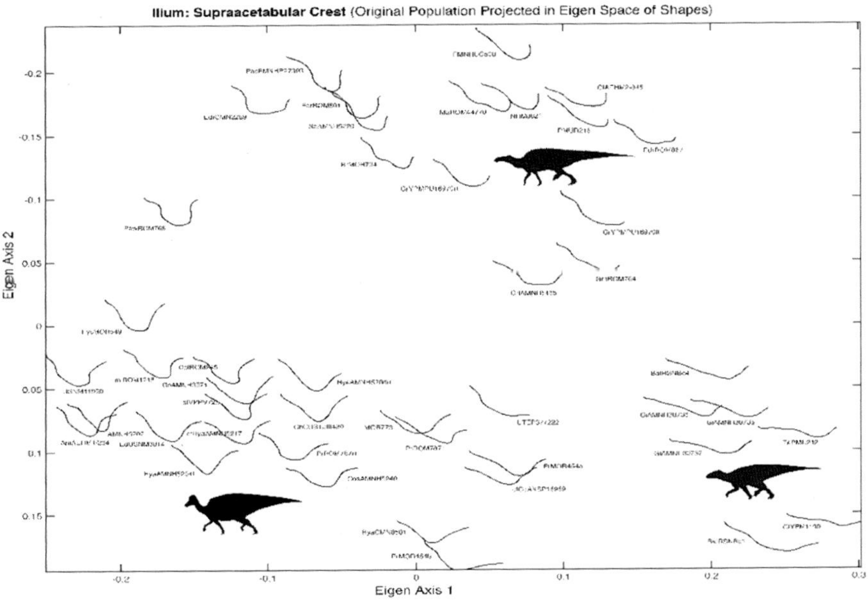

Fig. 9. Scatterplot of the PCA performed on the RAEC distance matrix for the post-acetabular process of the ilium of a sample of hadrosaurids and hadrosaurid outgroup taxa.

Pubis

Variation in the lateral profile of the iliac process of the pubis involved overall width and shape of the dorsal border (Fig. 10). Accordingly, lambeosaurines displayed a narrower and sub-rectangular process that contrasts with the broader and more rounded process of the saurolophines. Basal hadrosauroids exemplified by the *Bactrosaurus johnsoni* (IRSNB specimen; Fig. 10) take the saurolophine morphology to a greater extreme, i.e., being even broader and rounder.

The results also showed that the saurolophines may be distinct from the lambeosaurines in the lateral profile of the ischiadic process of the pubis. Specifically, the process of the lambeosaurines shows, in general, a more or less developed indentation along its ventral margin (Fig. 11).

The large pre-pubic process of the pubis is among the most informative elements of the pelvic girdle and the entire postcranium [1, 10, 12, and 19]. The lateral profile of the pre-pubic process represents a case in which it is not possible to recognize homologous points or landmarks among multiple specimens. This limitation is due to the continuous nature of the dorsal, anterior, and ventral margins of this osteological feature. Here, only those pubes that are complete or have experienced minor degrees of post-depositional damage during fossilization were included.

The results support characterization of most lambeosaurines by their 'fist-shaped' distal plates, in which the cranio-dorsal margin prominent and the ventral margin is gently directed cranio-ventrally (Fig. 12), in agreement with previous studies [10, 12, and 18]. A few lambeosaurine curves, however, show a more prominent cranio-ventral margin of the distal blade of the pre-pubic process (e.g., *Corythosaurus* sp., AMMH 3971). Such shapes overlap with the area of the morphospace occupied by a subset of brachylophosaurin (*Maisaura peeblesorum* and *Brachylophosauru canadensis*) and certain edmontosaurin (*Edmontosaurus regalis* and *Shantungosaurus giganteus*) saurolophines (Fig. 12). In general, saurolophines are characterized by cranio-ventrally oriented distal plates of the pre-pubic process. Differences lie in the depth and length of the proximal constriction ('pre-pubic neck') and the geometry of the distal plate (i.e., rounded versus sub-rectangular). Thus, a distinct subset of saurolophines occupies part of the dorsal region of the morphospace and is dominated the by the long proximal constrictions and sub-oval distal plates of species of *Edmontosaurus*, particularly *E. annectens* (Fig. 12). Longer proximal constrictions and sub-rectangular pre-pubic plates appear to characterize genera like *Gryposaurus* and *Prosaurolophus*, as previously noted by Prieto-Márquez and Salinas [24] and Prieto-Márquez [8].

Ischium

The results supported distinction of lambeosaurines from saurolophines on the basis of the morphology of the caudo-dorsal margin of the iliac process of the ischium. In agreement with previous studies [1,12], lambeosaurines share a 'thumb-shaped' process, in which the caudo-dorsal margin is recurved. This morphology is also present in hadrosaurid out-groups like *Gilmoreosaurus mongoliensis* (AMMH 30735) and *Bactrosaurus johnsoni* (SBDE 95E/26), to the extent that the processes of both groups could not be distinguished. In contrast, the iliac process of saurolophines lacks a recurved caudo-dorsal margin (Fig. 13).

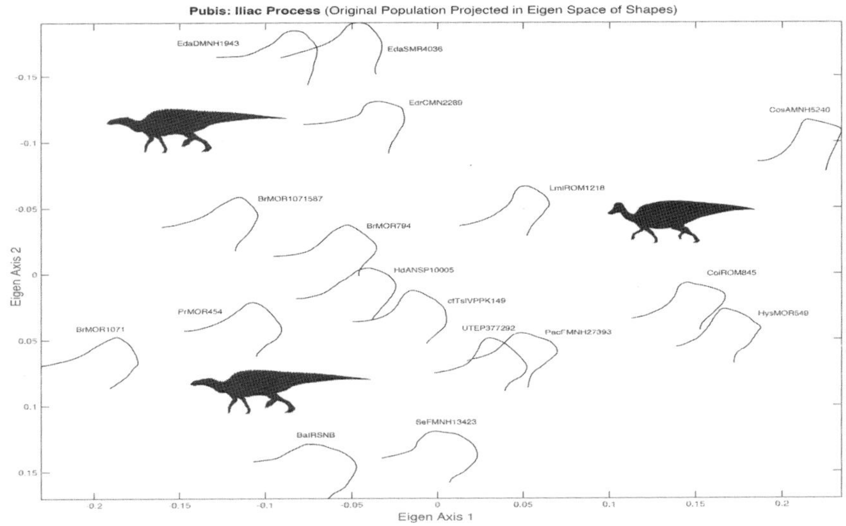

Fig. 10. Scatterplot of the PCA performed on the RAEC distance matrix for the pre pubic process of the pubis of a sample of hadrosaurids and hadrosaurid outgroup taxa.

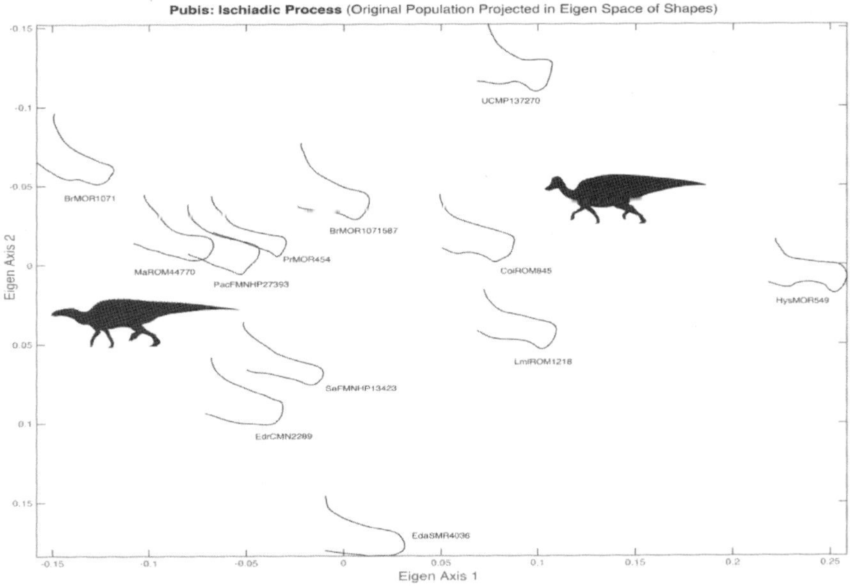

Fig. 11. Scatterplot of the PCA performed on the RAEC distance matrix for the iliac process of the pubis of a sample of hadrosaurids and hadrosaurid outgroup taxa.

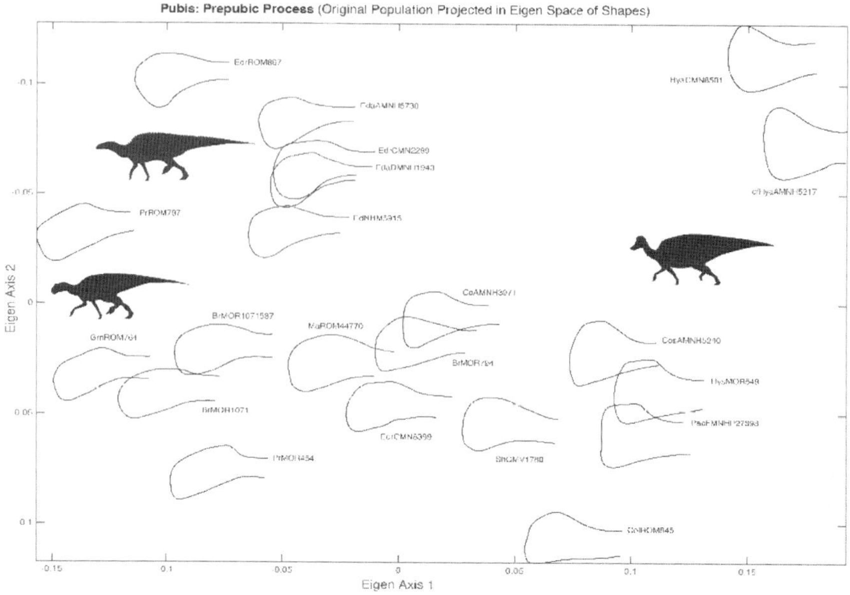

Fig. 12. Scatterplot of the PCA performed on the RAEC distance matrix for the ischiadic process of the pubis of a sample of hadrosaurids and hadrosaurid outgroup taxa.

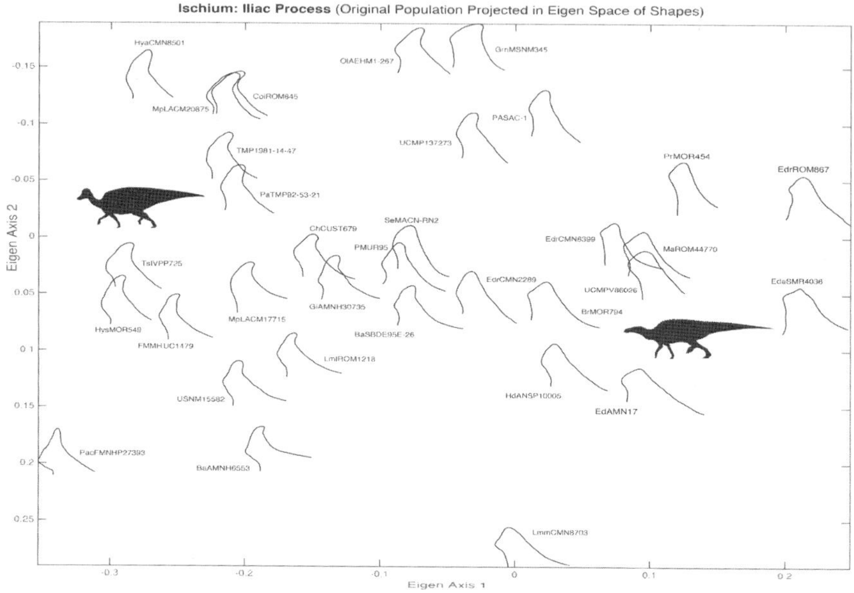

Fig. 13. Scatterplot of the PCA performed on the RAEC distance matrix for the iliac process of the ischium of a sample of hadrosaurids and hadrosaurid outgroup taxa.

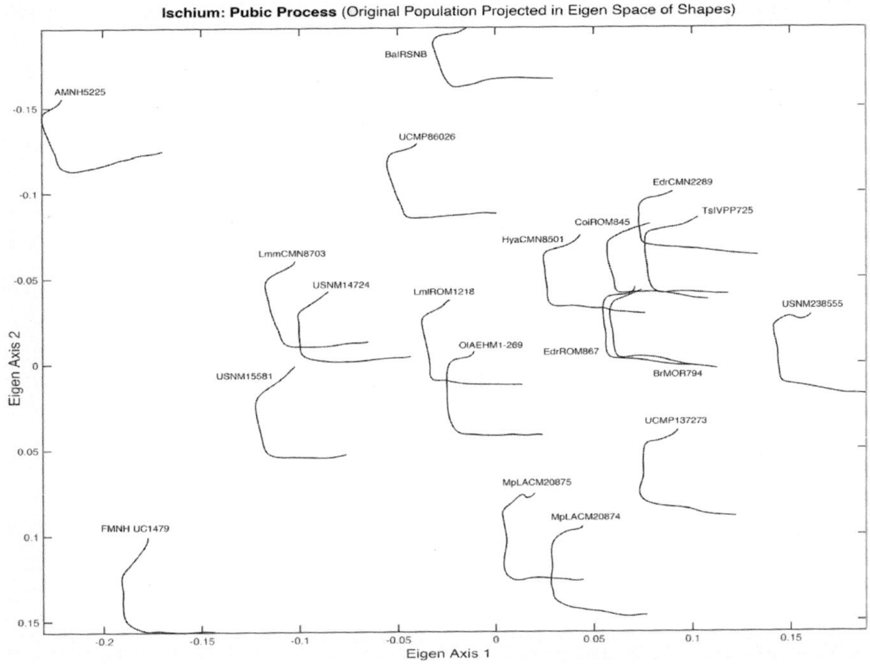

Fig. 14. Scatterplot of the PCA performed on the RAEC distance matrix for the pubic process of the ischium of a sample of hadrosaurids and hadrosaurid outgroup taxa.

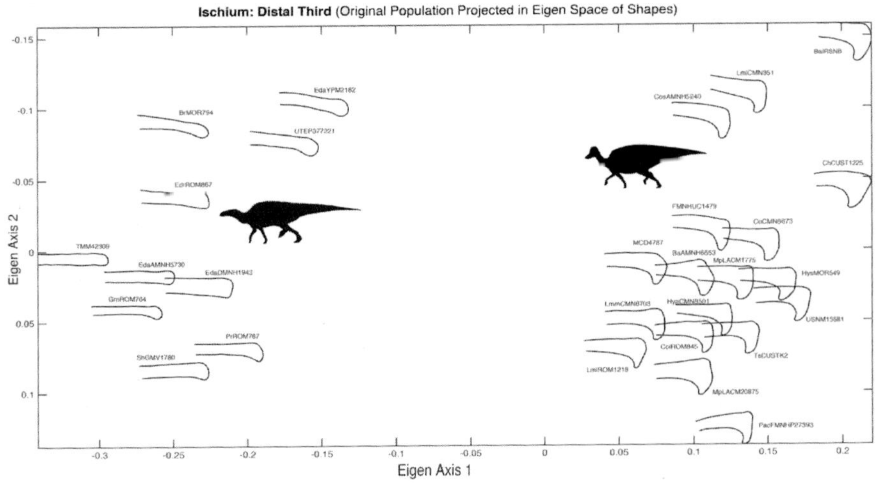

Fig. 15. Scatterplot of the PCA performed on the RAEC distance matrix for the caudal third of the distal process of the ischium of a sample of hadrosaurids and hadrosaurid outgroup taxa.

No pattern of variation was found in the pubic process of the ischium (Fig. 14). The sub-rectangular shapes of this structure appear scattered throughout the morphospace including overlapping variation of saurolophines, lambeosaurines, and outgroup hadrosauroid taxa.

Finally, the well-recognized [1, 12, and 18] distinction between the saurolophine and the lambeosaurine hadrosaurids in the morphology of the distal third of the ischium was also recovered. Thus, the lambeosaurines are characterized by a 'boot-like' distal process that extends ventrally and, often, also slightly cranially (Fig. 15). This morphology is also present in the basal hadrosauroids, as exemplified here by the presence of curves belonging to the hadrosauroid out-group *Bactrosaurus johnsoni* (IRSNB specimen) alongside those of the lambeosaurines. In contrast, the distal third of the ischiadic shaft of all saurolophines ends in a blunt margin, lacking a 'boot-like' distal process.

DISCUSSION AND CONCLUSIONS

Morphometric analysis of pelvic skeletal elements of hadrosaurid dinosaurs via RAEC indicates that this region of the postcranium contains various morphological characters useful for diagnosing the hadrosaurid taxa. However, such characters are primarily useful in distinguishing among higher taxonomic levels, mostly discriminating between the two major subclades of hadrosaurids, Saurolophinae and Lambeosaurinae, and between Hadrosauridae and closely related outgroup hadrosauroid species. Only a few structures are informative at lower taxonomic levels: the supra-acetabular crest of the ilium and pre-pubic process of the pubis can discriminate as far as the level of tribes, whereas only the latter may be able to inform on genera or species level.

Ongoing collection of additional hadrosaurid specimens, both at museum collections and from new field discoveries, should provide further testing for these conclusions, as well as providing for a more complete understanding of the utility of the pelvic and other postcranial bones of the hadrosaurids for taxonomy and for the inference of phylogenetic relationships.

ACKNOWLEDGEMENTS

We are grateful to the following individuals for access to specimens in their care: Mark Norell, Carl Mehling, Ted Daeschler, Daniel Brinkman, Peter Makovicky, John R. Horner, Michael K. Brett-Surman, Pacal Godefroit, Sandra Chapman, Paul Barrett, John Peel, Solweg Stuenes, Gerdard Plodowski, Olaf Vogel, Claudia Weißbrod, Bernard Battail, Philip Taquet, Tim Rowe, Jonathan R. Wagner, Lyndon Murray, Pamela Owen, Wann Langston, Luis Chiappe, Mark Goodwin, Patricia Holroyd, Kieran Shepherd, Margaret Feuerback, Robin Cuthbertson, Kenneth Carpenter, Kevin Seymour, Ian Morrison, David C. Evans, Robert R. Reisz, Brian Iwama, Cristiano Dal Sasso, and Simone Maganuco, Halska Osmólska, Karol Sabath, Xu Xing, Liu Liping, Liu Zhong Yun, Zhao Qin, Zheng Fang, Jia Cheng Kai, Zhang Zhijun, Jin Li-Young, Chen Jun, Li Tao, María del Carmen Perrillat, Claudia Inés Serrano Brañas, Marisol Montellano, Erandi García, René Hernández Rivera, José Manuel Padilla Gutiérrez, María del Rosario Gómez Núñez, Régulo

Zapata, Raúl Dante, Ruth Zúñiga, James Gardner, Alejandro Kramarz, Guillermo Salinas, Julia Desojo, Àngel Galobart, and Rodrigo Gaete. Thanks also to David Eberth for allowing the use of a photograph of the cast of MOR 764. This study was supported by the European Commission (Research and Innovation) through a Marie Curie Fellowship for Career Development, the Charlotte and Walter Kohler Charitable Trust, the National Science Foundation (EAR 0207744 and DBI 0446224 to G. Erickson) and a Visiting Scholarship from the The Field Museum.

REFERENCES

[1] Prieto-Márquez A (2010) Global phylogeny of Hadrosauridae (Dinosauria: Ornithopoda) using parsimony and Bayesian inference. *Zool J Linn Soc* **159**: 435-502.

[2] Lund EK and Gates TA (2006) A historical and biogeographical examination of hadrosaurian dinosaurs. *New Mex Mus Nat Hist Sci Bull* **35**: 263-276.

[3] Horner JR, Weishampel DB, and Forster CA (2004) Hadrosauridae. **In**: DB Weishampel, P. Dodson and H. Osmólska (Eds.), *The Dinosauria, Second Edition*, pp. 438-463. University of California Press, Berkeley.

[4] Lull RS and Wright NE (1942) Hadrosaurian dinosaurs of North America. *Geol Soc Am Special Papers* **40**: 1-242.

[5] Prieto-Márquez A and Wagner JR (2013) The 'unicorn' dinosaur that wasn't: a new reconstruction of the crest of *Tsintaosaurus spinorhinus* and the early evolution of the lambeosaurine crest and rostrum. *PLoS ONE* **8**(11): e82268.

[6] Evans DC and Reisz RR (2007) Anatomy and relationships of *Lambeosaurus magnicristatus*, a crested hadrosaurid dinosaur (Ornithischia) from the Dinosaur Park Formation, Alberta. *J Vert Paleontol* **27**: 373-393.

[7] Godefroit P, Shulin H, Tingxiang Y, and Lauters P (2008) New hadrosaurid dinosaurs from the uppermost Cretaceous of northeastern China. *Acta Pal Pol* **53**: 47-74.

[8] Prieto-Márquez A (2014) Skeletal morphology of *Kritosaurus navajovius* (Dinosauria: Hadrosauridae) from the Late Cretaceous of the North American south-west, with an evaluation of the phylogenetic systematics and biogeography of Kritosaurini. *J Syst Palaeont* **12**: 133-175.

[9] Prieto-Márquez A, Dalla Vecchia FM, Gaete R, and Galobart, A (2013) Diversity, relationships, and biogeography of the lambeosaurine dinosaurs from the European Archipelago, with description of the new aralosaurin *Canardia garonnensis*. *PLoS ONE* **8**(7): e69835.

[10] Brett-Surman MK (1975) The appendicular anatomy of hadrosaurian dinosaurs. Unpublished Master's Thesis, University of California.

[11] Chapman R and Brett-Surman MK (1990) Morphometric observations on hadrosaurid ornithopods. **In**: P Currie, and K Carpenter (Eds.) *Dinosaur Systematics, Approaches and Perspectives*, pp. 163-177. Cambridge University Press, Cambridge.

[12] Brett-Surman MK and Wagner JR (2007) Discussion of character analysis of the appendicular anatomy in Campanian and Maastrichtian North American hadrosaurids—variation and ontogeny. **In**: Carpenter K (Ed.) *Horns and beaks: ceratopsian and ornithopod dinosaurs*, pp. 135-169. Indiana University Press, Bloomington.

[13] Joshi SH, Klassen E, Srivastava A, and Jermyn I (2007) A novel representation for Riemannian analysis of elastic curves in R^n. *Proc IEEE Comput Soc Conf Comput Visual Pattern Recognition* 1-7.

[14] Joshi SH, Klassen E, Srivastava A, and Jermyn I (2007) Removing shape-preserving transformations in square-root elastic (SRE) framework for shape analysis of curves. *Proc IEEE Comput Soc Conf Comput Visual Pattern Recognition* **4679**: 387-398.

[15] Srivastava A, Klassen E, Joshi SH, and Jermyn IH (2011) Shape analysis of elastic curves in Euclidean spaces. *Pattern Analysis and Machine Intelligence, IEEE Transactions* **33** (7): 1415-1428.

[16] Srivastava A, Mio W, Klassen E, and Joshi SH (2003) Geometric analysis of continuous, planar shapes. *Proc IEEE Comput Soc Conf Comput Visual Pattern Recognition* 341-356.

[17] Klassen E, Srivastava A, Mio W, and Joshi SH (2004) Analysis of planar shapes using geodesic paths on shape spaces. *Pattern Analysis and Machine Intelligence, IEEE Transactions* **26**: 372-383.

[18] Joshi SH, Prieto-Márquez A, and Parker WC (2011) A landmark-free method for quantifying biological shape variation. *Biol J Linn Soc* **104**: 217-233.

[19] Prieto-Marquez A, Joshi SH, and Parker WC (2006) Morphometrics of the hadrosaurid pelvic girdle using a new method of shape analysis. *J. Vertebr. Paleontol.* **26** (Suppl. to 3): 112A–112A.

[20] Srivastava A, Joshi SH, Mio W, and Liu X (2005) Statistical shape analysis: clustering, learning, and testing. *Pattern Analysis and Machine Intelligence, IEEE Transactions* **27**: 590-602.

[21] Prieto-Marquez A, Chiappe LM, and Joshi SH (2012) The lambeosaurine dinosaur *Magnapaulia laticaudus* from the Late Cretaceous of Baja California, Northwestern Mexico. *PLoS ONE* **7**(6): e38207.

[22] Joshi SH and Srivastava A (2003) A geometric approach to shape clustering and learning. *Proc IEEE Workshop Stat Sig Proc*, 302-305.

[23] Joshi SH, Srivastava A, and Mio W, and Liu X (2004) Hierarchical organization of shapes for efficient retrieval. *Proc Euro Conf Comp Vis ECCV* **3023**: 570-581.

[24] Prieto-Márquez A and Salinas G (2010) A re-evaluation of *Secernosaurus koerneri* and *Kritosaurus australis* (Dinosauria, Hadrosauridae) from the Late Cretaceous of Argentina. *J Vertebr Paleontol* **30**: 813-837.

APPENDICES

Appendix I. Institutional Abbreviations

Each individual specimen of the sample is designated by a taxon abbreviation (see below) followed by the catalog number, which includes an acronym that refers to the repository institution. These acronyms are: AEHM, Amur Natural History Museum, Blagoveschensk, Russia; AMNH, American Museum of Natural History, New York, USA; ANSP, Academy of Natural Sciences of Philadelphia, Pennsylvania, USA; CMN, Canadian Museum of Nature, Ottawa, Canada; CUST, Changchun University of Science and Technology, Changchun, Jilin, China; DMNH, Denver Museum of Nature and Science, Denver; GDF, Gadofaoua; specimen cast accessed at the Muséum National d'Histoire Naturelle, Paris, France; GMV, National Geological Museum of China, Beijing, China; FMNH, The Field Museum, Chicago, USA; IRSNB, Institut Royal des Sciences Naturelles de Belgique, Brusels, Belgium; IVPP, Institute of Vertebrate Paleontology and Paleoanthropology, Beijing, China; LACM, Los Angeles County Museum of Natural History, Los Angeles, USA; MACN, Museo Argentino de Ciencias Naturales, Buenos Aires, Argentina; MCD, Museu de la Conca Dellà, Isona, Spain; MOR, Museum of the Rockies, Bozeman, Montana, USA; MSNM, Museo Civico di Storia Naturale di Milano, Milano, Italy; NHM, The Natural History Museum, London, UK; PASAC, Paleontological Association of Sabinas, Coahuila, Mexico; PMU, Museum of Evolution, Uppsala University, Uppsala, Sweden; ROM, Royal Ontario Museum, Toronto, Ontario, Canada; SBDE, Sino-Belgian Dinosaur Expedition collection, Institut Royal des Sciences Naturelles de Belgique, Brussels, Belgium; SM, Naturmuseum Senckenberg, Frankfurt, Germany; TMM, Texas Memorial Museum, Austin, Texas, USA; TMP, Royal Tyrrell Museum of Paleontology, Drumheller, Alberta, Canada; USNM, National Museum of Natural History, Washington DC, USA; UTEP, Centennial Museum at the University of Texas at El Paso, Texas, USA; YPM, Yale Peabody Museum of Paleontology, New Haven, USA; ZPAL, Instytut Paleobiologii, Polska Akademia Nauk, Warsaw, Poland.

Appendix II. Taxonomic Abbreviations

Each sampled individual specimen is designated by a taxon abbreviation, used here as a prefix to the catalog number. The sample of Hadrosauridae include members of both Saurolophinae and Lambeosaurinae. Included Saurolophine taxa are: Bb, *Barsboldia sicinskii*; Br, *Brachylophosaurus canadensis*; Ed, *Edmontosaurus* sp.; Eda, *Edmontosaurus annectens*; Edr, *Edmontosaurus regalis*; Gr, *Gryposaurus* sp.; Grl, *G. latidens*; Grn, *G. notabilis*; Ma, *Maiasaura peeblesorum*; Pr, *Prosaurolophus maximus*; Sa, *Saurolophus osborni*; Se, *Secernosaurus koerneri*; and Sh, *Shantungosaurus giganteus*. Lambeosaurine taxa include the following: Am, *Amurosaurus riabinini*; Co, *Corythosaurus* sp.; Coi, *C. intermedius*; Cos, *C. casuarius*; Ch, *Charonosaurus jiayinensis*; Lml, *Lambeosaurus lambei*; Lmm, *L. magnicristatus*; Hy, *Hypacrosaurus* sp.; Hya, *H. altispinus*; Hys, *H. stebingeri*; Mp, *Magnapaulia laticaudus*; Ol, *Olorotitan arharensis*; Pac, *Parasaurolophus cyrtocristatus*; Paw, *P. walkeri*; and Ts, *Tsintaosaurus spinorhinus*. Finally, outgroup taxa to Hadrosauridae included the following hadrosauroids: Cl, *Claosaurus agilis*; Ba, *Bactrosaurus johnsoni*; Gi, *Gilmoreosaurus mongoliensis*; and Ta, *Tanius sinensis*. The basal hadrosaurid *Hadrosaurus foulkii* is also included and abbreviated as 'Hd'.

The Direct Analysis of Digital Images (Eigenimage) with a Comment on the Use of Discriminant Analysis in Morphometrics

NORMAN MACLEOD[1]

Abstract

Contemporary shape analysis procedures are limited at present in their ability to represent, quantify, and summarize patterns of variation in any aspect of variation that cannot be represented by a set of landmarks or semilandmarks and/or features that are not present in every specimen across a population or sample. Yet, the problems presented to shape analysts for resolution fail to respect these rather stringent conceptual boundaries. An often-overlooked aspect of quantitative shape data inherent to all digital images is the pixel grid itself. For sets of images scaled to comparable pixel resolutions, this collection of quantitated brightness and colour values can be regarded as a set of semilandmarks that contain both spatial and feature-representational information. As such, the pixel grid can be considered to specify a 'shape' in the manner of a complex 3D (grayscale images) or as a 5D (colour images) surface and subjected to geometric morphometric analysis. Data analytic procedures necessary to implement the direct analysis of such digital image 'shapes' using standard geometric morphometric procedures are presented and discussed via reference to a hypothetical ladybird beetle dataset whose modes of variation are not well suited to traditional forms of shape analysis. Results indicate that all the formalisms associated with traditional shape analysis (e.g., summarisation of major patterns of form variation, visual portrayal of shape-based comparisons in an ordination space, empirical modelling of various locations within ordination spaces to inform either geometric and biological interpretation, *a priori* group discrimination, statistical testing of putative group differences) can be realised for the direct analysis of digital images. In addition, recent concerns that have been raised regarding the appropriateness of canonical variates analysis (CVA) as a useful procedure for shape data, and the proposal of between-groups principal components (BG-PCA) as an adequate alternative, are considered in light of the empirical results generated by the ladybird beetle dataset. So long as the ordination results of a CVA are tested statistically against expectations of reasonable null hypotheses (e.g., datasets drawn randomly from the same parent population), and so long as shape models are used to interpret the geometries of CVA ordination spaces, there seems no practical reason to prefer BG-PCA to CVA for the purpose of assessing group-structured shape variation hypotheses. When BG-PCA and CVA produce very different ordination patterns, the CVA result is likely to be the more useful especially insofar as it (1) utilises more information directly related to the group-discrimination problem (2) creates an ordination space more closely tied to the logic of the group-discrimination problem, (3) exhibits no insurmountable interpretive or statistical testing issues, and (4) usually achieves better group-discrimination results, than the former. Indeed, many of the same practical criticisms levelled at CVA are also true of BG-PCA, standard PCA, Procrustes PCA (= relative warps analysis), principal/partial warps analysis, and other geometric morphometric procedures when these are used to portray distinctions between *a priori*-defined groups. Precisely the same statistical procedures need to be implemented to assess and interpret the significance of the group-specific ordinations produced by these, alternative (and usually suboptimal), group-level data analysis procedures as are necessary for CVA.

[1] The Natural History Museum, Cromwell Road, London, SW7 5BD, UK, N.MacLeod9@gmail.com.

INTRODUCTION

Perhaps the most basic concept in the field of shape analysis is that of the variable. Variables are the observations we make, usually in the form of measurements, on the set of specimens that comprise a sample of some population of interest. Variables can come in many types and forms (e.g., nominal, ordinal, interval, ratio, etc.). But in all cases, for the purpose of describing and comparing the shapes exhibited by a set of specimens across a sample — not to mention drawing inferences about the population(s) from which the sample was collected — the variables that are used to describe the specimens' shape *are* the specimen; the only information any shape analysis procedure has access to. As such, it is of the utmost importance that the variables chosen to represent the specimens be appropriate, both to the specimens in question and to the hypotheses under consideration. If a set of variables does not meet these criteria in some reasonable and defensible manner it is likely that any results generated from their analysis will be compromised and/or (ultimately) questioned. For example, if one is interested in phenomena pertaining to the arrangement of component parts of the specimen relative to each other, and if the majority of those component parts are not located on the specimen's outline, it makes little sense to restrict data collection to the geometry of specimen outlines. Similarly, if one is interested in questions pertaining to the general form of the specimens, and if the general form is best represented by the specimen's outline, it makes little sense to restrict data collection to a small set of landmarks located within the outline.

Much of what has been written about shape sampling strategies [e.g., 1, 2, and 3] focuses on the advantages of one variable type over another in the context of particular (types of) applications. Nevertheless, all decisions regarding how to represent the aspects of shape variation individual investigators find interesting, important or useful involve compromises. In making these decisions it is important to remember not to approach them from the standpoint of an ideological commitment to one or another type of variable, or one or another approach to shape analysis, to the extent that the universe of data analysis possibilities is restricted needlessly at the outset of an investigation. In terms of biological shape analysis, the best advice is, where possible, to let the taxonomists who have experience in identifying specimens and/or placing them in categories provide guidance. Over the years they will have developed various 'rules of thumb' that guide their own qualitative assessments of patterns of morphological variation. It might be that a change in the manner in which forms are measured or otherwise assessed may produce results that are superior to these rules of thumb, especially when they are applied quantitatively and consistently. However, following the taxonomist's best advice in this area allows the shape analyst to access and benefit from this experience and, at the very least, has the advantage of providing a place to begin one's own consideration of this critical issue.

However, what if even after having taken this advice one remains unsure as to what parts of the morphology are important in terms of assessing similarity and/or difference relations among the specimens comprising a sample? It might be that the qualitative taxonomist's rules of thumb are arbitrary, difficult to quantify, or used inconsistently from specialist to specialist. Even more importantly, suppose one wants to include as much morphological information about ones specimens as possible and/or are uncomfortable with the idea of abstracting the specimens down to a (relatively) small set of variables *a priori*? Is there a way of handing this more generalized problem within the context of biological shape analysis?

To illustrate this all-too-often encountered problem consider the image dataset illustrated in Fig. 1. This small collection of ladybird beetle paintings serves to illustrate the full range of morphological variation in biological datasets. Among this representation of morphology, one sees copious variation in body shape, leg and antennae pose, as well as both colouration and colour texture of the elytra, pronotum and head shields. These sorts of variation patterns are typical of many morphological datasets — open just about any well-curated museum drawer; this is what will be seen. While these specimens may or may not incorporate all these types of variation, the methods developed here can be applied usefully to any images of biological specimens irrespective of what type of morphological variation they display. However, the most important point raised by this dataset is that none of the standard numerical shape analysis and/or morphometric tools can handle the problem of characterizing any but the most trivial similarity/dissimilarity relations illustrated by this sample.

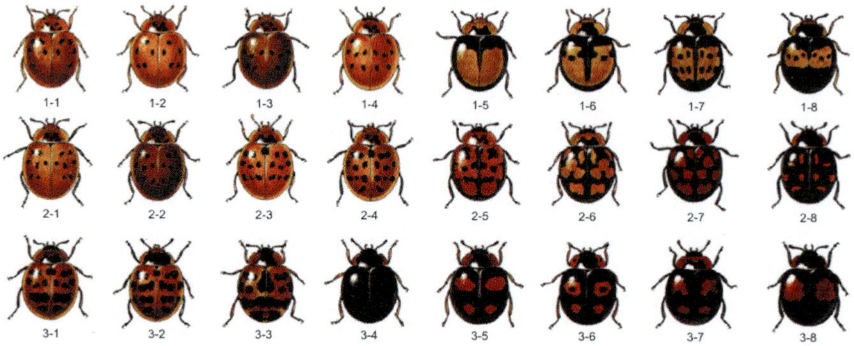

Fig. 1. Paintings of 24 Harlequin ladybird beetles (*Harmonia axyridis*, Family Coccinellidae). Note the complexity of variation in body form, colour, the distribution, numbers, sizes and shapes of spots. The complex nature of these variational modes makes this sample difficult to characterize using standard linear distance, landmark, or semilandmark variables. Copyright Scientific American Inc., images used with permission.

The patterns we see in Fig. 1 are too complexly structured and inconsistently distributed to be characterized adequately by sets of linear distances, landmarks or semilandmarks. Yet, in the absence of taking some sort of measurement, one cannot answer even the most basic questions about this sample. What is its mean form and colour pattern? Is morphological variation distributed continuously or discontinuously? Does the distribution of forms and colour blotches have a single or multiple modes? If the latter, how are those modes arranged relative to one another? Moreover, these are the easy, descriptive questions. If one wants to provide answers to more complex biological and/or evolutionary questions, such as how these patterns of morphological variation covary with environment, geography, ecology, behaviour, genotype, phylogeny, etc., an ability to use shape variables to compare each of these forms to one another quantitatively is crucial. However, if none of the tools, concepts, or methods currently regarded as being part of the shape analysis toolbox can accurately or adequately represent all these different modes of variation, must one conclude that shape analysis pertains only to some small subset of morphological data-analysis problems biologists encounter on a day-to-day basis? Is the morphometric approach unsuited to the majority of routine morphological data-analysis situations with which one is confronted? Is it really true that the best one can do in this case is shrug one's shoulders and go back to the qualitative inspection and appeals to 'authority' in addressing these questions posed above?

The standard morphometric response to a situation like this would be to either abandon a morphometric approach entirely or collect a small set of landmarks or outline semilandmarks and analyse these rather than the colour blotch patterns. After all, if there are no form-related features common to the blotch patterns in all specimens there would appear to be no basis on which to compare them. But this is obviously a false logic. Entomologists, non-specialist collectors, and even ladybird beetles' predator species are all able to make comparisons between these patterns. Indeed, in the case of Batesian mimicry, this capacity for comparison to an idealized form constitutes the very point of this intriguing evolutionary phenomenon! Accordingly, if entomologists and butterfly predators can make comparisons between morphologies like these, shape analysts and morphometricians should be able at least to make a stab at treating the same problem quantitatively. However, the question that then remains is how to best accomplish this.

MATERIAL AND METHODS

The current standard way of approaching the morphology-sampling problem is to find a series of topologically homologous, relocatable points across the forms of interest and record their coordinate positions. These points may mark the position of a singular structure of interest uniquely (landmarks) or be used in aggregate to quantify a boundary of interest (semilandmarks).

In either instance, the structures whose position(s) are described by the landmark or semi-landmarks coordinates are, in most cases (but see [4]), required to be present in all specimens in the sample.[2] Nevertheless, if the image of a specimen has been digitized, and if the sets of specimens have brought to a common size and orientation within each image frame, each specimen has already been subdivided into a series of topologically homologous coordinate points — the pixel grid (Fig. 2).

---

[2] Many discussions of the difference between landmarks and semilandmarks have argued that the former conform to the concept of biological homology. Unfortunately, this is an erroneous assertion. The concept of biological homology pertains to the identification of common derivation of tissues, organs, and support structural elements from a common ancestor, not to the derivation of unitary points on such structures from a common ancestor. Regardless of the type of comparison being made the matching criterion between points is always topological (or functional), never biological in the evolutionary sense of that term.

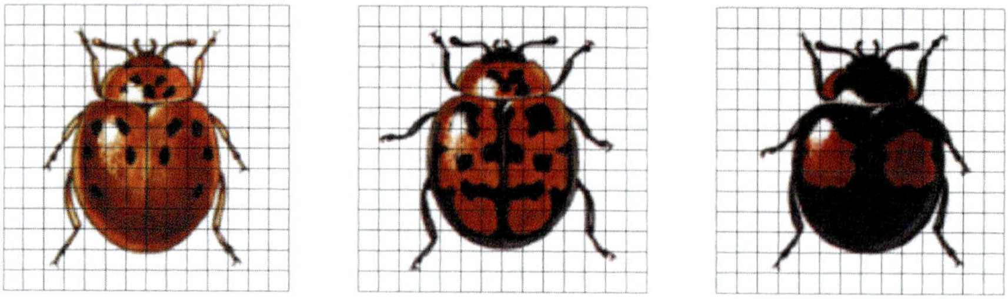

Fig. 2. Three-example representation of ladybird beetle morphology and ornamentation pattern with a superposed 14 x 14 pixel grid.

In order to appreciate the concept of the pixel grid as a 'shape', consider the grid as representing a 2D set of equally spaced semilandmark point-variables whose 'values' are the brightness (greyscale) or red, green, and blue (RGB) colour values assigned to each pixel. If these values are graphed by extending the height of the pixel up a hypothetical z-axis that represents the brightness/RGB colour scale the image can be regarded as a three dimensional (3D: greyscale or pooled RGB values) or five dimensional 5D: RGB values graphed separately) 'shape'. These image 'shapes' conform in all respects to the geometric morphometric concept of a configuration of points which represents the morphology of the specimen and this representation can be achieved irrespective of degrees of fine-scale correspondence among the geometries of the forms in question. So long as the dimensions and resolution of the pixel grid remain constant, and so long as the specimens are oriented in the grid in some reasonably consistent manner, this grid can be used to formulate comparable descriptions of these form geometries, including the colour and texture information irrespective of whether particular forms, colours and/or textures are present in all specimens comprising the sample.

Quantitative descriptions of specimen variation extracted in this manner can be organized into many formats. If only the outline of the form is needed, all pixels whose colour or grey level differs from that of the background can be assigned the same colour (usually black or white depending on the background).[3] If colour is not a parameter of interest *per se*, the grid can be set to sample only the grey-level values of the individual pixels comprising the image. If colour is of interest, the sampling grid can be used to extract the red, blue, and green (RGB) values of each pixel. Fig. 3 illustrates the effect various grid-sampling decisions have on the representation of a typical ladybird beetle form.

[3] In this case, the data would be more efficiently represented as a set of boundary outline semilandmark coordinates rather than a coordinate sampling grid.

What is remarkable about this comparison of sampling resolutions and colour formats is the degree of form, colour, and texture information content retained even at relatively low pixel grid resolutions. This suggests that, for the purpose of form, colour, and texture characterization, many pixels in a normal-resolution image (e.g., 72 pixels per inch) are redundant: the value of any particular pixel is much the same as the values of the pixels immediately adjacent. In mathematical terminology, this self-similarity is termed spatial autocorrelation. If possible, the spatial autocorrelation of raw morphological data should be reduced prior to data analysis so that the effective dimensionality of the data-analysis problem can be optimized.

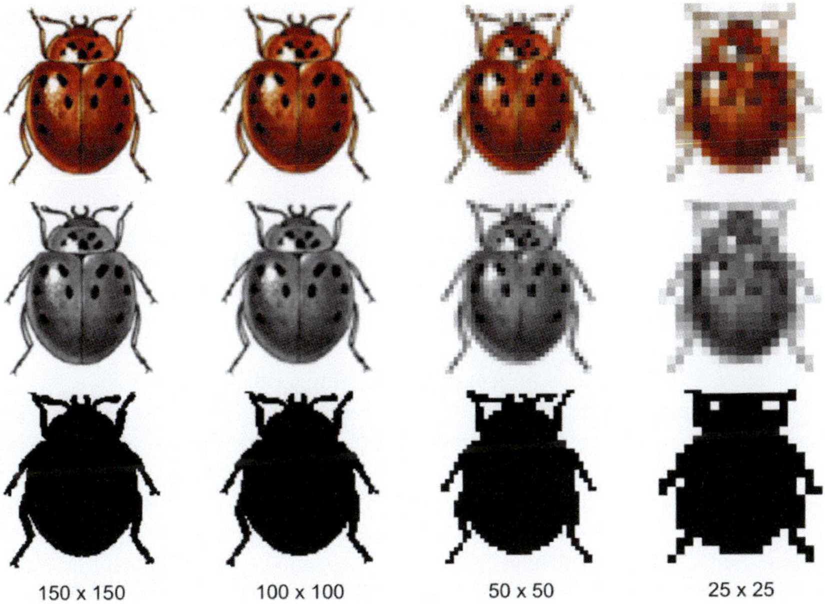

Fig. 3. Ladybird beetle morphology represented as four different pixel grid sampling resolutions and three different colour formats: 8-bit RGB values (upper row), 8-bit greyscale values (middle row), and binary (1-bit) values (bottom row). Note fidelity with which detail is retained even at low spatial resolutions.

As an initial step, it is possible to decrease the image resolution to the point where the number of self-similar pixels is minimized relative to the overall information content of the pixel grid. While there are algorithmic ways of accomplishing this minimization, for biological images it is usually recommended that one adopt an experimental approach to determining the spatial resolution and colour format required for each sample on a case-by-case basis. This recommendation reflects a personal belief that the analyst (or taxonomic specialist), rather than an algorithm, is usually best placed to determine which morphological features that need to be included in an image to ensure the down-sampled image set remains appropriate for the hypothesis test(s) under consideration.

Once the spatial and colour resolution necessary to represent the set of images has been established it is a simple matter to assemble the resultant pixel brightness or colour values into a mathematical description of the specimen. This is done by rearranging the matrix of pixel values into a single row of values in a standard data matrix in which the matrix rows contain all information collected from each specimen and the columns contain the variables. Each of the image pixels, then, becomes a variable and the colour or greyscale data the 'values' of those variables. Since each of the images processed in this manner is composed of the same number of variables (pixels), since each variable has a constant spatial relation to every other variable, and since the values each of these variables are of the same type and can adopt the same range of magnitudes, collectively these variables (pixels) form a mathematical space within which each of the specimens comprising the sample can be located. Specimens whose form, colour, and texture patterns are similar will lie close to each other in this space while those whose morphological attributes are distinct will lie at some remove from one another. This is precisely the same sort of inter-specimen representation achieved in a linear distance, landmark, or semilandmark-based shape analysis. Nonetheless, by using pixel values as the variables rather than coordinate point locations the resultant description preserves as much of the total morphological content of each image (and so each specimen) as possible as well as avoiding the necessity of making any *a priori* decisions about what aspect(s) of the morphology may, or may not, be important for resolving the problem at hand.

Relative to the more stringent data-matching criteria applied in a standard landmark or semilandmark analysis it might be said that the 'distances' separating images in this pixel space are inaccurate because the some aspects of the specimen may be represented by different pixels across the sample. In Fig. 1, it can be seen clearly that the orientations of the legs and antennae are inconsistent across these ladybird beetle paintings. Whether such inconsistencies are problematic depends on the issue being investigated (e.g., positional orientation of certain features may be the focus of the analysis). In such cases, it would clearly be advantageous to eliminate all extraneous sources of variation in the images prior to their analysis. However, even if such features are retained — as will be the case in the ladybird beetle analysis below — adjustment of the pixel-grid resolution will usually serve to minimize this source of spatial 'noise' relative to patterns of variation more germane to the phenomena under investigation. Moreover, various data modelling techniques (discussed below) are available for querying the patterns of variation summarized by various data-analysis procedures. These methods can be used to ensure the results of data-analysis procedures are not unduly biased by inclusion of obviously mismatched images. Also, once a better understanding of which aspects of variation within the images are associated with particular causal factors it may be possible to devise more formal landmark and/or semi-landmark-based analyses that will obviate the influence of feature mismatches entirely.

Use of the pixel grid as a basis for generalized form, colour, and texture assessment represents quite a flexible approach to the analysis of biological specimen morphology as well as one that incorporates many of the data-analysis features characteristic of more standard morphometric data analysis approaches. The conceptual link to semi-landmark-based outline and 3D surface analysis is straight forward [see 3, 5, and 6]. Nevertheless, even in the context of more traditional forms of landmark analysis, moving to a pixel-grid approach simply changes the focus of the analysis from the location of a small number of points (implicitly) embedded in a coordinate system to that of the total information content of the coordinate system itself. The price the data analyst pays for this change of focus involves having to deal with a much larger number of variables than would be the case in traditional landmark and semilandmark analyses and (usually) tolerance of small amount of feature mismatch across the image set. The benefit gained, however, is the ability to include much more morphological information that might be relevant to the questions being asked than would be the case under more traditional shape analysis sampling schemes.

An Example: Ladybird Beetles

To illustrate this procedure, consider the set of ladybird beetle images in Fig. 1 and ask whether this sample represents a continuously variable set of colour morphs (the null hypothesis) or whether ladybird beetles come in three basic types or morphs: orange beetles with black spots, black beetles with orange spots, and black beetles with red spots (the alternative hypothesis). Using a strictly Gestalt assessment of these morphologies the sample can be subdivided into these three colour-morph groups as listed in Table 1. Within this classification, specimens 5 and 6 in row 2 of Fig. 1 appear close to the intermediate condition between these groups with the former being slightly more black/orange and the latter slightly more black/red. However, whether the boundary between these putative groups is gradational or disjunct within the sample is not apparent.

Table 1. Putative Ladybird Beetle Colour Morph Assignments (by row and specimen number (see Fig. 1)

Orange w/ Black Spot Morphs		Black w/ Red Spot Morphs		Black w/ Orange Spot Morphs
1-1	2-3	2-5	3-7	1-5
1-2	2-4	2-7	3-8	1-6
1-3	3-1	2-8		1-7
1-4	3-2	3-4		1-8
2-1	3-3	3-5		2-6
2-2		3-6		

In preparation for analysis, one can reduce the spatial resolutions of the images that comprise Fig. 1 from 150 x 150 pixels to 32 x 32 pixels (Fig. 4). From inspection of these lower-resolution beetle images, one can see that, despite the total number of pixels representing each specimen having been decreased by 95 percent, these lower-resolution images preserve virtually all features of all the specimens observed in the original image set.

Accordingly, this operation has reduced the number of variables required to represent these specimens (= reducing their dimensionality) by preferentially eliminating redundant pixels (= reducing spatial autocorrelation) while suffering only a very minor loss of information content.

Fig. 4. Resample of the 24 Harlequin ladybird beetles (*Harmonia axyridis*, Family Coccinellidae) illustrations (see Fig. 1) using a 32 x 32 colour (RGB) pixel grid. This resampling protocol reduced each specimen's pixel number by circa 80% with little loss in the spatial resolution of features necessary for characterizing patterns of morphological variation.

Since these are colour images, and since, in this case, one wants to include an assessment of colour variation in the dataset, each specimen's image can be described completely by converting the pixel format to a 32 x 32 matrix of red, green, and blue (RGB) colour intensity values; an operation that results in the specification of 32 x 32 x 3, or 3,702 variables. This is a large and somewhat awkward dataset insofar as there are many more variables than specimens. However, it is a dataset that can be analyzed and such skewed data matrices are by no means uncommon, in data analysis generally or shape analysis in particular.

A covariance-based principal component analysis (PCA) of the pixel information in these images shows that 90 percent of the observed form, colour, and texture variation can be represented by 15 orthogonal components or axes of variation. This operation further reduces the effective dimensionality of the image measurement system from 3,702 variables to 15 (a reduction of 99.6%). This operation also drops the number of variables down below the number of specimens in the dataset. The number of specimens included in a morphometric dataset represents an important "curse of dimensionality" threshold [see 7, 8] of which more will be said.

Inspection of the subspace formed by the first few principal components of the pixel covariance matrix (Fig. 5) summarizes the major features of variation within the sample.

These components (Fig. 5A) account for the largest single share of observed form, colour, and texture variation (37.44%). But variation in this sample is arranged such that these axes portray only a minor share of the sample's variation overall. The distribution of the images in this 3D subspace suggests that broad-scale discontinuities in the pattern of form, colour, and/or texture variation do indeed exist. However, this ordination does not confirm the putative three-group model as representing the major modes of morphological variation in the sample accurately. In particular, the characters of the boundaries between these three putative morph-groups are not well defined. While the ordination of the dominantly orange morphs within this plane appears relatively gradational, the ordination of the dominantly black morphs appears highly structured.

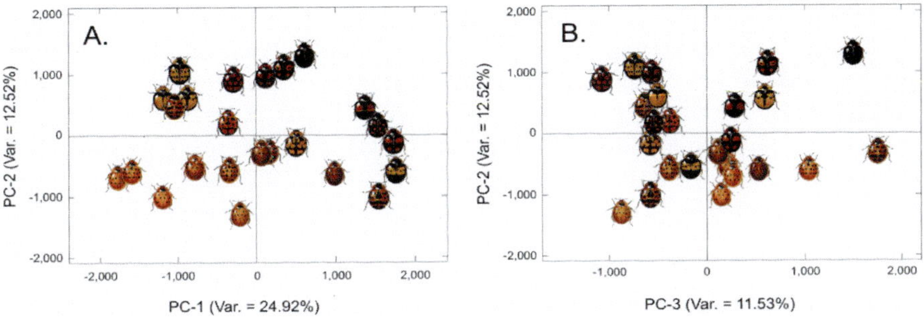

Fig. 5. Distribution of ladybird beetle form and colour texture similarity and difference relations on the first three principal component axes of the image covariance matrix. (A) the subspace formed by PC-1 (*x*-axis) and PC-2 (*y*-axis) and (B) the subspace formed by PC-3 (*x*-axis) and PC-2 (*y*-axis). Note the comparative lack of specimens that project to positions along the principal component axes. Lack of the empirical definition of axis form trends that might be afforded by such specimens makes it quite difficult to arrive at detailed geometric interpretations of the overall form space. See text for discussion.

Casual, qualitative inspection of this ordination pattern suggests that the predominant trend in morphological variation along PC-1 appears to contrast relatively narrow forms with large fields of orange pixels (low scores) with rotund forms exhibiting substantial, but localized, fields of red pixels (high scores). Intermediate positions along this axis tend to be characterized by either dusky orange morphs or morphs in which black and organ/red blotches occur in sub equal proportions. This PC-1 ordination differs distinctly with the morphological trend represented by PC-2 in which narrow forms characterized by orange pixels around the periphery of their elytra and pronotum shields (low scores) are contrasted with more rotund forms whose peripheral areas are increasingly characterized by black pixels (high scores). Within this overall ordination pattern three distinct modes of form, colour, and texture variation appear to be present: a group of bright orange beetles with small black spots (quadrant III), a group black beetles with either orange or red spots distributed over the entirety of the elytra (quadrants I and II), and a group of black beetles with orange or red spots distributed in a more localized manner over the elytra. Morphological discontinuities characterize the more extreme regions of the space between these modal groups whereas a few intermediate morphologies project to positions low on the PC-1 and PC-2 axes. Interestingly, the shape distinctiveness of these modal groupings is such that they subdivide the occupied space into quasi-equal sectors.

Interpretation of PC-3 (Fig. 5B) is, if anything, even more complex and ambiguous than the interpretations of PC-1 and PC-2. Along this axis, narrow and rotund forms are intermingled, as are the putative orange, black/orange, and black/red morph groups. The best that can be said, based once again on qualitative inspection of the projected image positions, is that black morphs characterized by relatively complex orange or red spot patterns (intermediate low scores) appear to be contrasted with black morphs characterized by relatively simple red and orange spot patterns (intermediate high scores) along PC-3. Patterns of variation at either extreme of this axis are very difficult to infer owing to the lack of beetle morphologies that occupy positions close to the PC-3 axis.

Given the complex character of image variation within this space, detailed interpretations of these axes are surprisingly difficult to devise. Unfortunately, this situation is by no means unusual in the context of a PCA. Whenever a dataset characterized by strong and sub equal contrasts among a small number (3-5) subordinate groupings is subjected to PCA the first few PC axes will tend to be aligned with positions between the groups rather than coming to rest in positions that reflect subgroup-level contrasts. This occurs because the eigenvector maximization procedure always searches for a global and mutually orthogonal solution to the problem of representing shape variation across the sample as a whole. In former times, a variety of strategies would be employed to rotate or deform the PC axis system to positions that were better aligned with modal variations within the dataset under the justification of better representing the inherent structure of these multivariate data (see [8]). However, as these methods distort representation of the inherent geometry of the shape distribution they have never been popular with the shape analysis community. Fortunately, alternative strategies exist for achieving precise interpretations of axes in the PC space.

Perhaps the most straightforward and useful of these is direct modelling of the coordinate positions in the PC space. MacLeod [9] presented a simple procedure for calculating geometric shape models at any position within a PC space that can be used with landmark, semilandmark, or pixel coordinate data [see also 10, 11, and 12]. In order to improve the interpretation of the ladybird beetle PCA space this procedure was used to calculate the series of equally spaced, along-axis shape models presented in Fig. 6.

Inspection of these models greatly clarifies interpretation of the major modes of variation in this dataset. Here one can see that PC-1 actually captures the distinction between orange morphs with numerous small black spots (extreme negative scores), through dusky orange morphs with larger black spots, to a median region in which forms composed of sub equal black and dusky orange regions occur and on to morphs characterized by a predominantly black body with a single pair of large red spots on the elytra and margins of the pronotum shield (extreme positive scores).

Orientation of the PC-2 axis is subtly different in that it contrasts bright orange morphs with a smaller number of larger black spots (extreme negative scores), passes through a median region composed of dusky orange morphs with small black spots and on to forms characterized by black bodies with four symmetrically placed red spots on the elytra, the anterior of which have black spots in their centres and elongate red spots at the margins of their pronotum shields (extreme positive scores).

Finally, PC-3 captures the distinction between forms characterized by black bodies with large, elongate orange spots whose axes are at right angles to the beetles' antero-posterior axis of symmetry (extreme negative scores), through a median region characterized by dusky orange morphs with black spots, and on to a region characterized by forms with black bodies and dusky orange stripes formed from the progressive amalgamation of orange spots (extreme positive scores).

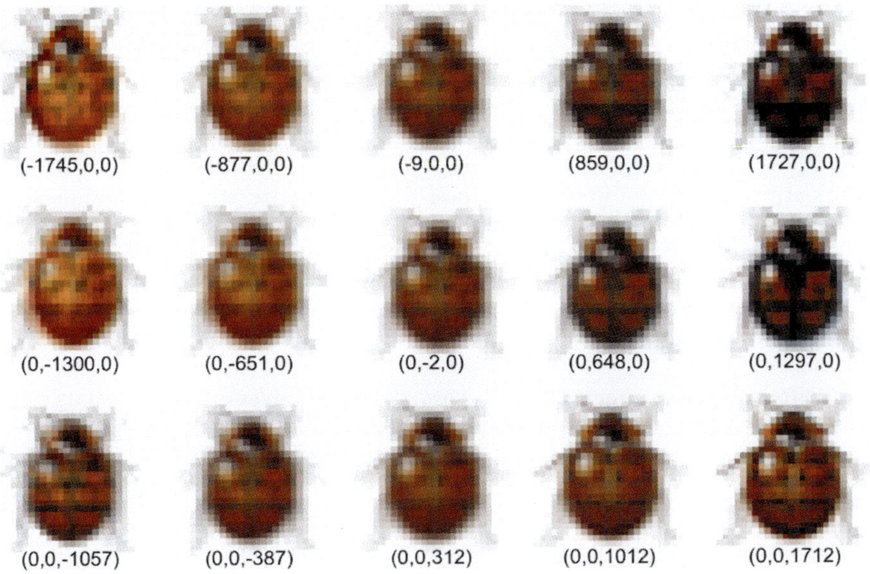

Fig. 6. Image models for the ladybird beetle dataset calculated at equally spaced coordinates along the first three principal component axes. See text for discussion. Numbers in parentheses refer to the coordinate positions at which each image model was calculated (see Fig. 5).

Close inspection of the PC-3 model set also reveals that, during the course of the transition from high negative to high positive scores a black band along the medial elytral margin changes colour and becomes orange. Naturally, the point is that, if one had not had access to these hypothetical image models to use as precisely located points of reference and comparison, it is very doubtful that such a detailed interpretation of the major dimensions of form variation could be deduced from a simple inspection of the actual specimen ordinations alone.

Although calculation of along-axis shape models allows more detailed and useful interpretations to be made of the PC ordination space, the interpretations offered above remain only semi-quantitative insofar as comparisons between the models calculated along each axis were done by eye.

An even more complete and nuanced picture of similarities of the differences among these models, can be assembled by calculating difference maps of the comparisons between these models (Fig. 7).

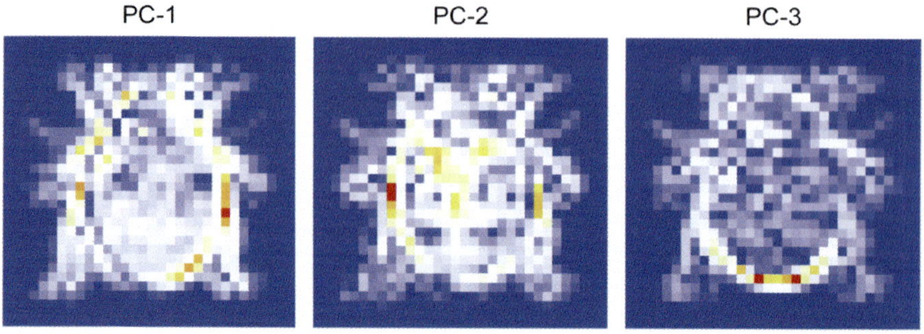

Fig. 7. Along-axis image model difference maps for the first three sets of principal component axis models (see Fig. 6). These plots summarize regions of the image at which the different proportions of change in pixel colour values are taking place as one moves along each of the principal component axes. In these plots, the individual pixels have been colour coded as follows: blue = no change, white = moderate change, yellow = moderately high change, orange = high change, red = very high change. Note how the different PC axes summarize changes in different regions of the form.

Difference maps compare (in this case) the RGB values of corresponding pixel locations across two or more images and assign a colour to the mapped pixel whose hue is based on the amount of change recorded at individual pixel locations. Typically, these maps employ a temperature metaphor to express their results visually. Under this convention regions of the image characterized by little of no change are signified by blue (= cool) and regions characterized by high levels of change by (progressively) yellow, orange, and red (= hot). This type of representation has the advantage of being a more objective, quantified, and detailed assessment of observed changes than the qualitative inspection of image differences.

However, it should be remembered that there is no simple relation between the absolute amount of pixel value change at any particular location and the biological significance of that change. The strength of the difference map approach is that the maps draw ones attention onto particular aspects of form comparisons, or regions of form change, that might be overlooked as our eyes try to process and pick out consistent patterns or trends in the image model results. Difference maps can be used either in an exploratory sense (e.g., to identify regions of high or low that we might want to pay special attention to in subsequent analyses) or as a means of testing specific hypotheses (e.g., if a generative hypothesis predicts that some sort of change will be localized in particular regions of the form), but should never be used on their own to interpret the biological significance of the form changes they record.

When the difference map approach is used to compare differences between extreme low and high scores, ladybird beetle models in each of the along-axis sequences (Fig. 7) they do pick out unique aspects of variation that were not readily apparent in the qualitative inspection of the models shown in Fig. 6. For example, there appears to be a general broadening of the elytra and pronotum shields along PC-1 with the elytral colour pattern variation differentially occurring in the posterior portion of the form. Along PC-2, this broadening is confined to the middle part of the elytra with substantial colour pattern variation occurring in the anterior elytral region. Along PC-3, moderate modes of colour pattern variation occur across the elytra accompanied by a slight concentration in the posterior portion of the pronotum shield. The predominant mode of form variation expressed along this axis, however, is a slight lengthening of the body that appears to be on the order of 2-3 pixels in dimensional aspect.

Testing Hypotheses Involving A Priori-Defined Groups

What about the original question regarding the distinctness of the putative orange, black-orange, and black-red morphs? Fig. 5 suggests (weakly) that the character of this transition might be gradational. Nevertheless, looking at a few projection planes through the original 15-dimensional PC space is rarely sufficient for determining the actual degree of separation between these morphs. If a separation between groups can be located along any of these axes, either singly or when employed in combination, then the groups are separable. Since one has neither the time, patience, nor software to inspect visually all possible geometries of points arising in a 15-dimensional space, some alternative procedure must be used to settle this question.

The classic way to handle this problem would be to either subject the raw data or a reduced set of PC scores to a discriminant analysis procedure such as canonical variates analysis (CVA). A CVA takes a multivariate dataset that has been partitioned into subgroups, and performs a series of standardized axis rotations and transformations, which result in the data for two or more groups being projected into a multivariate space in which each group centroid is maximally separated from every other group centroid relative to within-group dispersions [13]. This method usually cannot be applied to raw or Procrustes-aligned morphometric data because of computational difficulties that arise when attempting to invert the within-groups covariance matrix. Nevertheless, these computational difficulties can be solved (usually) by taking the raw data through an initial PCA and discarding the component axes that represent a statistically or biologically insignificant proportion of the observed variation.

Interestingly, the CVA procedure has been recently characterized as unsuitable in the context of shape analysis by Mitteröcker and Bookstein [14] for the following reasons:

1. Orientations of the CV axes are not orthogonal to each other in the space of the original variables used to calculate the CVA; as a result true Euclidean and/or Procrustes distances between specimens are not represented faithfully in the CVA ordination space.

2. If the number of variables greatly exceeds the number of specimens comprising a dataset CVA, one will always be able to find a linear combination of variables that separates groups even in cases of artificially generated data drawn from the same multivariate normal distribution.

3. The scores of objects projected into the CV space do not represent a linear transformation of the original variables, but a linear transformation of these variables after they have been transformed into a distorted space formed by an operation that renders the within-group covariance matrices spherical. Thus, modes of shape variation that optimally separate groups, are difficult to interpret as well as difficult to model properly.

As an alternative, Mitteröcker and Bookstein [14] advocate use of a 'between groups PCA' in which the group means are used to determine the orientations of a set of eigenvector axes and the data comprising the sample projected into this group mean-determined PCA ordination space.

To address Mitteröcker and Bookstein's [14] first point; certainly, the between-groups PCA (BG-PCA) is a simple procedure that does have the advantage of preserving true Euclidean and Procrustes distances more faithfully — though not exactly — in the resulting ordination space than a typical CVA would. If preservation of Euclidean or Procrustes distances is important for resolution of a particular hypothesis, then the CVA results should be used with caution. However, available shape modelling procedures (see above) enable any coordinate position within either the PCA space or the CVA space to be associated uniquely with a shape configuration. These models can be used to probe the character of the standard PCA space, the BG-PCA space, and/or the CVA space in order to develop quite a detailed understanding of the geometric character of these spaces quickly and easily regardless of their mathematical complexity (see an example below).

In addition, it must be remembered that the question posed in this example — and the question posed by most classification problems — does *not* require that Euclidean or Procrustes distances be preserved in the discriminant space. Resolution of these questions requires only that the multivariate space within which the pooled data reside be searched for a linear transformation of the data that optimizes group separation.

How well does BG-PCA optimize group separations? For datasets in which the number of objects exceeds the number variables of a BG-PCA will, in most cases, produce a result that is suboptimal in terms of group centroid separation within its discriminant space relative to the space calculated by a normal CVA. Thus, CVA would unquestionably be the better choice for these datasets, both in principle and in practice. Unfortunately, such large datasets are comparatively rare.

For more typical datasets, in which the number of variables greatly exceeds the number of objects, it is true that CVA may present a misleading picture of the true degree of the between-groups difference [14, 15] owing to the well-known 'curse of dimensionality' phenomenon (see [7]). Nevertheless, a preliminary PCA of the pooled dataset can be used to reduce the effective dimensionality of the data while preserving virtually all of its information content (see above). Once an otherwise large and unwieldy morphometric dataset has been pre-processed in this manner the results of a CVA ordination can be crosschecked for representational validity and interpretability using a wide range of standard procedures (see example below).

Fig. 8 compares results obtained by analyzing the ladybird beetle data using BG-PCA and CVA to maximize group differences for the putative three morph-group problem based on the scores of beetle images on the first 15 PC axes (the number needed to represent 90% of image variation within the pooled image sample).

As expected, the standard CVA of the PCA scores produced the more definitive result. Indeed, these two results disagree with one another with regard to their respective assessments of the distinctiveness of group separations. An analyst basing their interpretation on the BG-PCA result would conclude that morphological variation between black-red and black-orange morphs was continuous with a morphological discontinuity existing between these and the orange morphs. Analysts basing their interpretation on the CVA result would conclude morphological variation was discontinuous among all three putative morph groups.

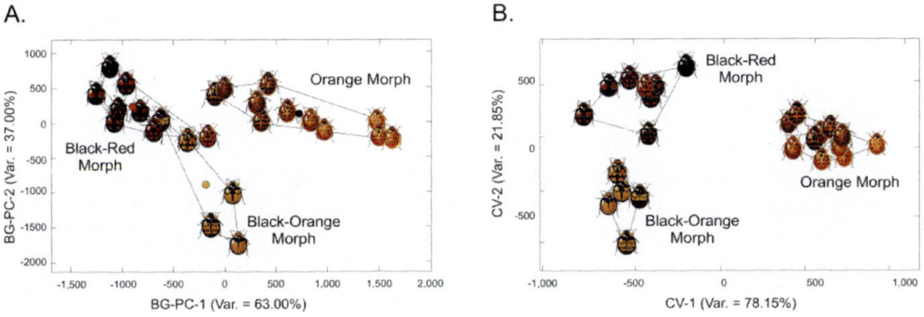

Fig. 8. Comparison of results obtained from a BG-PCA (A) and CVA (B) of the a reduced set of PCA scores (first 15 eigenvectors) that in aggregate, represent 95% of the information content of the original ladybird beetle image set. Note the superior between-groups discrimination offered by the CVA. See text for discussion. Small filled circles in A mark the positions of the group centroids that served as the data used to define the BG-PCA discriminant space.

For the purposes of both classification, and of determining the character of morphological variation among beetles within this dataset, the CVA result is clearly better resolved.

Both results are correct insofar as both are based on correct calculations. However, these ordination spaces portray different aspects of the sample's structure. The BG-PCA space represents an assessment of the major axes of group centroid positions. Such a space is an accurate representation of this — and only of this — aspect of the data. As no information related within-group variation was used to define this space, the BG-PCA ordination cannot be relied upon to portray anything definite about the distinctiveness of groups or patterns of within-group variation accurately. The group range patterns displayed in Fig. 8A are best regarded as circumstantial by-products of an analytic result that was only attempting to represent the major directions of variation among group centroids, not to maximize between-groups differences relative to within-groups dispersions.

It is mathematically possible to project any set of comparable data into such a space. But by projecting the positions of the entire image dataset into this space, and then using this projected space to assess relations between groups there is a serious danger of producing an ordination patterns that does not represent the true structure of the data set and one which is oriented in a non-optimal position with regard to between-groups similarity and difference. Moreover, the 'curse of dimensionality' problem that Mitteröcker and Bookstein (14) cite as a fatal flaw in the use of CVA on morphometric datasets is clearly present in this space as well, albeit in a different — though no less problematic — manner. As the dataset that defines the space shown in Fig. 8A included 15 variables, but only 3 objects, the BG-PCA result is a representation of variation based on an extremely underdetermined (and therefore unstable) system of synthetic (mean) observations.

To my mind the CVA result (Fig. 8B) is by far the more informative and robust characterization of true group differences. Information from both these factors was used to define the space within which the ladybird beetle data are portrayed. The linear transformations inherent in the mathematics of CVA (see [13]) cannot force groups that do not overlap in the space of the original variables to overlap. Mitteröcker and Bookstein's [14] discussion alludes obliquely to statistical methods that might be used to test the CVA discrimination space for statistical significance, but neglects to either discuss or reference the use of these procedures explicitly. The fact is, any suspicion that an apparent effect along these lines has influenced the CVA ordination result can be tested quantitatively using a variety of standard statistical resampling procedures (see [16] and Fig. 9).

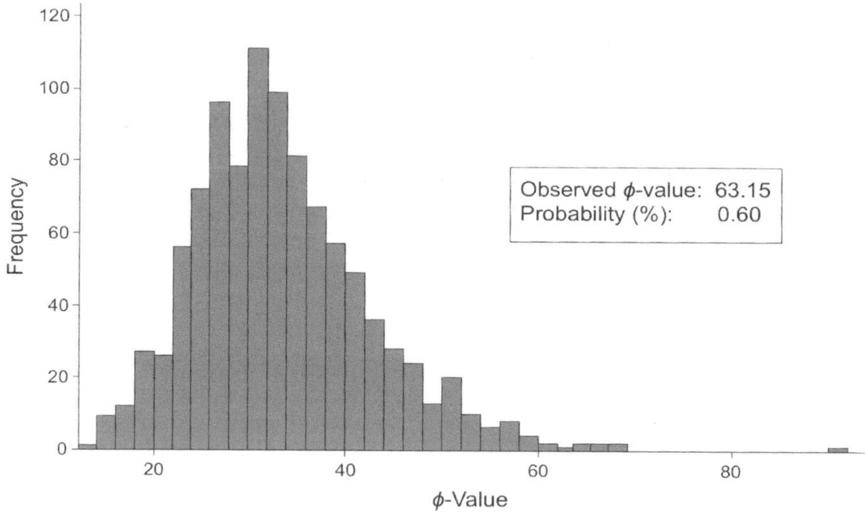

Fig. 9. Results of a Monte Carlo simulation test of the difference between putative colour morph mean vectors (as assessed by the log-likelihood ratio test, see [16]) for the CVA result shown in Fig. 8. The distribution of ϕ-values was determined by creating 1,000 pseudo replicate datasets of sizes and covariance structures identical to that of the ladybird beetle dataset but with variable values all drawn from a normal distribution with identical means and variances, submitting these pseudo replicate datasets to CVA and recording the resultant ϕ-values. This result indicated that there is less than a 0.6% probability that the observed ϕ-value for the ladybird beetle CVA result (63.15) could be generated via random collection of the dataset from a single underlying population. Bootstrap resampling strategies can also be used to evaluate the statistical significance of any CVA result.

With regard to concerns over the interpretability of CVA ordination results, as noted earlier, the best way to approach this problem is not via study of axis loadings or by calculation of along-axis form models since the relevant between-groups contrasts are not well aligned with the traces of either the BG-PCA or CVA axes.

But, as every coordinate location in the BG-PCA and CVS ordination spaces corresponds to a unique configuration of (in this case) pixel colour values it is a relatively easy matter to reconstruct the images that would be located along any trajectory through either space and inspect these transformations to understand the geometric nature of the ordination spaces themselves as well as the biological significance of the between-group discriminations. The fact that the CVA axes are not necessarily oriented orthogonally to one another in the space of the data used to calculate the CVA space, or scaled in an inconsistent manner along the CV axes in question, neither prevents, inhibits, nor greatly complicates the calculation of these models or their interpretation (see Fig. 10).

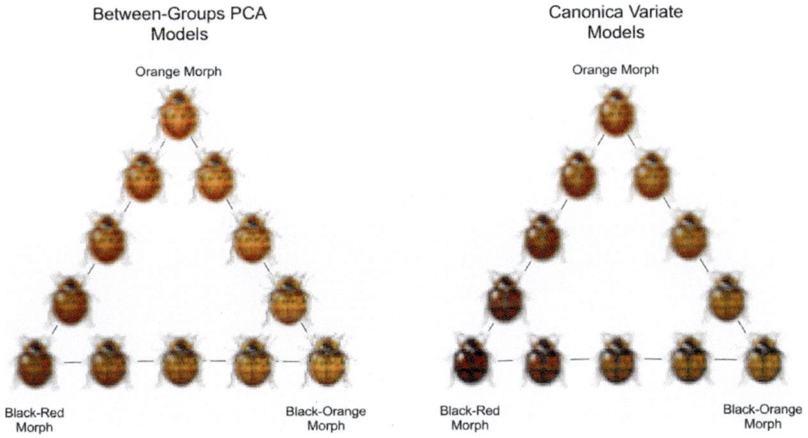

Fig. 10. Image models for sets of equally spaced coordinate points in the BG-PCA and CVA ordination spaces illustrating the morphological transition between the each putative morph group. Mean group images are shown at the triangle apices. Note the overall similarity between these model sets in terms of their ability to illustrate and probe the fine-scale aspects of the ordination spaces shown in Fig. 8 and greater degree of between-groups distinction afforded by the CVA model set.

Both BG-PCA and CVA have their merits and their limitations when it comes to visually portraying between-group similarities and differences. If the purpose is to achieve an optimal group assignment decision or to test hypotheses concerning a morphological distinctions referenced to an optimal between-groups discrimination space, and if the size of the sample is large relative to the number of variables being used to define the discriminant analysis, a CVA analysis of PCA scores should return an appropriately robust result. Alternatively, if the purpose is to achieve between-groups discrimination while preserving a high degree of correspondence in the ordination result to the fundamental Euclidean or Procrustes character of the data, and/or if the size of the sample is very small relative to the number of variables being used to define the discriminant space, a BG-PCA should return an acceptable result. Both approaches are subject to the 'curse of dimensionality. Accordingly, both should always be checked for conformance to a null model of between-groups variation whenever they are being used to portray group differences. Perhaps the best way of conceptualizing the relation between these two results is that the BG-PCA result is similar to what would be seen if we could view the 15-dimensional PCA space in an orientation that best separated these three putative morphs whereas the CVA result (in this case) allows us to perform a similar operation and, at the same time, hold a mathematical magnifying glass up to the critical transition interval.

DISCUSSION AND CONCLUSIONS

While the results described do not prove that ladybird beetles come in three varieties, they do show that the data encoded in digital images can be used to test specific hypotheses regarding the character of variation within a morphologically complex sample of organisms even if one cannot reasonably specify the positions of relocatable landmarks or boundary outlines across all specimens comprising the sample. While the eigenimage technique mimics the Gestalt approach to image analysis that has been employed by traditional taxonomists for (literally) millennia, it offers considerable advantages in terms of sensitivity, accuracy, consistency, reproducibility, and speed. This 'image analysis' technique represents a conceptual link between morphometrics and approaches to computer vision that can have widespread application across the biological sciences, and perhaps across the natural sciences (see [6]).

Of course, this is not the first investigation that has attempted to use digital images directly as input to a morphometric analysis. Most research in this area to date has been focused on the issue of human face recognition. Human faces are considered a difficult image identification problem largely because human faces — even the face of the same individual — can be quite variable depending on lighting, expression, age, hair style, skin colour (e.g., tanning), and various epistylistic features (e.g., distribution of facial hair, glasses). The human visual system is very good at identifying the faces of individuals despite variation in all these factors.

In biology, it is comparatively rare that recognition needs to be performed at the level of the individual. However, copious evidence is now emerging that humans are not especially good at identifying a wide range of biological objects and placing these into even broadly defined taxonomic categories consistently [17, 18, 19, 20, 21, and 22]. The development of semi-automated and fully-automated generalized approaches to the problem of taxonomic identification based on visual characteristics represents an outstanding problem in biology, one that research into shape analysis procedures will be required to solve, but one that most biological shape analysts have yet to consider in any detail [23].

Interestingly, most approaches to the face-recognition problem among computer vision specialists, as most approaches to contemporary morphometric analyses, have focused on summarizing the information content of an image based on identification of a relatively small number of features (e.g., position and size of eyes, nose, mouth, and head outline) and use these to construct a model of the face(s) in question. Early attempts to achieve face recognition relied on trained observers marking these features on images by hand (e.g., [24, 25]) though automated pattern recognition algorithms were developed for this purpose in the 1970's [26, 27, and 28].

By the late 1970's, it was suspected that the distribution of these features alone was insufficient to duplicate the routine performance of the human face-recognition system [29]. Nevertheless, this approach remains the one most commonly employed in automated face recognition contexts. Throughout the 1980's and 1990's, experiments based on artificial neural networks and template-matching strategies were undertaken, all with limited success (e.g., [30, 31, 32, 33, 34, and 35]). The 'eigenface' approach ([36, 37]; see also [38]) was the first to use a generalized PCA-based strategy to summarize the information content of entire images of faces in a reduced multidimensional space without presupposing which features in the frame might be of greater or lesser importance in solving the recognition problem. This approach is close to the eigenimage method outlined earlier, but differs from it in several important respects.

Eigenface analysis begins with a training set of face images, which have been captured digitally and processed to conform to a relative high-resolution pixel-sampling grid (originally 256 x 256 pixels, see [37]). Pixel information from each image is then transformed so that each face is represented by a column vector in a high-dimensional variable space (originally 65,536 variables) and the covariance matrix for these images calculated. Since the sample size will almost always be considerably less than the number of variables used to calculate the face space, a PCA was used to reduce the dimensionality of this variable set while suffering no loss of information content.

Under the eigenface procedure, the dimensionality of this 'face space' is reduced further by using only eigenvectors corresponding to the m largest eigenvalues (corresponding to the m most distinctive faces) to define a reduced face space. This step loses information but so long as m is sufficiently large (e.g., 40-50) the loss should be minor.[4] Each individual face (or set of faces from the same individual) is then projected into the reduced face space so that they form implicit (single face images) or explicit (multiple images of the same face) identification domains. This space is defined by the first m eigenvectors of the image covariance matrix whose loadings represent a series of empirically determined pixel (= feature) weights. Identification of unknown images is achieved by projection of the processed unknown image into the reduced face space and determining which known face lies closest to it. Boundaries of implicit recognition regions can be erected about each training-set face so that faces not included in the training set (or aberrant training set faces) can be identified and treated separately.

The eigenimage method outlined above is similar to the eigenface approach in the context of using a PCA-based approach to reduce the dimensionality of the image space, but goes beyond the typical scope of the eigenface technique in that it encourages adjustment of the effective resolution of the images used in the analysis prior to PCA processing to achieve additional dimensionality reduction.

---

[4] Turk and Pentland used modelling procedures to reconstruct individual faces based on the eigenvectors of the reduced face space will little noticeable loss of critical information content [36, 37].

The Turk and Pentland [36, 37] eigenface procedure recommended a minimum resolution of 256 x 256 pixels for digital representations of the human face, but failed to provide an explanation or justification for these parameters. Most likely visual inspection of images at this resolution 'looked' right in terms of the images' ability to represent fine details of facial structure. As shown in Fig. 4 though, considerable reductions in the image resolution can be made to most images of biological specimens and/or structures without engendering significant loss of representational detail. Such reductions increase the efficiency of eigenimage analyses by lowering the computational overhead. By placing an emphasis on flexibility in terms of setting the raw image resolution means that the analyst remains free to set the image resolution to whatever value is felt is necessary to address the problem at hand and encourages an experimental approach to the setting of this critical parameter.

In addition to adjustment of the image set's resolution, the eigenimage procedure differs from the eigenface procedure in encouraging the images to be aligned with one another so as to remove extraneous sources of variation from the scene. This alignment can be accomplished in various ways. In the case of the ladybird beetle example, the illustrations of beetle morphology were already aligned. Framing of the drawings at the centre of a pixel-constrained field-of-view was accomplished via direct image processing. Alternatively, the mathematical Procrustes method (e.g., [39, 40]) could be used to correct the region of interest within a field-of-view for inconsistencies in size, position, and rotation relative to a reference form (usually the mean shape or mean image) with the background being automatically blanked once these transformations had been completed. In addition, irregular regions of interest can be accommodated under the eigenimage procedure by specifying a (usually hand-drawn) region of interest within the frame and transforming the pixel brightness/colour values from a rectilinear to an polar coordinate system with the latter being interpolated to a consistent number of pixel values (see [37] for an example; also [41, 42, and 43]).

While this procedure produces images that do not 'look' like the specimens or features in question, virtually all the information content of the original image is present in this highly processed format and remains available for subsequent analysis and/or modelling procedures to operate on. The ability to use such images as the basis for an eigenimage analysis makes the point that automated image recognition procedures do not need to rely on representations of objects, specimens, features, etc., that make visual sense to human taxonomists in order to be useful and effective. Such procedures also do not need to confine themselves to keying on traditional taxonomic features in order to subdivide the image space appropriately in order to achieve image identification. Indeed, systems constructed on the principles employed in eigenface and/or eigenimage analysis should be seen as generalized pattern-recognition engines capable for formulating their own rules for segmenting the face/image space according to empirical principles. Consequently, such systems are capable — to a greater or lesser extent — of basing image identifications on novel features or characteristics of the images that they 'discover' during the course of their calculations.

Finally, the eigenimage procedure differs from Turk and Pentland's eigenface procedure in utilizing PCA primarily for dimensionality reduction, but then addressing the between-groups discrimination problem via secondary analyses of the PC scores such as BG-PCA or CVA. Discriminant analysis strategies can be based on PCA (e.g., the SIMCA procedure, [44]) and have the advantage of preserving the natural geometries of the data. It is suspected that this issue lies close to the heart of Mitteröcker and Bookstein's [14] advocacy of BG-PCA in the context of morphometric studies. However, because PCA is tied fundamentally to the simple representation of the major directions of variation within a pooled dataset, and not to the separation of *a priori*-defined groups, the performance of PCA-based procedures in the context of group separation problems will always be suboptimal in terms of testing the group-separation hypothesis. This contrast in performance is seen clearly in the comparison of BG-PCA and CVA results for the ladybird beetle dataset. Both the degree of between-groups separation and the geometric specificity of the discriminant spaces were notably lower in the BG-PCA result relative to that achieved by CVA (see Fig. 8).

As stated above, Mitteröcker and Bookstein [14] were correct to call attention to the interaction between CVA and dimensionality relations within the dataset when used to analyze underdetermined data matrices. This method is so powerful it can find differences among samples where none exists statistically. However, this does *not* mean those differences do not exist numerically. For example, a CVA of a partitioned set of data artificially constrained such that each subgroup has the same mean but different variances will result in calculation of an ordination space in which the group means project to exactly the same point and whose within-group dispersions overlap entirely. If group distributions are not separable, CVA will not separate them (see [13]). A variety of statistical methods are available to estimate expected levels of between-groups separation for any partitioned dataset under the null model that the samples were drawn from the same distribution and so represent nothing more than idiosyncratic sampling issues which would be expected if sample sizes are low relative to variable numbers. This is a statistical issue that arises any time group-level data are portrayed in a CVA space. However, this is not an issue that is in any way unique to CVA. Precisely the same phenomenon can be present in the ordination patterns that result from a BG-PCA (Fig. 8A), a standard PCA, a Procrustes PCA (= relative warps analysis), and/or any other multivariate data analysis procedure whose results can be portrayed in an ordination space.

Questions regarding the statistical validity of apparent group separations portrayed in multivariate spaces should not be answered by implying that some ordination spaces are susceptible to this issue while others are immune from such considerations and can be interpreted at face value. Rather, this must always be addressed through application of statistical testing procedures such as those already described.

After reformulations in the late 1980s and 1990s based around the analysis of landmark data, the subsequent history of the development of shape analysis theory and practice has exhibited a strong theme. One of extension to incorporate alternative types of data and application of modern shape analysis methods to an ever greater range of interesting problems throughout the biological, and now extending into the physical (see [6]), sciences. This should be considered a normal trajectory of development for the field of shape analysis and underscores the unusually broad scope of shape-analysis problems across the natural sciences.

In this chapter procedures were described that can be employed to relax the traditional constraint of using morphometric methods to only analyze a relatively small set of quantified positions and landmarks and/or semilandmarks collected from a sample of forms and answer the question of the extent to which geometric morphometric methods can be applied to the analysis of images themselves in the form of pixel brightness and/or colour values. Conceptually, this problem inverts the standard shape analysis sampling situation of tracking the positions of landmarks/semilandmarks moving across the physical space defined by a sample of forms, but ignoring all data other than those the analyst has decided to track *a priori*, to one of tracking all aspects of morphological variation simultaneously as recorded by the changing pixel brightness/colour values within a common, fixed, pixel grid.

This strategy avoids the need to make any *a priori* decisions regarding which aspects of the forms present in a sample, are important for resolving the shape analysis question at hand and the manner by which these features (e.g., points, structures, characters, boundaries, outlines) are represented in the shape analysis data matrices. In addition, owing to the fact that pixel brightness/colour values enable those viewing the image to infer that the 3D structure of the objects that have been imaged with this strategy represent a way of incorporating aspects of 3D information into a shape analysis investigation that can be conducted quickly, easily, and with minimum expense.

The example analysis of patterns of variation displayed by a sample of ladybird beetle drawings here shows that this eigenimage procedure is capable of accurately assessing aspects of morphological variation that would be very difficult to capture adequately using traditional approaches to shape analysis, thus facilitating the testing of hypotheses that, at present, cannot be addressed by contemporary shape analysis procedures. These hypotheses involve both the assessment of major patterns of variation within the sample of specimen images and hypotheses related to the level of morphological distinction existing between *a priori*-defined groups.

In addition, shape-modelling procedures developed originally to facilitate interpretation of the ordination spaces created through the use of standard geometric shape analysis procedures, can be applied in the context of eigenimage analysis for the same purpose. Indeed, the utility of these modelling procedures is even more obvious in the context of an eigenimage analysis than in traditional shape analysis contexts owing to the greatly increased geometric complexity of the data being analyzed.

Finally, a minor controversy exists within the field of shape analysis in general, and morphometrics in particular, regarding the most appropriate data analysis strategy to use to assess the character of separations between *a priori*-defined groups. Comparisons between eigenimage-based discriminant results obtained using the BG-PCA and CVA indicate that the latter produce more definitive results owing mostly to the fact that CVA utilises more information relevant to the group discrimination problem than BG-PCA. Concerns about the use of CVA in morphometric contexts raised recently by Mitteröcker and Bookstein (14) are valid, but only in those instances in which the raw ordination results of a CVA are presented without any subsequent statistical testing of the level of group separation that would be expected under various null hypotheses. When appropriate statistical tests are applied to the ordinations that result from a CVA analysis, these statistical results can be used to assess the significance of the group separations achieved and so curb over enthusiastic interpretations. Regardless, it should be understood that these concerns do not apply only to the ordination results of a CVA. The same types of tests can — and should — be used to assess the significance of any ordination analyse that involves questions pertaining to group separation(s).

Eigenimage analysis represents a generalization of the semilandmark-based eigenshape approach to the analysis of morphological data. It effectively bridges the gap that exists currently between shape analysis and computer vision (e.g., eigenfaces) and might represent a means through which these presently distinct research communities can interact with one another to their mutual benefit; especially in the much-needed area of automated specimen identification. While eigenimage results can be used as the basis for shape analysis hypothesis tests on their own, they can also be used in an exploratory context, to identify regions and/or characteristics of specimens that might benefit from more detailed consideration using either traditional shape analysis, or refined and more specifically focused eigenimage data analysis strategies.

ACKNOWLEDGEMENTS

I would like to thank Peter Lestrel for requesting that I write up these results, which were originally presented verbally as part of the Third International Symposium on Biological Shape Analysis (14-17 June 2013, University of Tokyo, Japan) for inclusion in the present volume and for being an understanding editor when this ms was submitted slightly behind the deadline agreed originally.

REFERENCES

[1] Bookstein FL (1991) *Morphometric Tools for Landmark Data: Geometry and Biology*. Cambridge University Press, Cambridge.

[2] Lestrel PE (1997) *Fourier descriptors and their applications in biology*. Cambridge University Press, Cambridge.

[3] MacLeod N (1999) Generalizing and extending the eigenshape method of shape visualization and analysis. *Paleobiology* **25**: 107-138.

[4] Brown, CM, Arbour JH, and Jackson DA (2012) Testing of the effect of missing data estimation and distribution in morphometric multivariate data analyses. *Syst Biol* **61**: 941-954.

[5] MacLeod N (2008) Understanding morphology in systematic contexts: 3D specimen ordination and 3D specimen recognition. **In**: Q Wheeler (Ed.) *The New Taxonomy* CRC Press, Taylor & Francis Group, London.

[6] MacLeod N, Krieger J, and Jones KE (2013) Geometric morphometric approaches to acoustic signal analysis in mammalian biology. *Hystrix* **24**: 116-125.

[7] Bellman RE (1957) *Dynamic Programming*. Princeton University Press, Princeton, New Jersey.

[8] Jöreskog KG, Klovan JE, and Reyment RA (1976) *Geological Factor Analysis*. Elsevier, Amsterdam.

[9] MacLeod N (2007) Groups II. *Palaeont Assoc Newsletter* **65**: 36-49.

[10] MacLeod N (2009) Form and shape models. *Palaeont Assoc Newsletter* **72**: 14-27.

[11] MacLeod N (2012) The center cannot hold II: elliptic Fourier analysis. *Palaeontological Association Newsletter* **79**: 29-42.

[12] MacLeod N (2013) Semilandmarks and surfaces. *Palaeont Assoc Newsletter* **83**: 37-51.

[13] Campbell NA and Atchley WR (1981) The geometry of canonical variate analysis. *Syst Zool* **30**: 268-280.

[14] Mitteröcker P and Bookstein FL (2011) Linear discrimination, ordination, and the visualization of selection gradients in modern morphometrics. *Evol Biol* **38**: 100-114.

[15] Bookstein FL (2002) Creases as morphometric characters. **In**: N MacLeod and PL Forey (Eds.) *Morphology, Shape and Phylogeny* Taylor & Francis, London.

[16] Manly BFJ (1997) *Randomization, Bootstrap and Monte Carlo Methods in Biology* Chapman Hall, London.

[17] Colquhoun WP (1959) The effect of a short rest pause on inspection efficiency. *Ergonomics* **2**: 367-372.

[18] Colquhoun WP (1982) Biological rhythms and performance. **In**: WB Webb (Ed.) *Biological Rhythms, Sleep, and Performance* Wiley, Chichester.

[19] Culverhouse PF, Williams R, Reguera B, Herry V, and González-Gils S (2003) Do experts make mistakes? *Marine Ecol Prog Ser* **247**: 17-25.

[20] Culverhouse P (2007) Natural object categorization: man vs. machine. **In**: N MacLeod (Ed.) *Automated Taxon Recognition in Systematics: Theory, Approaches and Applications* CRC Press, Taylor and Francis Group, Boca Raton, Florida.

[21] MacLeod N, Benfield MC, and Culverhouse PF (2010) Time to automate identification. *Nature* **467**: 154-155.

[22] Culverhouse PF, MacLeod N, Williams R, Benfield MC, Lopes RM, and Picheral M (2013) An empirical assessment of the consistency of taxonomic identifications. *Marine Biol Res* **10**: 73-84.

[23] MacLeod N (2007) *Automated Taxon Identification in Systematics: Theory, Approaches and Applications*. CRC Press Taylor & Francis Group, London.

[24] Bledsoe WW (1966) The model method in facial recognition. Panoramic Research Inc., Pal Alto, CA, *Research Report, August* 15.

[25] Bledsoe WW (1966) Man-machine facial recognition. Panoramic Research Inc., Palo Alto, CA, *Research Report August* 22.

[26] Goldstein AJ, Harmon LD, and Lesk AB (1971) Identification of human faces. *Proc IEEE* **59**:78.

[27] Harmon LD (1971) Some aspects of recognition of human faces. **In**: OJ Grusser and R Klinke (Eds.) *Pattern Recognition in Biological and Technical Systems* Springer Verlag, Berlin.

[28] Fischler M and Elschlanger RA (1973). The representation and matching of pictorial structures. *IEEE Trans Comput* **c-22**: 67-92.

[29] Carey S and Diamond R (1977) A computer vision system for the identification of individuals. *Science* **195**: 312-313.

[30] Kohonen T and Lethtio P (1981) Storage and processing of information in distributed associative memory systems. **In**: GE Hinton and JA Anderson (Eds.) *Parallel Models of Associative Memory* Erlbaum, Hillsdale, New Jersey.

[31] Cannon SR, Jones GW, Campbell R, and Morgan NW (1986) A computer vision system for the identification of individuals. *Proc IECON* **1**: 347-351.

[32] O'Toole AJ, Millward RB, and Anderson JA (1988) A physical system approach to recognition memory for spatially transformed faces. *Neural Networks* **1**: 179-199.

[33] Burt P (1988) Smart sensing within a Pyramid Vision Machine. *Proc IEEE* **76**: 139-153.

[34] Kohonen T (1984) *Self-Organization and Associative Memory*. Springer-Verlag, New York.

[35] Fleming M and Cottrell G. (1990) Categorization of faces using unsupervised feature extraction *Proc IJCNN* **90**: 2.

[36] Turk M and Pentland A (1991) Face recognition using eigenfaces. *Comp Vision Pattern Recog Proc CVPR '91* (IEEE)_DOI: 10.1109/CVPR.1991.139758.

[37] Turk M and Pentland A (1991) Eigenfaces for recognition. *J Cognitive Neurosciences* **3**: 71-86.

[38] ONeill M (2007) *DAISY*: a practical tool for semi-automated species identification. **In**: N MacLeod (Ed.) *Automated Taxon Recognition in Systematics: Theory, Approaches and Applications* Taylor and Francis, London.

[39] Rohlf FJ (1990) Rotational fit (Procrustes) methods. **In**: FJ Rohlf and FL Bookstein (Eds.) *Proceedings of the Michigan Morphometrics Workshop* The University of Michigan Museum of Zoology, Special Publication No. 2, Ann Arbor.

[40] Rohlf FJ and Slice D (1990) Extensions of the *Procrustes* method for optimal superposition of landmarks *Syst Zool* **39**: 40-59.

[41] MacLeod N, O'Neill MA and Walsh SA (2007) A comparison between morphometric and artificial neural net approaches to the automated species-recognition problem in systematics. **In**: G Curry and C Humphries (Eds.) *Biodiversity Databases: Techniques, Politics, and Applications* CRC Press, Taylor and Francis Group, Boca Raton, Florida.

[42] MacLeod N (2010) Alternative 2D and 3D form characterization approaches to the automated identification of biological species. **In**: PL Nimis and R Vignes Lebbe (Eds.) *Tools for Identifying Biodiversity: Progress and Problems* Edizioni Università di Trieste, Trieste, Italy.

[43] MacLeod N (in press) A comparison of alternative form-characterization approaches to the automated identification of biological species. **In**: A Hamilton and E Pierson (Eds.) *Patterns in Nature* University of California Press, Berkeley, California.

[44] Wold S (1976) Pattern recognition by means of disjoint principal component models. *Pattern Recog* **8**: 127-139.

Robust Generalized Superimposition Methods: A Comparison Using 3D Facial Images

KATLEEN DANIELS[1, 2] DIRK VANDERMEULEN[1, 2],
PAUL SUETENS[1, 2] and PETER CLAES[1, 2]

Abstract

Superimposition is a popular technique to separate scaling, position and orientation differences from true differences in shape represented as landmark configurations. The well-known generalized Procrustes superimposition (GPS), however, is affected by outliers (Pinocchio-effect) and assumes variation of the landmarks to be homoscedastic and uncorrelated, which influences the correctness of the superimposition. This chapter proposes and investigates two robust superimposition methods, which are generalizations (superimposing more than two landmark configurations) of two ordinary (superimposing two landmark configurations) superimposition methods from recent literature: Outlier Process (from the framework of dysmorphometrics) and Scaled Mixture. In their generalization and in contrast to the GPS approach, the landmarks are not assumed to be homoscedastic. Furthermore, both methods are robust against the Pinocchio-effect. While the Outlier Process Generalization (OPG) explicitly introduces the concept of outliers and assumes the inliers to follow a normal distribution, the Scaled Mixture Generalization (SMG) assumes the displacements of the landmarks to follow a student-t distribution, which is outlier-tolerant. In a first test set-up the methods are tested in their ability to recover a known covariance structure (containing 6 to 20 landmarks), based on perturbed configurations. Additionally, a database of 469 facial images is used, on which 7,150 3D quasi-landmarks are established using an Anthropometric Mask. After generalized superimposition, different quasi-landmarks appear to have a difference in variance, confirming the importance of a separate displacement distribution per landmark. Both OPG and SMG are able to detect these differences. In a last test set-up, an artificially created Pinocchio aspect, containing large outliers at the nose, is added to the database. This allows the investigation of the outlier detection capability of the two methods. The OPG performs better than the SMG in estimating a known covariance structure and is able to correctly and explicitly delineate the region of atypical facial variation in the face of Pinocchio.

INTRODUCTION

Morphometrics is the field that is concerned with the analysis of biological shape and shape variation [1]. Typically, form is defined as size and shape, independent of position and orientation [2]. In geometric morphometrics, shapes are represented by homologous landmark configurations [3, 4], where homologous landmarks are defined as points of correspondence (homology) on an object that match between and within populations [5]. In order to perform a geometric morphometric analysis on biological shape, the objects under observation, represented as homologous landmark configurations, should be brought into the same shape space. For this purpose, superimposition can be used to eliminate scaling, position and orientation differences [6], such that only true shape differences remain. Superimposition of two landmark configurations is referred to as ordinary superimposition, while the superimposition of more than two configurations is called generalized superimposition.

[1] Medical Image Computing, ESAT/PSI, Department of Electrical Engineering, KU Leuven, Herestraat 49 bus 7003, 3000 Leuven, Belgium, katleen.daniels@esat.kuleuven.be.
[2] iMinds-KU Leuven Future Health Department.

Given a set of N shapes, the current practice is to use a generalized Procrustes superimposition (GPS) [7, 8]. This method uses a least-squares (LS) approach, which minimizes the sum of squared differences between corresponding landmarks [9]. However, in a least-squares criterion some assumptions are implicitly made about the variance and covariance of the individual landmarks. Indeed, all landmarks are assumed to follow a standard normal distribution on their displacements [10], which means that their variance is the same in all directions (isotropic), that they move independently from each other, and that the variance is identical across all landmarks. These assumptions are quite strict and might generally not apply in actual biological shape variation. Furthermore, the GPS is influenced by the Pinocchio effect, when dealing with localized but substantial shape differences [10]. This is because in a least-squares superimposition all landmarks are considered equally important due to the equal variance assumption, which causes the discrepancy of the excessive displacements (considered to be outliers) to be smeared out over all the other landmarks. This is undesired since the resulting residuals will then not reflect real variation in shape. The solution to this conundrum is the use of robust superimposition methods [10].

Siegel and Benson [11] were the first to recognize the problem of LS superimpositions in the context of outliers and introduced the Resistant Fit (RF), a "repeated median resistant fitting algorithm". However, in contrast to LS, this method is time-consuming with increasing number of landmarks as well as being heuristic, which implies a lack of explicit criterion being optimized. Much more recently, Mechelke and Habeck [12] were looking for means to fit optimally two structures containing regions that differ largely, as encountered in proteins undergoing large-scale deformations or conformational changes. They proposed a probabilistic method in which the shape differences are considered to follow a 'scale mixtures of Gaussians', or more specifically a student-t distribution, which is a heavy-tailed distribution (in contrast to the standard normal distribution used in LS).

This approach, referred to as Scaled Mixture (SM) in this chapter, as such implicitly enables outlier-tolerant superimpositions [12]. In 2012, Claes *et al.* [10] suggested the explicit detection of shape differences as outliers compared to an appropriate norm. Through the introduction of a latent variable coding for anomalies or outliers, they extended the probabilistic description of LS landmark superimpositions with outlier processes. In the remainder of this chapter, this technique will be referred to as the Outlier Process (OP).

Although starting from different underlying frameworks, both SM and OP converge in theory to a weighted least-squares approach, allowing them to be robust by assigning lower weights to outliers. However, in their current implementation, both methods assume all the landmarks in a configuration to follow the same perturbation distribution: a normal and a student-t distribution in the case of OP and SM respectively. However, biologically it is quite possible for different landmarks within a configuration to fluctuate at different levels of variances. Outliers can then be defined relative to the variance of that particular landmark. For the estimation of a separate variance per landmark, a database of multiple landmark configurations is needed (generalized superimposition).

In this work, the OP and the SM method are extended to allow for the estimation of a separate distribution per landmark, leading to the Outlier Process Generalization (OPG) and the Scaled Mixture Generalization (SMG) respectively. In concordance with previous work of the authors, the focus will be on the superimposition of a spatially-dense 3D landmark configuration of human faces. A first validation test starts from a known covariance structure and examines how well the superimposition methods are able to recover this covariance structure, based on perturbed configurations. In a first instance, the test set-up will work with spatially-sparse synthetic data as developed by Walker [13]. In a second instance, to generate synthetic data more allied to biologically spatially-dense facial data, the eigenvalues of the face dataset after principal component analysis are inserted into the synthetic test set-up. In a second validation test, the dataset of human faces is superimposed and the dissimilarity in variance across the face is visualized for the two proposed methods. Finally, in a last validation, the capability of the methods to detect an anomaly in the dataset is investigated.

MATERIALS AND METHODS

Databases Utilized

Faces

Three-dimensional (3D) facial surface scans of 469 healthy individuals of mixed ancestry, gender and age were recorded using a 3dMD Facial^{TM} stereophotogrammetric system (3dMD Inc., Atlanta, Georgia, USA). All individuals in the database were mapped with an Anthropometric Mask (AM) using a non-rigid surface registration, resulting in a spatially dense set of quasi-landmarks [14, 15]. This dense representation is desired in order not to overlook relevant salient shape information. As a result, each of the faces in the database available is represented by 7,150 homologous 3D-points, referred to as landmark configurations.

Synthetic data

Synthetic 3D landmark configurations of 6 to 20 landmarks are created as explained by Walker [13]. The influence of four factors is investigated: (i) the number of landmarks, or landmark count (LC); (ii) the magnitude of standardized covariance within the covariance matrix, or correlation level (CL): (iii) the number of coordinates with excessive variance relative to the variance of the other coordinates, or Pinocchio level (PL): and (iv) the magnitude of excessive variance, or variance magnitude (VM) [13]. First, a covariance matrix is constructed, after which random perturbed landmark configurations are obtained using this covariance structure. Note that the higher the VM, the higher the relative variance of the landmarks with excessive variance, and not the higher the variance of all landmarks in general. Also consider that at PL = 0, all landmarks are giving excessive variance, resulting in none of the landmarks having excessive variance due to rescaling. Increasing PL decreases the number of landmarks having excessive variance, and finally, at PL = 1, only one landmark was given excessive variance. For each of the variables under investigation (LC, CL, PL and VM), 100 different synthetic datasets of 100 landmark configurations are superimposed, and the resulting dissimilarity measures as defined later in this chapter, are averaged.

Synthetic data with facial eigenvalues

The synthetic data, as obtained by Walker [13], only generates a limited set of significant principal components. Indeed, a Parallel Analysis (data not shown here) determines that there are only three. A scree plot [16] can be used to investigate the amount of variance explained by the different principal components. Scree plots of covariance structures are shown in Figs. 1a and 1b for a CL of 1 and 0 respectively. Respectively 3 and 5 PCs are enough to explain more than 95% of the variance in the configurations. The Correlation Level (CL) influences the correlation between the landmarks and thus influences the number of significant principal components (PC): a high/low CL corresponds to high/low correlation between the landmarks and a lower/higher number of PCs. However, the influence of this factor by the value ranges given in [13] is rather limited.

It would be interesting to construct synthetic covariance structures that are closer to biologically meaningful covariance structures such as the one observed in the facial data. This is done by replacing the fairly random construction of an initial covariance matrix, in which the eigenvalues of the covariance structure are assigned, as done by Walker [13], with an insertion of the eigenvalues of the database of faces. Doing so, a more biologically relevant, but synthetic nevertheless, dataset is obtained. Landmark configurations (40,000) of 1,000 3D-points are created, in order to get relatively dense representations of shape. VM and PL were varied as before, but CL remained unchanged as this parameter only influenced the definition of the synthetic eigenvalues.

In order to retrieve facial eigenvalues the facial landmark configurations were reduced to 1000 uniformly selected landmarks across the facial surface. Then a generalized Procrustes superimposition (GPS) was performed, followed by a Principal Component Analysis (PCA) providing the desired eigenvalues. Subsequently, Parallel Analysis (PA) was performed determining 34 significant PCs. A scree plot of a resulting set of test configurations is depicted in Fig. 1c. In this case, more than 10 PCs are necessary to explain 85% of the variance in the configurations. In other words: a more diverse, uncorrelated covariance structure is obtained, which will be used as a reference that the different generalized superimposition methods should estimate as well as possible. For these tests, only PL and VM are varied and the results are averaged for 50 different datasets.

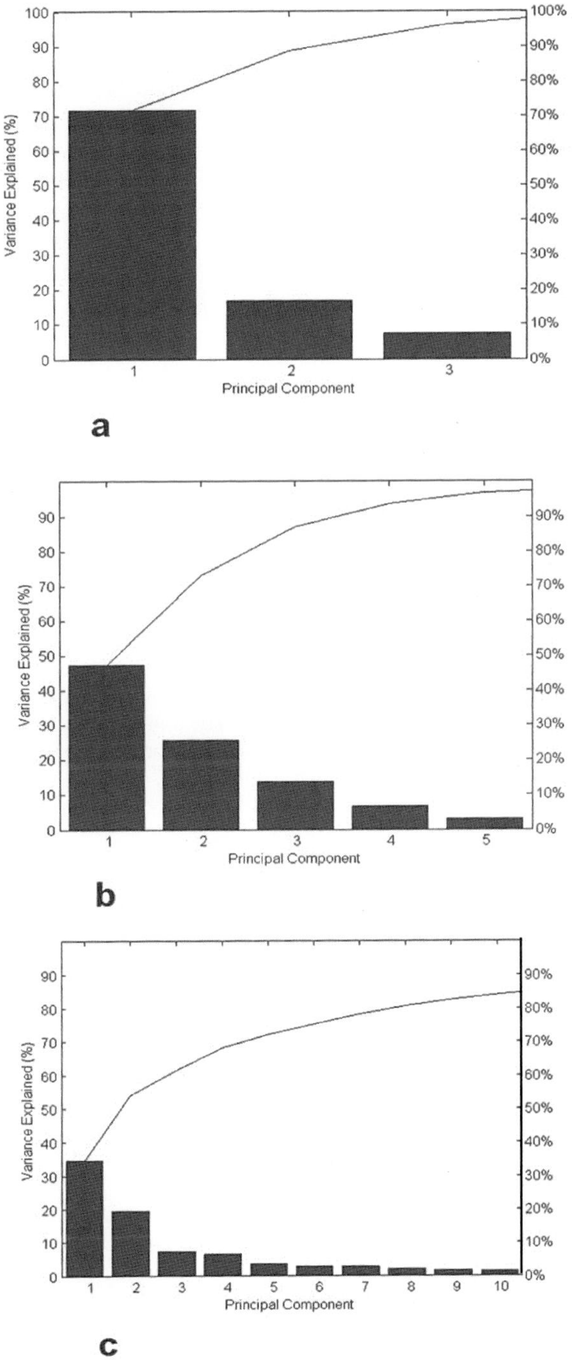

Fig. 1. Scree plots of the (a) synthetic data with $CL = 1$ and (b) $CL = 0$, and (c) of the synthetic data with facial eigenvalues.

Methodology

Outlier process generalization

The Outlier Process Generalization (OPG) method is an extension of the OP method by Claes *et al.* [10]. The extension lies in the fact that, instead of a single normal distribution for the entire face, a separate normal distribution per landmark is estimated. OPG considers two factors when superimposing landmark configurations in a weighted LS fashion: (i) the expected variance of a particular landmark (which is allowed to be different for different landmarks in a configuration), and (ii) the belief that this landmark is an inlier or not, taking into account the variance of that landmark. The practical layout of the algorithm is as follows:

1. Least Squares superimposition of all landmark configurations onto a randomly chosen landmark configuration [6].

2. Determine the average landmark configuration.

3. Determine or update the inlier and outlier distribution per landmark. The inliers for a landmark i are considered to be normally distributed $N(0,\sigma_i)$, with σ_i the standard deviation of the displacement of the *ith* landmark [10]:

$$\sigma_i = \sqrt{\frac{\sum_j^N w_{ij} d_{ij}^2}{\sum_j^N w_{ij}}} \tag{1}$$

where w_{ij} and d_{ij} correspond respectively to the weight and the displacement of the *ith* landmark in the *jth* configuration, and N is the number of configurations available. The parameters of the outlier distributions per landmark are then updated according to [10]:

$$\lambda_i = \frac{1}{\sqrt{2\pi}\sigma_i} \exp\left(-\frac{1}{2}\kappa^2\right) \qquad \forall\ landmark\ i \tag{2}$$

Changing κ changes the estimated amount of inliers versus outliers. By setting $\kappa = 2$, landmarks are considered outliers as soon as their displacement exceeds twice the standard deviation σ_i for that landmark i.

4. Update the weights for the landmarks over all landmark configurations:

$$w_{ij} = \frac{b_{ij}}{2\sigma_i^2} \tag{3}$$

The belief b_{ij} of landmark i in the *jth* configuration to be an inlier is defined as the probability to be an inlier divided by the total probability, under the current transformation $\tilde{T}$ [10]:

$$b_{ij} = \frac{N(d_{ij}|0,\sigma_i)}{N(d_{ij}|0,\sigma_i) + \lambda} \tag{4}$$

5. Perform a weighted Least Squares superimposition of all landmark configurations onto the average.

6. Update the weighted average landmark configuration A:

$$p_{iA} = \frac{\sum_{j=1}^{N} b_{ij}\, p_{ij}}{\sum_{j=1}^{N} b_{ij}} \qquad \forall\ landmark\ i \tag{5}$$

   where p_{ij} are the 3D coordinates of the *ith* landmark in the *jth* configuration. Since the average is weighted, outliers do not affect it.

7. If the value of the objective function has improved: go back to step 3, where the objective function is given by:

$$\min \sum_{i=1}^{L} w_{ij} E_{ij}^2 \qquad \forall\ face\ j \tag{6}$$

The factor $\frac{1}{2\sigma_i^2}$ of the weights allows those regions exhibiting small variations to have a tighter fit. The factor b_{ij} is introduced because of the uncertainty of a particular landmark to be an inlier. Landmark configurations will be superimposed based on the variance of the landmarks and the probability of the landmarks to be an inlier or an outlier, taking into account the typical variation of that respective landmark only. The pseudo code for this algorithm can be seen in Appendix. I.

Scaled mixture generalization

Similar to OPG, SMG is an extension of the method developed by Mechelke and Habeck [12]. While SM considers one heavy-tailed student-t distribution for the entire landmark configuration, SMG allows for a separate student-t distribution per landmark over all configurations. Due to the heavy tails of the student-t distribution, the method is outlier tolerant. An overview of the algorithm is as follows:

1. Initialize the weights randomly: pick random values from the uniform distribution on the open interval (0,1).

2. Determine or update the rigid transformation: perform a weighted Least Squares superimposition using the weights w_{ij}.

3. Determine or update the parameters α_i and β_i of the heavy-tailed distribution by fitting a Gamma distribution to the distribution of the weights, for every landmark i [12].

4. Update the weights by replacing them with their expected value, taking into account the parameters α_i and β_i and the displacements d_{ij} [12]:

$$w_{ij} = \frac{2\alpha_i + 3}{2\beta_i + \left\|d_{ij}\right\|^2} \tag{7}$$

   where α_i and β_i are the scale and the shape parameters of the student-t distribution for the *ith* landmark.

5. If the weights have changed, go back to step 2.

Notice that in contrast to the OPG approach, SMG does not explicitly mark inliers from outliers. Outliers, however, are given a lower weight due to their high displacement d_{ij} in the denominator of the weights w_{ij}. The pseudo code for this algorithm can be found in Appendix II.

Comparative Study

Comparing covariance structures

As explained by Walker [13], covariance structures obtained after superimposition can be compared with the known covariance structure, from which the synthetic dataset was constructed, and the similarity between these two indicates the ability of the superimposition method to reconstruct a known covariance structure. Both OPG and SMG are used for superimposition of the synthetic dataset. For comparison reasons, the generalized Procrustes superimposition is used as well. For each approach, the discrepancy between the recovered and the reference covariance structure is assessed using different dissimilarity or similarity scores, assessing different aspects of the structure differences.

In a first instance, two moments following the work of Claes [17] are computed: variance-covariance scale and variance-covariance orientation. For the difference in these moments between two covariance structures, a non-pivotal distance-based statistic (D-statistic) is used as a measure for the dissimilarity. The first statistic measures the scale difference or the difference in central moment. This test assesses the difference in magnitude of variance using Euclidean distances of samples to their sample mean [18]. The second statistic examines the orientation difference or the difference in mixed moment. This can be considered as a difference in variance orientation or correlation. For this measure, a principal component analysis (PCA) on the superimposed landmark configurations was performed first. Afterwards, a Parallel Analysis (PA) was used to determine the number of significant principal components (PC) K. This was done to exclude irrelevant information, without throwing away the relevant variation. Finally, a set of distance measures D_k for k = 1,...,K based on the principal angles between subspaces was computed as a measure for the difference in variance-covariance orientation between two covariance structures:

$$D_k = \sqrt{k - \sum_{i=1}^{K} (\cos\theta_i)^2} \tag{8}$$

where θ_i for i = 1,...,K are the principal angles and K is the number of significant principal components (17).

In a second instance, as done by Walker [13], the Pearson Product Moment Correlation (PPMC) is used to compare the estimated covariance structures (given a certain superimposition) with the reference covariance structure, which is an overall measure of similarity [13]:

$$PPMC = \frac{\sigma_{12}}{\sigma_1 \sigma_2} \tag{9}$$

where σ_{12} is the covariance between the off-diagonal elements of the two covariance matrices to be compared, and where σ_1^2 and σ_2^2 are the variance of the off-diagonal elements of the first and second covariance matrix respectively.

In a third and last instance, the RV coefficient [19] was used, which is a kind of correlation metric between two covariance matrices, related to the PPMC:

$$RV = \frac{trace(\Sigma_{12}\Sigma_{21})}{\sqrt{trace(\Sigma_{11}\Sigma_{11})trace(\Sigma_{22}\Sigma_{22})}} \tag{10}$$

where Σ_{12} is the covariance matrix of the two covariance matrices 1 and 2, and where $trace(\Sigma_{11}\Sigma_{11})$ defines the variance of the first covariance matrix 1.

Visualizing the different distributions per landmark

The database of faces was superimposed using both OPG and SMG, and the different displacement distributions found per landmark i were investigated by looking at the standard deviation σ_i of the normal distribution for OPG, and by looking at the scale parameter α_i for the student-t distribution in the case of SMG. This was done to confirm or reject the importance of the use of a separate distribution per landmark.

Detection of Pinocchio

In order to investigate whether the methods are able to detect anomalies, the Pinocchio effect was added to the database. Pinocchio was constructed by taking the average face from the dataset and virtually increasing its nose length. This way, homology was automatically obtained. Afterwards, the landmarks of Pinocchio were colour-coded according to the belief factor b_{ij} for the case of OPG, and according to the weights w_{ij} for the case of SMG. Since the nose of Pinocchio displays excessive variance compared to normal variation in the nose, the belief factor and the weights for these landmarks is expected to be small. In contrast, landmarks in the remainder of Pinocchio are expected to have high belief factors and high weights, since Pinocchio has the average face when the nose is excluded.

RESULTS

Comparing Covariance Structures

The results of the comparison between the estimated and the reference covariance structure for the synthetic datasets are presented in contour plots, displaying Landmark Count (LC) on the x-axis and Pinocchio Level (PL) on the y-axis. Because the Correlation Level (CL) had little effect, results were averaged over all values of CL. Such contour plots are compared for every combination of VM x superimposition method. Different levels of VM are displayed in different rows of contour plots, different methods in different columns. The results are shown in Figs. 2 to 5.

The number of maximum components used to investigate the variance-covariance orientation was determined in a separate study using PA. The resulting average amount of significant PCs K was 3 for all LCs. In Fig. 4, D_k with $k = 3$ is shown, because this measure incorporates all significant PCs. Since the PPMC and the RV coefficient are measures of correlation, a higher value (more red in the contour plots) corresponds to a larger similarity and therefore represents a better result. The other metrics are measures of dissimilarity and thus lower values (more blue) present a better result.

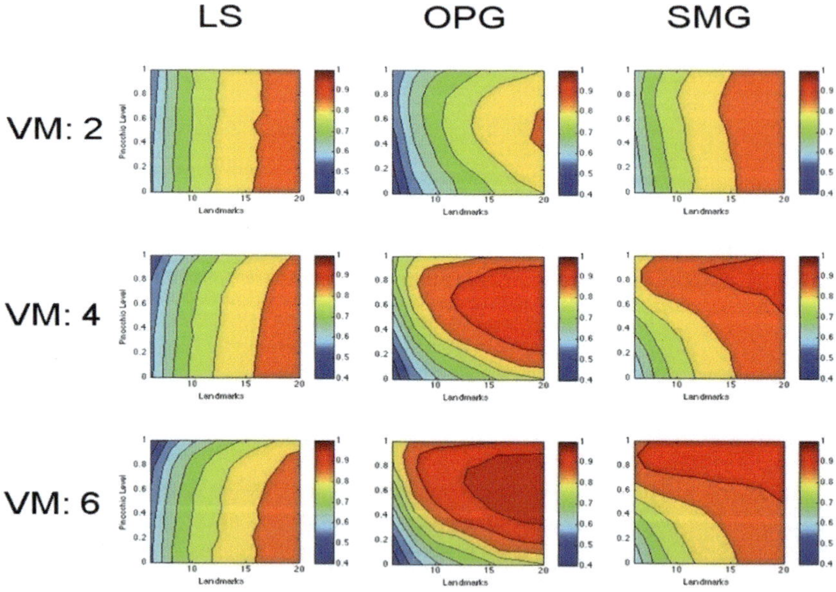

Fig. 2. Results of the synthetic test set-up using the Pearson Product Moment Correlation (PPMC): the influence of LC (x-axes), PL (y-axes), VM (rows) and the method used (columns) on the ability to estimate the covariance structure is investigated. Red values indicate a good result, while blue values indicate a bad result.

In general, OPG outperforms SMG and SMG outperforms LS, according to all metrics. However, at VM = 2, the performance of both LS and SMG appears to be slightly better than that of OPG. A reason might be that in this case the excessive variance is only twice as high as the regular variance, but OPG only considers landmarks to be outliers as soon as their variance exceeds κ times the normal variance. Since $\kappa = 2$, the advantage of OPG is very limited when VM = 2. For small values of PL (PL < 0.3), LS also performs a little better than the OPG method. This corresponds to most of the landmarks having excessive variance, but this excessive variance being relatively less prominent because the average variance stays the same. SMG performs well if a limited set of landmarks (high PL) is given a high excessive variance (high VM). For OPG, there appears to be an optimum in the value of PL (between 0.5 and 0.80). Since the performance of all methods increases with LC, it is expected that the performance will increase even more for even higher values of LC.

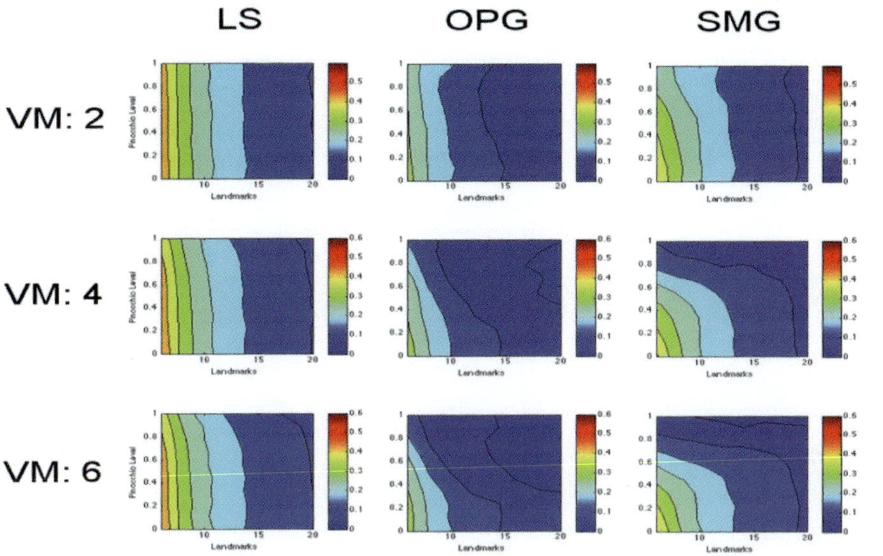

Fig. 3. Results of the synthetic test set-up using variance-covariance scale differences: the influence of LC (x-axes), PL (y-axes), VM (rows) and the method used (columns) on the ability to estimate the covariance structure is investigated. Blue values indicate a good result, while red values indicate a bad result.

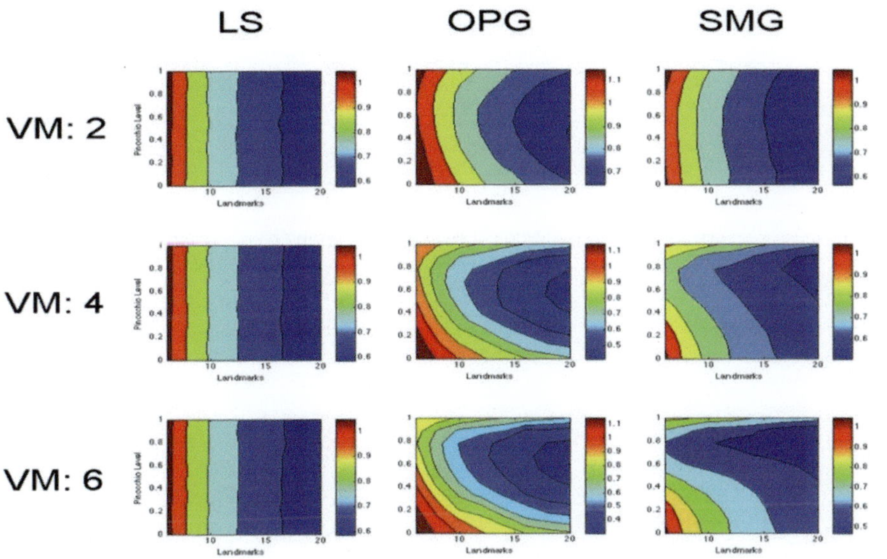

Fig. 4. Results of the synthetic test set-up using variance-covariance orientation differences: the influence of LC (x-axes), PL (y-axes), VM (rows) and the method used (columns) on the ability to estimate the covariance structure is investigated. The value of D_k with $k = 3$ and $K = 3$ is shown here. Blue values indicate a good result, while red values indicate a bad result.

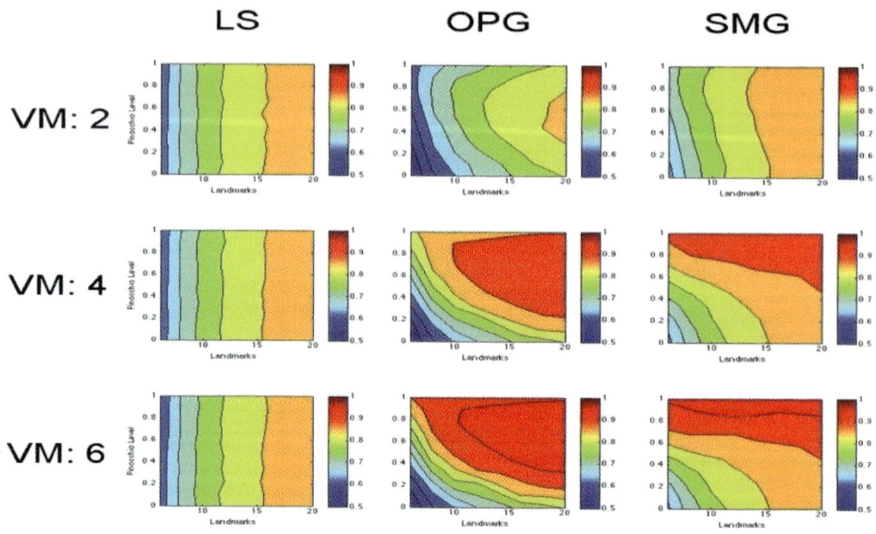

Fig. 5. Results of the synthetic test set-up using RV coefficient: the influence of LC (x-axes), PL (y-axes), VM (rows) and the method used (columns) on the ability to estimate the covariance structure is investigated. Red values indicate a good result, while blue values indicate a bad result.

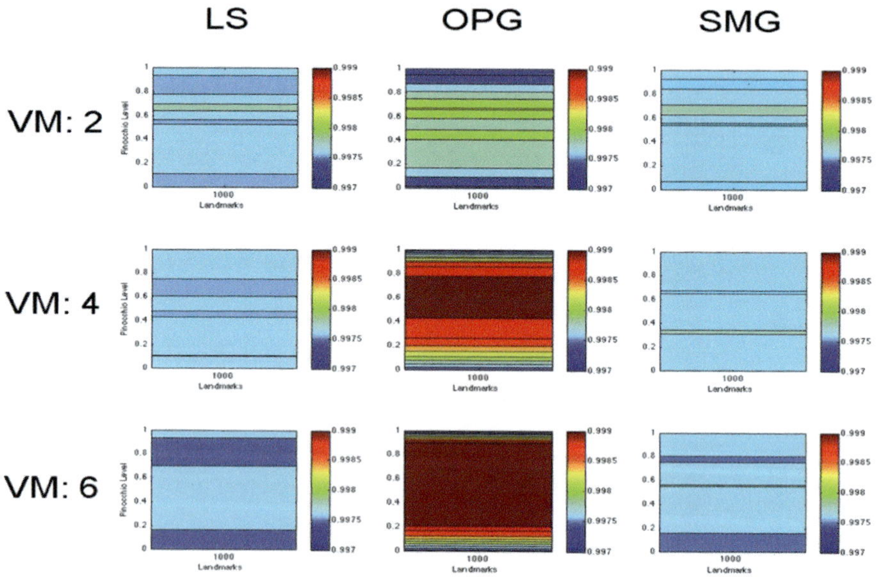

Fig. 6. Results of the synthetic test set-up using PPMC when the eigenvalues of the biological database are inserted, the influence of PL (y-axes), VM (rows) and the method used (columns) on the ability to estimate the covariance structure is investigated. The number of landmarks (x-axes) is fixed to 1,000. Red values indicate a good result, while blue values indicate a bad result.

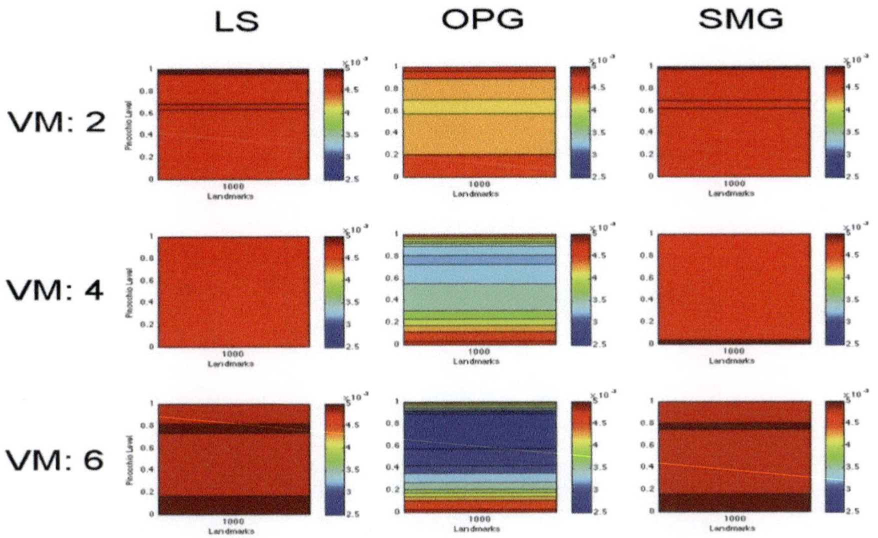

Fig. 7. Results of the synthetic test set-up using variance-covariance scale difference when the eigenvalues of the biological database are inserted: the influence of PL (y-axes), VM (rows) and the method used (columns) on the ability to estimate the covariance structure is investigated. The number of landmarks (x-axes) is fixed to 1,000. Blue values indicate a good result, while red values indicate a bad result.

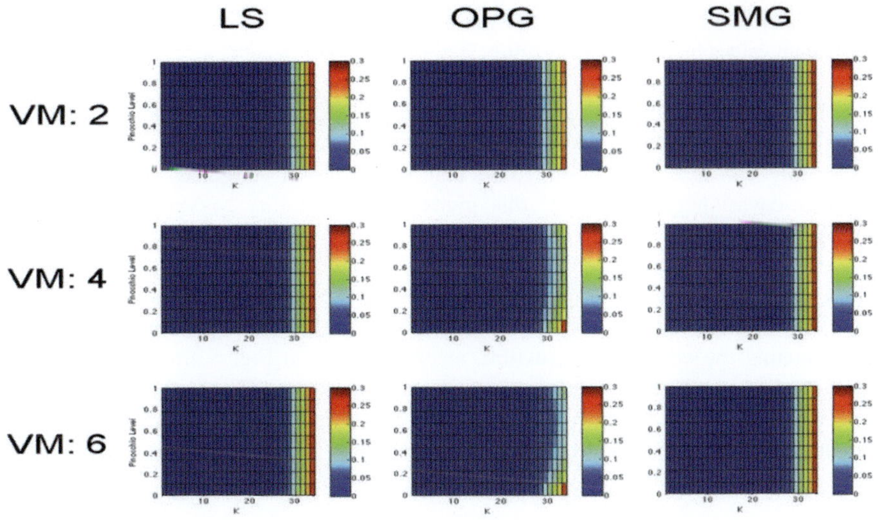

Fig. 8. Results of the synthetic test set-up using variance-covariance orientation difference when the eigenvalues of the biological database are inserted: the influence of PL (y-axes), VM (rows) and the method used (columns) on the ability to estimate the covariance structure is investigated. The values of D_k with are shown with k varied on the x-axes and $K = 34$. The number of landmarks is fixed to 1,000. The more blue the better the results.

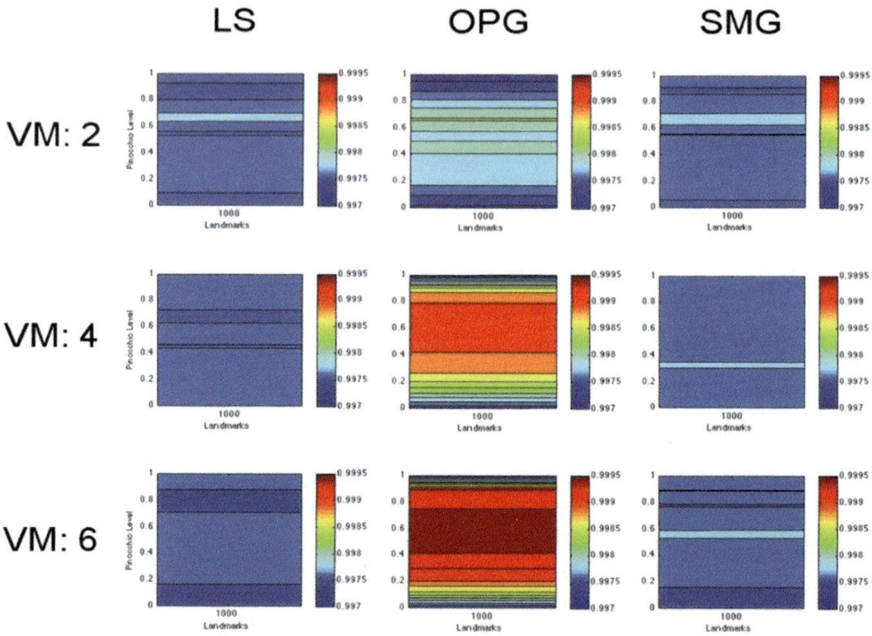

Fig. 9. Results of the synthetic test set-up using the RV coefficient when the eigenvalues of the biological database are inserted: the influence of PL (y-axes), VM (rows) and the method used (columns) on the ability to estimate the covariance structure is investigated. The number of landmarks (x-axes) is fixed to 1,000. Red values indicate a good result, while blue values indicate a bad result.

The results for the synthetic data using facial eigenvalues are shown in Figs. 6 to 9. Notice that the x-axis gives the number of landmarks, which remains constant since only configurations with 1,000 landmarks are investigated. In the case of variance-covariance orientation, this gives the opportunity to vary the k of the distance measure D_k on the x-axes instead of the number of landmarks, as done in Fig. 8.

Overall, OPG performs best in this test setup. Since higher values for PPMC and for RV (more red) indicate better results, the OPG method does an excellent job in estimating the covariance structure, as can be seen in Fig. 6 and Fig. 9. On the level of variance-covariance scale (Fig. 7) and orientation (Fig. 8) the OPG method performs best, because the difference with the reference covariance structure is smaller here (more blue indicates a better result). In Fig. 8, it can be seen that the OPG method results in the smallest difference measures D_k for higher values of k.

Visualizing the Different Distributions Per Landmark

In the case of differences in variance per landmark, OPG and SMG will be capable of finding these differences. Indeed, when performing a generalized superimposition of the database, both methods indicate that these differences are present when investigating facial shape. Figures 10a and 10b show the Anthropometric Mask (AM), in which each landmark i is colour-coded according to the estimated standard deviation σ_i of the normal distribution for the OPG approach and according to the estimated scale factor α_i of the student-t distribution for the SMG approach respectively.

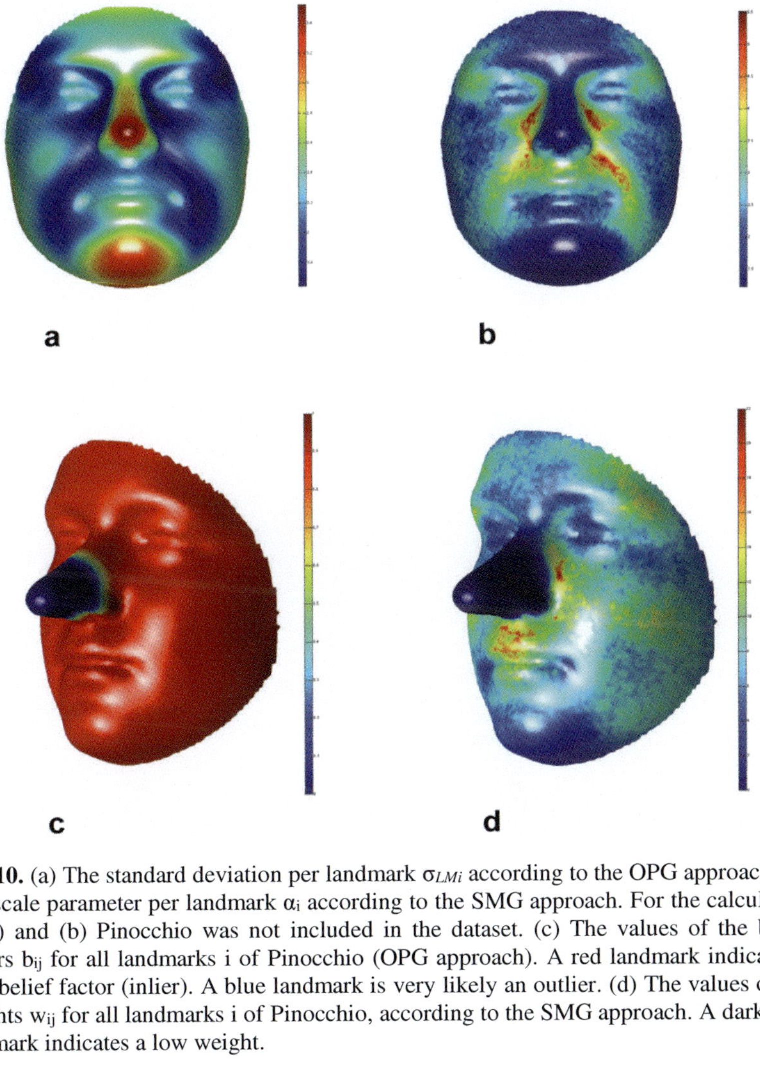

Fig. 10. (a) The standard deviation per landmark σ_{LMi} according to the OPG approach. (b) The scale parameter per landmark α_i according to the SMG approach. For the calculation of (a) and (b) Pinocchio was not included in the dataset. (c) The values of the belief factors b_{ij} for all landmarks i of Pinocchio (OPG approach). A red landmark indicates a high belief factor (inlier). A blue landmark is very likely an outlier. (d) The values of the weights w_{ij} for all landmarks i of Pinocchio, according to the SMG approach. A dark blue landmark indicates a low weight.

Both the results of OPG and SMG generate similar feedback. According to OPG, areas such as the orbits are relatively stable (low σ_i), while the chin and the nose exhibit relatively large displacements. According to SMG, the displacements in areas such as the orbits and in areas next to the nose follow a quite concentrated student-t distribution with low tails (high α_i). Displacements in areas such as the nose and the chin are assumed to be less concentrated and exhibit larger tails. For clarification, these distributions are also visualized in Fig. 11.

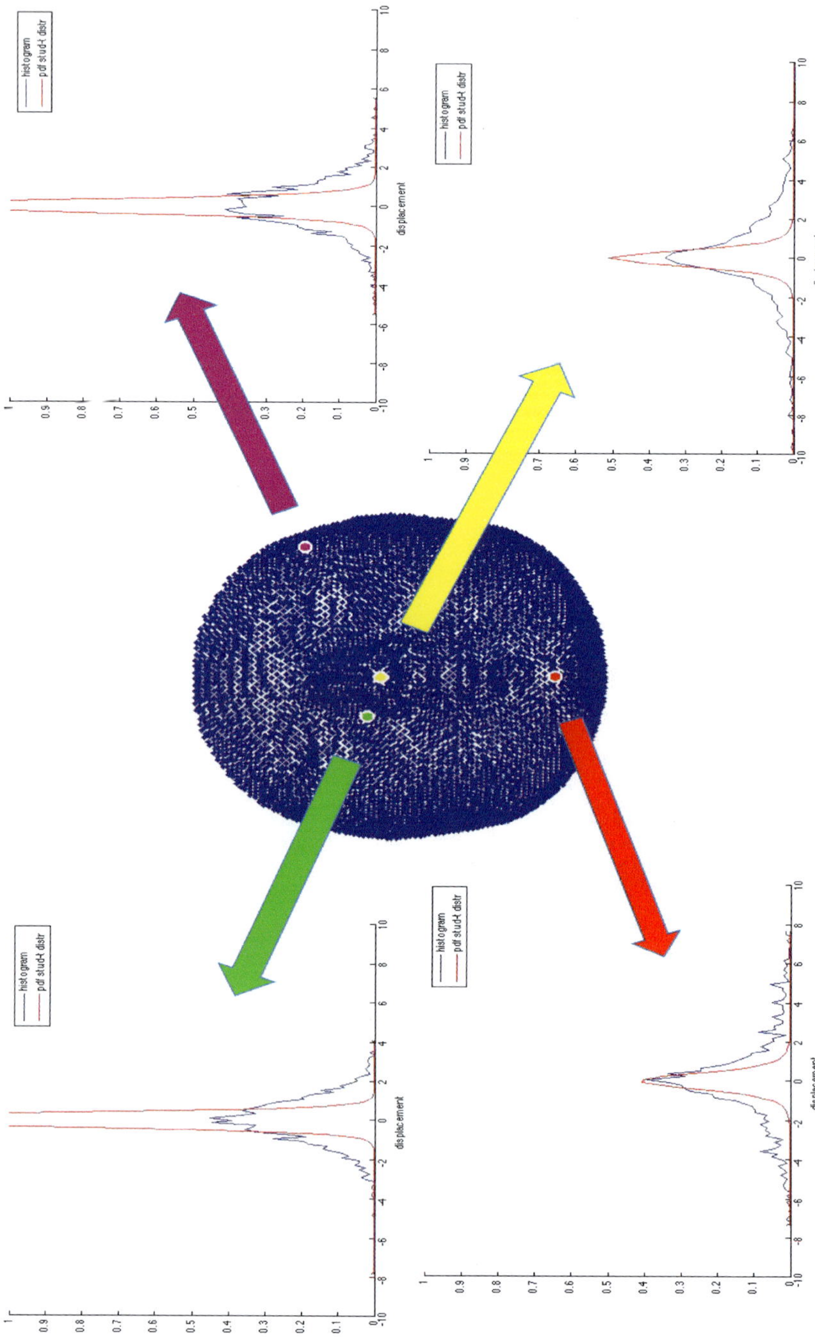

Fig. 11. Distribution of the displacements (pooled over the three dimensions) according to their student-t distributions.

Detection of Pinocchio

The landmarks of Pinocchio were colored according to their belief of being an inlier, i.e. according to the belief factor b_{ij}. As can be seen in Fig. 10c, the non-typical variation in Pinocchio's nose is detected: b_{ij} is virtually zero in the nose. Because the model of normal facial variation is build using the available database, this database is not expected to manifest low average belief factor in certain regions.

As expected, the average b_{ij} has a value of ± 0.8 across the entire landmark configuration. When comparing this result with the part of the face of Pinocchio that is not atypical (i.e., everything except the nose), these landmarks are expected to have a high belief value, since Pinocchio has an average face if his nose is not taken into account. Indeed, many of the values in Pinocchio's face have a high belief factor (± 0.88, red region), compared to the average belief factor for the entire database.

SMG is also able to reject the anomalies of Pinocchio, by giving the according landmarks a low weight w_{ij}. Figure 10d indeed shows that the anomalies in the nose of Pinocchio are detected. However, this detection is implicit compared to OPG.

DISCUSSION AND CONCLUSIONS

A geometric morphometric approach to biological shape analysis involves defining homologous landmarks across observations, followed by a generalized superimposition, needed to bring landmark configurations into a common frame of reference. In the search of a robust superimposition approach, LS is considered inadequate since it introduces a Pinocchio effect when dealing with substantial but localized atypical variations, and therefore the resulting displacements will not represent true shape differences. Although OP and SM try to deal with anomalies in a mathematically based way, they do not take the difference in variance throughout the landmark configuration into account. Both methods are based on only a single displacement distribution for the entire landmark configuration. However, for example, in the case of the human face, it is not unlikely that landmarks in the nose follow a different distribution than for example the landmarks in the forehead. If biologically, the variance of the landmarks in the nose is larger than that of landmarks in the forehead; normal variations in the nose can be considered outliers, while abnormalities in the forehead can go undetected. Therefore, this work proposed two methods that compute a model of variance per landmark, which can then be used to implement a more robust generalized superimposition. The Outlier Process Generalization (OPG) makes a distinction between inliers and outlier and assumes a separate inlier normal distribution per landmark. The Scaled Mixture Generalization (SMG) assumes each landmark to follow a separate student-t distribution, which is outlier-tolerant.

OPG and SMG indeed detect differences in variance throughout the face, making the application of these methods justified and superior to the application of Outlier Process and Scaled Mixture. By visualizing the different distributions per landmark, it can be seen that OPG and SMG come to similar conclusions regarding the variations in certain regions: variance in the nose and the chin is generally larger. However, keep in mind that both methods assume a different type of distribution for the displacements, and that they handle outliers in a different way: OPG explicitly introduces an outlier distribution while SMG uses outlier-tolerant distributions.

Regardless, the introduction of a separate distribution per landmark appears to be worthwhile, since displacement distributions are considerably different throughout the face.

During the test setups, OPG achieved better results to reconstruct known covariance structures. SMG was less successful and a reason for this might be the assumption of a heavy-tailed distribution for the displacements. Initially this was developed to superimpose geometrically similar substructures in proteins. However, this assumption might not be optimal when dealing with e.g. variation in facial shape. A normal instead of a heavy-tailed distribution is assumed in the original Procrustes superimposition, which has been extensively used in the analysis of biological shape so far. Similarly, OPG assumes a normal distribution as well for the inliers and as such, it is a more straightforward extension of the Procrustes superimposition.

As indicated by its ability to detect the anomaly of Pinocchio, OPG is an adequate approach to perform a robust generalized superimposition. The power of OPG lies in the fact that it can detect differences in variance of landmark displacements across the face and additionally detect excessive displacements compared to this normal variance per landmark. Moreover, OPG can explicitly detect outliers by investigating the belief factors, allowing this technique to be used to delineate regions of interest when for example planning surgery.

During the test set-up of Walker [13], outliers are not defined in the same way as in assumptions of OPG. Walker defined outliers as landmarks possessing excessive variance compared to the other landmarks. In other words, when a landmark is given excessive variance, this landmark experiences excessive variance in each landmark configuration of the dataset. This is in contrast to Pinocchio as part of the dataset: here Pinocchio's nose has excessive variance, whilst the remainder of the faces have "normal" noses. In other words, a particular landmark or group of landmarks within a single landmark configuration exhibits excessive variance compared to the average variance of that particular landmark in the complete dataset, which is the assumption of OPG. Hence, the capacity of OPG is expected to be larger than the one tested in the synthetic test set-up.

In the comparative study, a strong correlation between the performance and the number of landmarks was observed. This stimulates the use of spatially dense landmark configurations when comparing faces.

It should be noted that the generalizations proposed here, are still simplifications of reality, namely, these methods do not take into account anisotropic displacements, nor covariance between landmarks. Future work should focus on this.

ACKNOWLEDGEMENTS

This work is financially supported by KU Leuven's Concerted Research Action GOA/11/006.

REFERENCES

[1] Rohlf FJ (1990) Morphometrics. *Ann Rev Ecol and Syst* **21**: 299-316.

[2] Penrose L (1952) Distance, size and shape. *Annals of Eugenics* **17**: 337-343.

[3] Klingenberg CP, Barluenga M, and Meyer A (2002) Shape analysis of symmetric structures: quantifying variation among individuals and asymmetry. *Evolution* **56**: 1909-1920.

[4] Slice DE (2007) Geometric morphometrics. *Annu. Rev. Anthropol* **36**: 261-281.

[5] Dryden IL and Mardia KV (1998) *Statistical Shape Analysis*. John Wiley & Sons, New York.

[6] Rohlf F and Slice D (1990) Extensions of the procrustus method for the optimal superimposition of landmarks. *Syst Zool* **39**: 40-59.

[7] Sneath PH (1967) Trend-surface analysis of transformation grids. *J Zool* **151**: 65-122.

[8] Rohlf FJ (1999) Shape statistics: Procrustes superimpositions and tangent spaces. *J Classification* **16**: 197-223.

[9] Mitteroecker P and Gunz P (2009) Advances in geometric morphometrics. *Evol Biol* **36**: 235-247.

[10] Claes P, Daniels K, Walters M, Clement JG, Vandermeulen D, and Suetens P (2012) Dysmorphometrics: the modelling of Morphological abnormalities. *Theoret Biol Med Modelling* **9**: 1-39.

[11] Siegel A and Benson R (1982) A robust comparison of biological shapes. *Biometrics* **38**: 341-350.

[12] Mechelke M and Habeck M (2010) Robust probabilistic superposition and comparison of protein structures. *BMC Bioinformatics* **11**: 13.

[13] Walker JA (2000) Ability of geometric morphometric methods to estimate a known covariance matrix. *Syst Biol* **49**: 686-696.

[14] Claes P (2007) *A robust statistical surface registration framework using implicit function representations: Application in craniofacial reconstruction*. **In**: Faculteit ingenieurswetenschappen, department Elektrotechniek, afdeling PSI Leuven: K.U.Leuven, Belgium.

[15] Claes P, Walters M, and Clement J (2012) Improved facial outcome assessment using a 3D anthropometric mask. *Int J oral maxillofac surg* **41**: 324-330.

[16] Cattell RB (1966) The scree test for the number of factors. *Multivariate Behav Res* **1**: 245-276.

[17] Claes P, Walters M, Shriver MD, Puts DA, Gibson G, Clement JG, Baynam G, Verbeke G, Vandermeulen D, and Suetens P (2012) Sexual Dimorphism in Multiple Aspects of 3D Facial Symmetry & Asymmetry defined by Spatially-dense Geometric Morphometrics. *J Anat* **221**: 97-114.

[18] Anderson MJ (2006) Distance-Based Tests for Homogeneity of Multivariate Dispersions. *Biometrics* **62**: 245-253.

[19] Robert P and Escoufier Y (1976) A unifying tool for linear multivariate statistical methods: the RV-coefficient. *Applied Stat* **25**: 257-265.

APPENDICES

Appendix I. Pseudo Code for of the OPG Algorithm (refer to text)

```
A = random initial landmark configuration
B = all other landmark configurations
wij = 1, for all landmarks i and all landmark configurations j
for each landmark configurations j in B
        superimpose B(j) onto A using Least Squares
end
B = all initially superimposed landmark configurations
A = mean(B)
for each landmark configurations j in B
        determine the displacements dij
end
for each landmark i
        update the standard deviations σi
        update the belief factors bij
        update the weights wij
end
change = true
while change
        for each landmark configurations j in B
                superimpose B(j) onto A using weighted Least Squares using weights wij
        end
        A = weighted mean of B, using belief factors bij as weights
        for each landmark configurations j in B
                determine the displacements dij
        end
        for each landmark i
                update the standard deviations σi
                update the belief factors bij
                update the weights wij
        end
        if objective function does not improve
                change = false
        end
end
```

Appendix II. Pseudo Code for the SMG Algorithm (refer to text)

```
A = random initial landmark configuration
B = all other landmark configurations
wij = rand(), for all landmarks i and all landmark configurations j
for each landmark configurations j in B
        superimpose B(j) onto A using weighted Least Squares using weights wij
end
B = all initially superimposed landmark configurations
A = mean(B)
for each landmark i
        update the parameters αi and βi of the heavy-tailed distribution
end
for each landmark configurations j in B
        determine the displacements dij
        update the weights wij
end
change = true
while change
        for each landmark configurations j in B
                superimpose B(j) onto A using weighted Least Squares using weights wij
        end
        A = mean(B)
        for each landmark i
                update the parameters αi and βi of the heavy-tailed distribution
        end
        for each landmark configurations j in B
                determine the displacements dij
                update the weights wij
        end
        if the weights wij have not changed
                change = false
        end
end
```

A PLS Regression Framework for Spatially-dense Geometric Morphometrics to Analyze Effects on Shape and Shape Characteristics: Applied to the Study of Genomic Ancestry and Sex on Facial Morphology

PETER CLAES[1, 2], KATLEEN DANIELS[1, 2], DIRK VANDERMEULEN[1, 2], PAUL SUETENS[1, 2] and MARK D. SHRIVER[3]

Abstract

Shape-regression is an important technique in geometric morphometrics for investigating the effects of various independent variables on biological morphology represented using landmark configurations (dependent variables). Spatially-dense (in contrast to the traditional sparse) landmark configurations, typically cover the complete shape with thousands of landmarks such that salient features of the shape are not overlooked. Furthermore, as proposed in this chapter, spatially-dense configurations allow for the computation and representation of shape variation in terms of local shape characteristics such as curvature, area, and normal displacement. One challenge in using spatially-dense data is the large number of correlated dependent variables in comparison to the number of observations, leading to model instability when using an ordinary least squares regression. This problem has been addressed by using the more advanced technique of partial least squares regression (PLSR), which uses the correlation between the dependent variables for model stabilization, and have investigated genomic ancestry and sex in relation to 3D facial morphology. Briefly, the effect on facial morphology with respect to a particular landmark is measured as the magnitude or Euclidean distance of its displacement in 3D space. The effect-size or strength of the relationship is reported as the variance explained by the PLSR model. Statistical significance is tested under permutation for multivariate regressions. While the effect and effect-size provide insight into which parts of the face are being affected it fails to summarize and convey how facial characteristics are changing. Therefore, the PLS regression framework has been expanded to analyze local effects on normal displacement, curvature, and area. The results are in agreement with general expectations for differences in facial shape due to sex and genomic ancestry. The incorporation of normal displacement, curvature, and area provide additional and valuable biological insight and feedback of the effects on facial morphology due to sex and genomic ancestry.

INTRODUCTION

Shape regression, or a multivariate regression of shape variables, which are invariant to position, orientation and size [1], on another variable is a useful technique to investigate the effect of a single factor on the morphology of biological organisms [2]. In the particular framework of geometric morphometrics, shape variables consist of superimposed landmark coordinates [3], which are defined as a point of correspondence (homology) on an object that matches between and within populations [4]. Traditionally, only a sparse set of specific landmarks that have developmental, functional, structural, or evolutionary significance [5] are used. In practice, one selects a set of anatomical points that are believed to capture relevant aspects of shape variation [3]. As a result, they are almost exclusively located manually on shape per shape independently, which introduces a perceptual and subjective bias in the choice of landmarks. However, it has been discussed and

[1] Medical Image Computing, ESAT/PSI, Department of Electrical Engineering, KU Leuven, Herestraat 49 bus 7003, 3000 Leuven, Belgium, peter.claes@esat.kuleuven.be.

[2] iMinds-KU Leuven Future Health Department.

[3] Department of Anthropology, Pennsylvania State University, University Park, PA, USA.

shown that different landmark choices can lead to contrasting study outcomes [6]. Furthermore, in the example of human faces, only a limited amount of landmarks can be chosen owing to the lack of anatomically discrete features in regions of the face, such that salient features of facial shape are overlooked [7, 8].

Several extensions to spatially-denser sets of landmarks have been proposed. For example, besides using "true" landmarks reflecting homologous points, more automatically indicated points defined by relative locations (for example "the point of highest curvature"), or defined relative to and based on other landmarks (for example "halfway between eye corners"), can be used [9]. Better known in geometric morphometrics are the *semi- or sliding-landmarks* as proposed in [10]. Here additional landmarks are often first sampled equidistantly in between indicated landmarks, e.g., and are then allowed to slide along the curve or surface until some measure of shape difference is minimized [2]. Alternatively, following previous work, a spatially-dense set of landmarks can also be obtained using an Anthropometric Mask (AM) and a mapping technique [6, 11, and 12]. This provides a spatially-dense set of *quasi-landmark* indications, which are essentially obtained using a non-rigid surface registration (mapping) of a predefined template (the AM). This template defines the amount of landmarks used, their relative location in regard to each other and the connectivity in between with triangles.

Besides differences in the way additional landmarks are identified and called or referred to, all spatially-dense representations of shape share the same advantages and disadvantages. The main advantage is in the complete coverage of shape, which obviously has interesting scientific implications. Furthermore, using the right resolution of landmarks and defining triangular connectivity in-between them (as defined in an AM e.g.), real and synthetic shapes (simulated through shape regression e.g.) can be displayed as "picture-alike" images thanks to appropriate computer graphics techniques. This has clear practical applications as well, such as the creation of synthetic perceptual stimuli (e.g. [13]), planning of surgical treatment [11,12, and 14] or forensic reconstruction of faces for victim identification purposes [15, 16]. Another advantage, shown in related work in the current proceedings [17], is that an increased amount of landmarks lead to better superimposition results. This is not unimportant, considering the need for correctly superimposed landmark coordinates prior to any analysis in geometric morphometrics.

The main disadvantage of spatially-dense shape representations, however, is that the large number of shape variables easily and practically always exceeds the number of cases or observations. Therefore popular statistical methods may become unreliable, especially those involving an inversion of the landmark variance-covariance matrix [2, 6]. In the context of shape regression, an ordinary multiple least squares regression suffers from model instability and therefore alternative and computationally feasible regression techniques are required. Finally, the use of spatially-dense representations of shape is limited so far and some possible advantages are still unexplored. For example, a local coordinate frame per landmark based on the surface normal estimated from neighbouring landmarks and their connectivity can be established such that local shape characteristics and changes therein can be computed.

In this work a partial least squares regression (PLSR [18]) framework is proposed for spatially-dense geometric morphometrics and it is applied to the investigation of sex and genomic ancestry in relation to 3D facial morphology. PLSR utilizes the correlations between the dependent variables (facial shape variables) for model stabilization and does not require the independent variables (sex and genomic ancestry) to be linearly independent [19]. In addition it is proposed to look at three local shape characteristic changes (LSCC), (stretching/shrinking, convex/concave changes and inward/outward movement) and incorporate them into the same PLSR framework for purposes of statistical inference. This provides a means of translating effects on facial morphology from the abstract shape space into both visual representations and into the terms useful for the clinical, forensic, and anthropological descriptions of faces.

MATERIALS AND METHODS

Population Sample

592 Research participants from three West African/European admixed populations contributed DNA and 3D facial images. They were collected in the United States (N = 154, State College, PA, Williamsport, PA, and The Bronx, NY); Brasilia, Brazil (N = 191); and Cape Verde (N = 247, São Vicente, and Santiago), all under a Penn State University Internal Review Board (IRB) approved research protocol titled, "Genetics of Human Pigmentation, Ancestry and Facial Features." Height, weight, age, self-reported ancestry, and sex were collected by survey. DNA was collected both with buccal cell brushes and using finger-stick blood on four-circle Whatman FTA cards (Whatman, Florham Park, NJ). Ancestry informative markers (AIMs) were used to estimate individual genomic ancestry from DNA [20]. 68-AIM ancestry estimates were generated using ADMIXMAP.

3D Image Acquisition and Processing

3D images composed of surface and texture maps were taken using the 3dMDface system (3dMD, Atlanta, GA). Participants were asked to close their mouths and hold their faces with a neutral expression for the picture. Images were then exported from the 3dMD Patient software in OBJ file format and imported into a scan cleaning program for cropping and trimming, removing hair, ears, and any dissociated polygons.

An AM (7,150 quasi-landmarks) was non-rigidly mapped onto the original 3D images and their reflections [11, 21], which were constructed by changing the sign of the x-coordinate. This established homologous spatially-dense quasi-landmark (Q-L) configurations for all original and reflected 3D images. Following previous work [6, 12], faces were decomposed into their components of symmetry and asymmetry, where the component of asymmetry was dropped. This was done as deviations from bilateral symmetry are thought to be the effects of developmental noise and/or environmental factors rather than genes [22].

A principal component analysis (PCA) was performed on the superimposed and symmetrized quasi-landmark configurations. The PCs corresponding to the last two percent of the variance observed in the complete dataset were deleted, since they typically correspond to insignificant variance due to random errors or artefacts arising from the scanning or imaging and mapping processes. Finally, superimposed, symmetrized and cleaned quasi landmark-configurations are reconstructed from the first set of PCs explaining 98% of the total variance.

Partial Least Squares Regression

The superimposed, symmetrized and cleaned quasi landmark-configurations were regressed on sex and genomic ancestry using PLSR. The effect on a particular quasi-landmark was measured as the magnitude or Euclidean distance of its displacement in 3D space. The effect-size or strength of the relationship was reported as the variance explained by the PLSR model (R^2). The partial effects (one variable independent from the others in a multiple regression) were coded in the partial regression coefficients. The partial effect-sizes were reported as the partial R^2 values obtained from a reduced regression model. This reduced model is the regression model for a single independent variable after statistically removing the effect of all the other independent variables onto both the single independent variable itself and the dependent variable [23]. Statistical significance of both multiple and partial effects are tested under permutation for multivariate regressions [24]. Here, the respective multiple and partial R^2 values are used as test-statistics with 10,000 permutations. For significance of the partial effects, permutation is performed under the reduced model [23].

Localized effects and effect-sizes per quasi-landmark were visualized using heat maps, while localized significance per quasi-landmark was plotted as a significance map using binary colours coding for being significant (yellow) or not (green) according to a p-value < 0.001. Additionally, shape transformations across the range of independent variable values provide visual changes illustrating the effect on facial morphology. Per independent variable, two shape transformations were constructed from the average face in the direction of the regression-line. This at -3 and +3 times the standard deviation in the case of sex and genomic ancestry.

Local Shape Characteristic Changes

While the effect (R) and effect-size (R^2) analysis illustrated which quasi-landmarks were being affected, these failed to usefully communicate how the quasi-landmarks are changing and what is happening to the facial morphology as a consequence. In order to further illustrate the effects, we defined three local shape characteristic changes (LSCC). The effect on these LSCCs was tested within the same PLSR framework under permutation. Each LSCC was measured between the two shape transformations from the average face and served as an observed test-statistic. LSCCs were either obtained as the difference/ratio between measured features on both shape transformations or as a directed change from one shape transformation to the other. Under each permutation of independent variables, the shape transformations were created again and the LSCC under permutation was compared to the observed value. This process is repeated 10,000 times and the number of times the permuted values are greater than or equal to the observed values divided by the total number of permutations provides an empirical p-value for a one-sided test (H1 + or −) of the null that there is no significant LSCC. In all cases, however, the direction in which the change in shape characteristic occurred between the two shape transformations was irrelevant, such that the absolute permuted LSCC values was compared to the absolute observed value, generating the *p-value* for the two-sided test (H2). Similar to above, for the significance of partial effects of sex and genomic ancestry, permutation was performed under the reduced model.

The following categories of shape features and directed changes were used in this work:

- *Curvature:* The signed mean curvature in each quasi-landmark was used where a negative and positive sign indicate a concave and convex local shape, respectively. A curvature of zero indicates a locally flat shape. A curvature-based LSCC was obtained by taking the difference between corresponding curvature measurements on both shape transformations. These types of LSCCs provide insight whether or not facial shape is changing in aspects of flatness, concavity and convexity.
- *Area:* The average area of all triangles of the mesh structure in which a quasi-landmark participates as a vertex was used to summarize the local area in each quasi-landmark. An area-based LSCC was obtained by taking the negative log ratio between corresponding area measurements on both shape transformations. These types of LSCCs provide insight whether or not facial shape is locally stretching or shrinking.
- *Normal Displacement:* A directed displacement was measured as the signed magnitude of the positional change of a quasi-landmark in space from the first to the second shape transformation. The displacement was projected onto the direction of the normal through the quasi-landmark in the first shape transformation. These types of LSCCs provide insight whether or not facial shape is locally changing inwards or outwards.

RESULTS

PCA run on the superimposed and symmetrized quasi-landmark configurations of the panel of 592 participants resulted in 44 PCs that together summarize 98% of the total variation in the face space. To examine the effect of excluding lower PCs one first reconstructed actual quasi-landmark configuration from the 44 PCs only and compared these to the original remapped face. One found that the average root mean squared error (RMSE) was as small as 0.2 mm per quasi-landmark.

The effect size and statistical significance per quasi-landmark along with alternate shape transformations for sex is depicted in Fig. 1. This figure also shows the effects of sex in terms of area, curvature and normal displacement. The effects observed here are consistent with the effects found in a recent study on sexual dimorphism in facial symmetry and are primarily on the supraorbital ridges, nose, cheeks, mandible, and midface [6]. Using the new local shape characteristic changes (LSCC) described in this chapter, additional insights into the sexual dimorphism of the face can be made. Males exhibit a larger nose, chin, mandible, upper lip, philtrum, inner upper canthic region, and supraorbital ridges, while having a smaller midface and smaller eyes in terms of surface area. Curvature differences mainly occur in the orbital regions and around the mouth, with the nasal bridge and supraorbital ridges standing out as showing the most significant differences in local curvature. An outward movement of the entire nose, chin, supraorbital ridges, and philtrum and an inward movement of the cheekbones, cheeks and eyes are seen moving from the female to male transformed face.

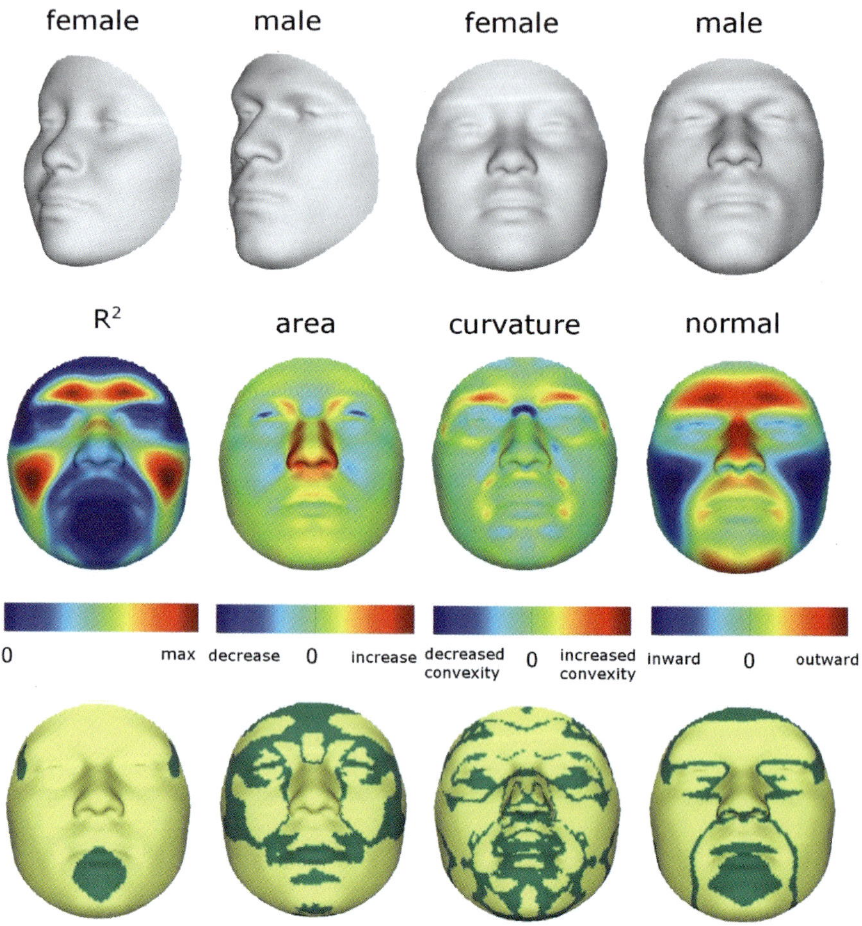

Fig. 1. Transformations and heat maps showing how face shape is affected by sex are displayed here. The top row of each panel shows the shape transformations for −3 and +3 times the standard deviation for sex. The bottom row shows the R^2 (proportion of the total variation in each quasi-landmark), and the three local shape characteristic changes: area ratio, curvature difference, and normal displacement. The statistical significance under permutation (t = 10,000) is shown in the bottom row, where yellow indicates significant and green not ($p < 0.001$).

The effect-size and statistical significance per quasi-landmark along with alternate shape transformations for genomic ancestry are depicted in Fig. 2. As for sex, this figure also shows the effects in terms of area, curvature and normal displacement. The effects on the West African/European axis of ancestry mainly involve changes in the nose, lips, chin, mandible and supraorbital ridges. In terms of surface area, the European transformed face shows larger paranasal tissues and inner canthic regions, and a larger midface and chin. The European transformed face also shows smaller lips, philtrum, alaenari and nares as well as a smaller central forehead and smaller eyes. The main curvature differences are located at the nasal bridge, nasal ridge, supraorbital ridges, columella, philtrum, and chin, all of which show greater convexity in the European transformation than in the African transformation. An outward movement of the nasal bridge, nasal ridge, supraorbital ridges, chin, and mandible and an inward movement of the alaenari, lips, perioral region, cheeks, and orbital regions and are seen moving from the African to the European transformed face.

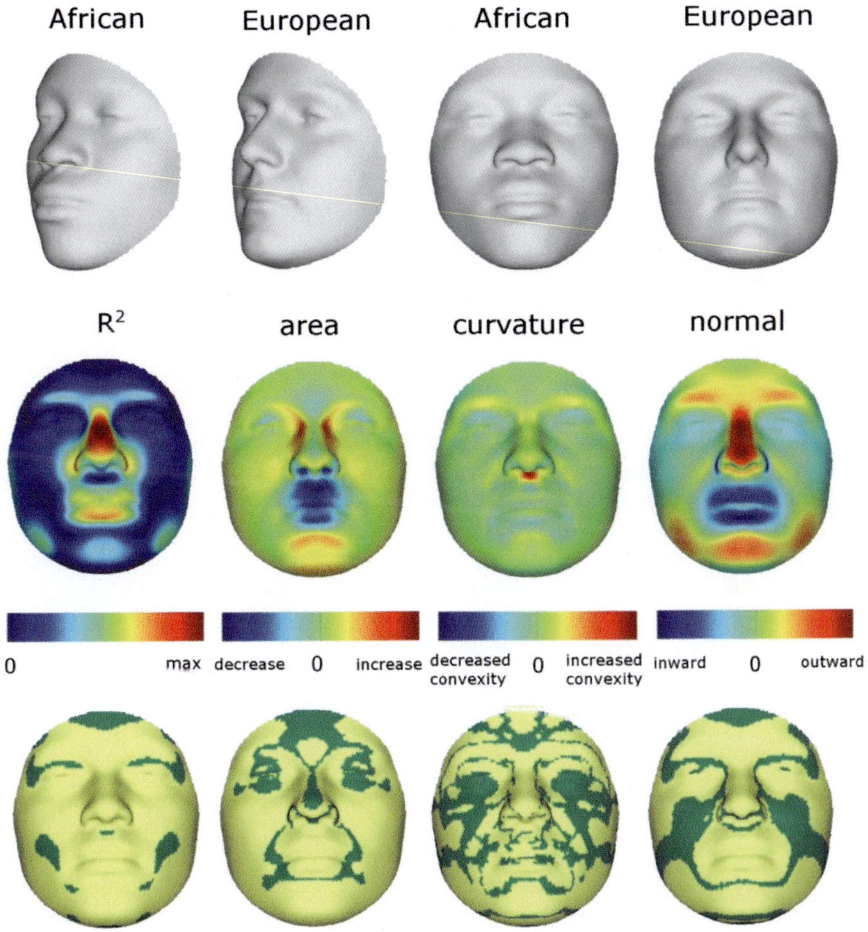

Fig. 2. Transformations and heat maps showing how face shape is affected by genomic ancestry are shown here. The top row of each panel shows the shape transformations for −3 and +3 times the standard deviation for genomic ancestry. The bottom row shows the R^2 (proportion of the total variation in each quasi-landmark), and the three local shape characteristic changes: area ratio, curvature difference, and normal displacement. The statistical significance under permutation (t = 10,000) is shown in the bottom row, where yellow indicates significant and green not ($p < 0.001$).

DISCUSSION AND CONCLUSIONS

Sex and genomic ancestry have clear effects on facial morphology, the results of which are interesting for a variety of reasons. Foremost among these is the fact most people are quite familiar with the facial effects of these variables. Given that quasi-landmark mapping, PCA, and PLS regression are abstract and relatively complex statistical methods, it is encouraging to observe familiar results for variables like sex and ancestry. Observing the shape transformations in Figs. 1 and 2, for example, one can clearly recognize which faces result from transformations in the male, female, European and African directions. Because, sex and genomic ancestry, have just been used to illustrate the methodological applicability, a full review of sexual dimorphism for example is outside the scope of this discussion. Instead, the reader is referred to a previous publication on the topic [6].

The ability to perceptually recognize and evaluate the results is a strong advantage of working with spatially-dense shape representations. In combination with shape characteristics and changes therein, the feedback from the results is augmented, enabling objective and clear descriptions of the effects observed. It has been demonstrated that three different types of local shape characteristics; curvature, area and normal displacement are involved. These are interesting examples to augment the analysis of shape in a regression framework. Changes in intuitive aspects such as stretching/shrinking, convex/concave changes and inward/outward movement can be exposed and quantified.

The regression framework presented was designed to deal with spatially-dense data. In these scenarios, typically a large number of variables are given compared to the number of observations, which makes traditional techniques, like ordinary multiple regression unstable. Instead the more advanced technique of PLS regression was used in combination with permutation based significance testing. Although computationally intensive, model stability and computational feasibility is still guaranteed. With the increased computing power available over the last decade, these types of analyses are accessible today.

ACKNOWLEDGEMENTS

We thank the participants in this study, without whom none of this research would have been possible. This investigation was supported by grants from Science Foundation of Ireland Walton Fellowship (04.W4/B643) from the National Institute Justice (2008-DN-BX-K125). The work was also supported by the Flemish Institute for the Promotion of Innovation by Science and Technology in Flanders (IWT Vlaanderen), the Research Program of the Fund for Scientific Research - Flanders (Belgium) (FWO), and the Research Fund of the KU Leuven.

REFERENCES

[1] Penrose LS (1954) Distance, size and shape. *Annals of Eugenics* **18**: 337-343.

[2] Mitteröcker P and Gunz P (2009) Advances in Geometric Morphometrics. *Journal of Evolutionary Biology* **36**: 235-247.

[3] Slice D (2007) Geometric Morphometrics. *The Ann rev Anthrop* **36**: 261-281.

[4] Dryden IL and Mardia KV (1998) *Statistical Shape Analysis.* John Wiley & Sons, New York.

[5] Richtsmeier JT, Deleon VB, and Lele SR (2002) The Promise Of Geometric Morphometrics. *Am J Phys Anthrop* **45**: 63-91.

[6] Claes P, Walters M, Shriver MD, Puts DA, Gibson G, Clement JG, Baynam G, Verbeke G, Vandermeulen D, and Suetens P (2012) Sexual Dimorphism in Multiple Aspects of 3D Facial Symmetry & Asymmetry defined by Spatially-dense Geometric Morphometrics. *J Anat* **221**: 97-114.

[7] Thomas CDL (2005) Three-Dimensional Quantification of Facial Shape. **In**: JG Clement and MK Marks (Eds.) *Computer-Graphic Facial Reconstruction.* Elsevier Academic Press, London.

[8] Thomas CDL, Claes P, Shaweesh AI, and Clement JG (2011) Three-Dimensional Facial Shape Archetypes For Identification and Diagnosis. **In**: PE Lestrel (Ed.). *Biological Shape Analysis: Proc 1st Int Symp Biol Shape Anal* (ISBSA) World Scientific Pub, Singapore.

[9] Smeets D, Claes P, Vandermeulen D, and Clement JG (2010) Objective 3D Face Recognition: Evolution, Approaches and Challenges. *Forensic Sci Int* **201**: 125-132.

[10] Bookstein FL (1991) *Morphometric tools for landmark data: geometry and biology*. Cambridge University Press, Cambridge.

[11] Claes P, Walters M, and Clement JG (2012) Improved Facial Outcome Assessment using a 3D Anthropometric Mask. *Int J Oral Maxillofac Surg* **41**: 324-330.

[12] Claes P, Walters M, Vandermeulen D, and Clement JG (2011) Spatially-dense 3D facial asymmetry assessment in both typical and disordered growth. *J Anat* **219**: 444-455.

[13] Hill H, Claes P, Crocoran M, Walters M, Johnston A, and Clement JG (2011) How different is different? Criterion and sensitivity in face-space. *Frontiers Psych, Percept Sci* **2**: 1-14.

[14] Claes P, Walters M, Gillett D, Vandermeulen D, Clement JG, and Suetens P (2013) The normal-equivalent: a patient-specific assessment of facial harmony. *Int J Oral Maxillofac Surg* **42**: 1150-1158.

[15] Claes P, Vandermeulen D, De Greef S, Clement JG, Willems G, and Suetens P (2010) Bayesian estimation of optimal craniofacial reconstructions. *Forensic Sci Int* **201**: 146-152.

[16] Claes P, Vandermeulen D, De Greef S, Willems G, Clement JG, and Suetens P (2010) Computerized craniofacial reconstruction: Conceptual framework and review. *Forensic Sci Int* **201**: 138-145.

[17] Daniels K, Vandermeulen D, Suetens P, and Claes P (2013) Robust generalized superimposition methods: a comparison using 3D facial images. **In**: PE Lestrel (Ed.) *Biological Shape Analysis: Proc 3rd Int Symp Biol Shape Anal* (ISBSA) World Scientific Pub, Singapore.

[18] Abdi H (2003) Partial Least Squares (PLS) Regression. **In**: M Lewis-Beck, A Bryman, and T Futing (Eds.) *Encyclopaedia of Social Sciences Research Methods*. Sage, Thousand Oaks (CA).

[19] Tøndel K, Indahl UG, Gjuvsland AB, Vik JO, Hunter P, Omholt SW, and Martens H (2011) Hierarchical Cluster-based Partial Least Squares Regression (HC-PLSR) is an efficient tool for metamodelling of nonlinear dynamic models. *BMC Syst Biol* **5**: 1-17.

[20] Halder I, Shriver MD, Thomas M, Fernandez JR, and Frudakis T (2008) A panel of ancestry informative markers for estimating individual biogeographical ancestry and admixture from four continents: utility and applications. *Hum Mutation* **29**: 648-658.

[21] Claes P (2007) *A robust statistical surface registration framework using implicit function representations: Application in craniofacial reconstruction.* **In**: Faculteit ingenieurswetenschappen, department Elektrotechniek, afdeling PSI. PhD Thesis, K.U. Leuven, Belgium.

[22] Palmer AR (1994) Waltzing with Asymmetry Is fluctuating asymmetry a powerful new tool for biologists or just an alluring new dance step? *Biosci* **46**: 518-532.

[23] Anderson MJ and Legendre P (1999) An empirical comparison of permutation methods for tests of partial regression coefficients in a linear model. *J stat comput simul* **62**: 271-303.

[24] Anderson JM (2001) Permutation tests for univariate or multivariate analysis of variance and regression. *Canadian J Fisheries Aquatic Sci* **58**: 626-639.

A Study of the Mandible in 3D: It's Utility for Clinical Comparisons

REENA KHULLAR[1], PETE E. LESTREL[2], WON MOON[3] and CHARLES A. WOLFE[4]

Abstract

The traditional approach for characterizing the biological form consists of linear distances, angles, and ratios; known as the conventional metrical approach (CMA). Distances presumably measure size while angles are a measure of shape. In spite of the widespread use of CMA, this approach suffers from limitations that include, the fact that the measurements as commonly used are too sparse to actually measure the precise form of complex and irregular morphologies encountered in the biological sciences, and that it suffers from the limitation that these measurements form a disconnected dataset and cannot adequately describe the boundary or shape of biological forms. This arises because linear measurements precisely apply only to regular objects and were never intended for the measurement of irregular structures like the human skull. Alternative methods such as Fourier Descriptors (FDs) have been successfully applied to describe the boundary of biological forms in 2-space ($\mathbb{R}^2$). Although, such measurements in $\mathbb{R}^2$ are highly useful and have produced useful data, there remain limitations with such studies. Namely, that the overwhelming majority of biological information resides in $\mathbb{R}^3$. The quantitative representation of the biological form in 3D continues to present challenges. This study is a continuation of our research to quantify the mandibular morphology in 3-space. A sample of 10 normal cases was utilized for comparison with two cases exhibiting hemifacial microsomia (HM). Five steps were required: (1) 3D CT scans of the mandible were obtained, and the boundary outline digitized utilizing Dolphin software from Dolphin Imaging and Management Solutions; (2) once the 104 points had been digitized, the x, y, and z coordinates were exported from the Dolphin software into a format acceptable to EFFA, a routine that generates FDs; that is, curves in 3D, (3) since superimposition is necessary for clinical comparisons; this required the calculation of a centroid and the bounded surface area. A special routine, Surface Evolver, was required for this procedure; (4) once the centroid and surface area are available it is a comparatively straightforward procedure to superimpose the mean of the control mandibles (n=10) onto the mean of the HM clinical cases (n=2); (5) finally, the EFFA-generated 3D curves were superimposed using both area-standardization and Procrustes superimpositions and displayed graphically, allowing for a visual assessment of the differences between controls and abnormals in a clinical setting.

INTRODUCTION

The Conventional Metrical Approach

The conventional metrical approach (CMA) involves measurement of the biological form using lines, angles and ratios. This approach is widely used to make linear measurements for regular geometric objects due to its ease of use, and its assumed representation of forms [1]. When irregular biological structures like the skull are measured using this approach, its various limitations become evident, as can be illustrated with an example of the cranial base (Fig. 1).

[1] Section of Orthodontics, UCLA School of Dentistry, Los Angeles, CA 90024-1668, USA, drrkhullar@gmail.com.
[2] Formerly, Section of Oral Biology, UCLA School of Dentistry, Los Angeles, CA 90024-1668, USA.
[3] Section of Orthodontics, UCLA School of Dentistry, Los Angeles, CA 90024-1668, USA.
[4] C. Wolfe Software Engineering, 13376 Dronfield Avenue, Sylmar, CA.

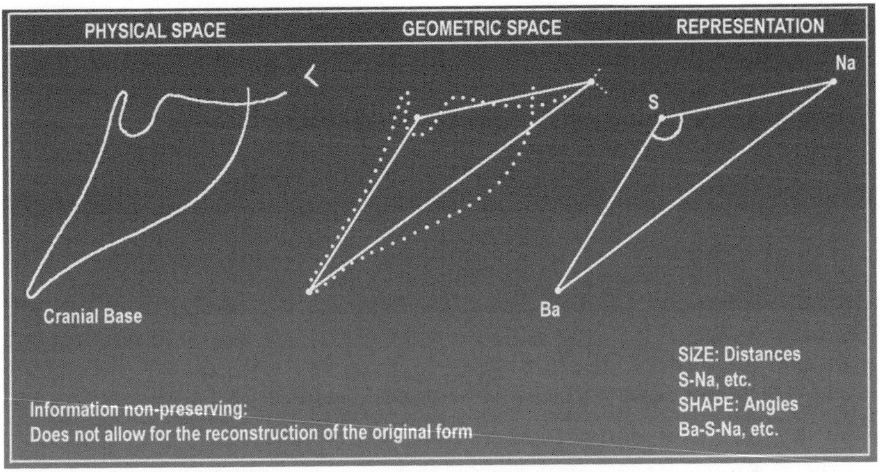

Fig. 1. The cranial base depicted with three points (N-S-Ba) represented by a triangle (modified from a figure originally developed by Read [23].

In Fig. 1, the cranial base can be represented by an outline formed by the points *Nasion* (Na - a point at the anterior limit of the frontonasal suture), *Sella* (S - a point representing the midpoint of the pituitary fossa), and *Basion* (Ba - the most inferior posterior point of the occipital bone at the anterior margin of the Foramen Magnum). Using CMA, lines are drawn to join the points Na, S and Ba, resulting in the shape of a triangle. This triangle, although providing some information about the cranial base, is not a true representation of the cranial base itself. Hence, there is a substantial loss of information of the cranial base in those areas not covered by the lines drawn. This continues to be a major limitation of CMA. Utilizing CMA to measure the cranial base, results in the loss of the actual physical representation of the shape of the cranial base itself.

Specifically in the field of orthodontics, CMA is utilized extensively to make cephalometric measurements from 2D x-rays, such as a lateral cephalogram and a frontal cephalogram. Several points are marked on the cephalograms and measurements are made from those points. These cephalometric measurements are then used as an aid in diagnosis and treatment planning. However, the human head is a 3D biological form. Hence, the approach of using the CMA once again results in loss of information in areas, which cannot be measured by lines, angles or ratios.

Moreover, some points marked on the cephalogram are not stable, and this variation in the position of the points can result in differences in the measurements themselves, as illustrated by the following example. Consider the points S and Na. In the Steiner cephalometric analysis [2], the S-Na plane is used as a reference plane and all measurements are made from that plane. However, the points S and Na are not stable. Thus, any variation in the points S and Na, either vertically or antero-posteriorly, can result in differences in the measurements, as well as differences in the interpretation of the of the positions of the upper and lower jaws. This highlights another limitation of 2D cephalometrics. These differences in measurements are discussed and displayed elsewhere (Figs. 2 – 5 in [16]).

Hemifacial Microsomia

Hemifacial microsomia (HM) is a congenital disorder in which the lower half of one side of the face is underdeveloped, and does not grow normally; most commonly, the ears (aural), the mouth (oral) and the mandible (jaw) are affected. This condition is also known as Goldenhar syndrome, brachial arch syndrome, facio-auriculo-vertebral syndrome, oculo-auriculo-vertebral syndrome, or lateral facial dysplasia. There are several clinical variations of this disorder. Since the growth of one side of the face is affected, it results in different degrees of facial asymmetry. In these cases, lateral cephalogram and frontal cephalograms are taken to determine the degree of asymmetry of the affected side, and then compared to the normal side. As discussed above, the limitations of CMA once again become evident during this process of making cephalometric measurements using lines, angles and ratios. Moreover, the overlapping of the right and left sides of the face in a lateral cephalogram makes it extremely difficult to gauge the variation from the norm. Since the two lower borders of the mandible can be seen in a lateral cephalogram, the clinician has to decide which one to choose in order to make the cephalometric measurements; that is, the right side, the left side, or the average between the two. Each decision will result in different cephalometric measurements. Clearly, an alternative procedure would be highly desired. This has been recently accomplished by joining a new 3D technique based on cone-beam tomography with a 3D mathematical approach; namely, a fitting function, derived from Fourier analysis (Fig. 2).

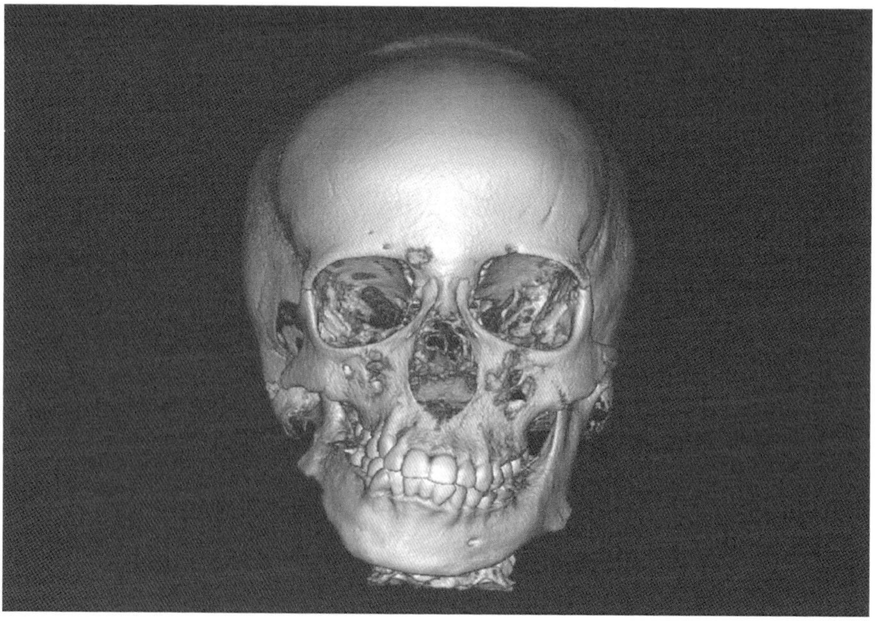

Fig. 2. Displayed here is a hemifacial microsomia case with the mandibular deviation on the right hand side, which affects the shape of both the maxilla and the mandible.

Cone-Beam Computed Tomography

The advent of cone-beam computed tomography (CBCT) has enabled the clinician to view sliced images of the biological form, specifically the human skull, by using divergent x-rays. With CBCT, scans as sliced images can now be viewed in all three planes. These scans, generated from the CBCT, are from a NewTom scanner[5]. A single scanned image of the skull is all that is required. CBCT allows for virtual treatment planning in 3D. This also aids in diagnosis and treatment planning of the right and left sides of the skull as separate elements, as is needed in patients with HM. With this technique, pre- and post-treatment superimpositions can also be done effectively, which allows the clinician to assess post-treatment stability [3]. In spite of this new technology to view the biological form in 3D, there is no methodology to quantify this vast amount of data in 3-space. Therefore, this data has to be converted into multiple 2D images, and then CMA again has to be used for quantification of this data. This raises the question of how to best quantify this data in 3D, and this leads to elliptical Fourier functions (EFFs).

Elliptical Fourier Functions

Fourier descriptors (FDs) [4, 5, and 6], a technique widely used in engineering, pattern recognition, and image processing, represents the boundary outline or shape as a segmented aspect of an image. Elliptical Fourier Functions (EFFs), a specific type of FD, transforms the data from the spatial domain into the frequency domain Refer to [7] for a basic description of these terms derived from harmonic analysis. EFFs can be effectively applied to very irregular forms of the type encountered in biology. EFFs used here were developed by Kuhl and Giardina [8].

EFFs allow for any number of points along the boundary outline with either equal or unequal spacing between any two points. This is in contrast to conventional Fourier analysis, which requires that both the number of points have to be as a power of two and the spacing between points has to be even. Differences between EFFs and conventional Fourier methods (including the Fourier transform are briefly alluded to in [16] but reference should be made to [7] for a detailed discussion of Fourier methods. A set of DOS routines have been developed by the authors PEL and CAW over the years called EFFA[6], which are used to compute the EFFs from the digitized data, generate size and shape data, shape only data, test the goodness of fit between the actual data and the Fourier function as residuals, and provide data files for visualization. Files generated by this package include the residuals file (.RES), the predicted point files (.P, .PA), and the harmonic coefficient files (.H, .HA).

EFFs have been used to describe: (1) shape changes in the maxillary (MX) cleft palate in 2D [9], (2) shape of the human mandible [10, 11, 12], (5) the complete craniofacial complex [13], (6) assessment of skeletal jaw relationships [14], (7) shape changes in cranial base due to age [15], (8) boundary of the mandible in 3D [16], etc.

In this study, EFFs were applied to describe the boundary of the mandible (MD) in normal cases (n = 10). Subsequently, EFFs were also applied to the HM cases

---

[5] NewTom Scanner Manufacturer is QR srl, Verona, Italy, Website: http://www.newtom.it/home.

[6] Available at a nominal cost from plestrel@earthlink.net. A newer Microsoft Windows version, MLmetrics, is under development.

(n = 2) to describe the boundary outline of the mandible for this abnormality. Various superimpositions were then carried out using the mean of the normal cases and HM cases. These superimpositions (normal and HM) were then used to visually assess the variation of the HM compared to the norm. A primary goal of this study was to evaluate the clinical utility of EFFs as a method that facilitates the visualization of morphological changes in the mandible.

MATERIALS AND METHODS

Study Subjects

A sample (n = 10) of controls was comprised of patients from the Orthodontic Clinic at UCLA. Criterion for subject selection included normal permanent dentition, absence of abnormalities and availability of an existing CBCT scan [16]. CBCT scans were also available for the HM cases (n = 2). These two cases were affected on the right side.

Boundary of the Mandible using EFFs

In a previous study, a dry skull mandible was used to create a mandibular outline with a brass wire to test the positioning of the eventual points to be used to describe the outline [16]. The boundary was then converted into a set of 104 closely spaced points that accurately represented the outline of the mandible in 3D. To digitize these points, Dolphin software [7] was utilized. Each of the medical CBCT scans was then loaded into the Dolphin software and the set of 104 points digitized. Each point consisted of a set of x, y, and z co-ordinates, all of which were compiled into a file, identified as a digitized 3D Dolphin file (Appendix I).

From Dolphin to EFFA

The digitized 3D Dolphin file was then modified to be acceptable as 3D input data to EFFA. Several changes had to be made in the process. The file name was reduced to fit the 8.3 DOS convention. All digitized files had to be renamed without the three-character extension. For example, the file AE_REDolphinfile.txt was reduced to AE with no 3-character extension as required for the otherwise equivalent EFFA digitized file. The format for this newly generated digitized EFFA file is shown in Appendix II. The structure of this file consists of one long physical record without line breaks or carriage returns. It contains multiple logical records, each 88 characters long. This is a 3D file with the coordinates of 4 points (x, y, z) on a line with each point taking up 21 characters; that is, 7 characters per coordinate (x, y, z). Thus, 4 points take up 84 characters on each line. The remaining four characters of the total of 88 characters are reserved for the logical record number. This digitized file has 29 records of which the first three are header records (Appendix II).

Computing the Area of a Bounded Curve in 3-space ($\mathbb{R}^3$)

The area and centroid of any irregular form in 2D can be computed using calculus [17] although a line integral may be required [18]. However, to compute the area/centroid of a bounded outline in 3D (the mandible here); that is, from a curve in

---

[7] Dolphin Imaging and Management Solutions, a Patterson Technology, USA 9200 Eton Ave, Chatsworth, CA 91311, www.dolphinimaging.com.

3-space ($\mathbb{R}^3$), there has not been an approach available until very recently. Computing this 3D area is essential because all forms are standardized to a constant area to control for differences in size. While for irregular forms in 2-space ($\mathbb{R}^2$) computation of the area and centroid are straightforward, for the $\mathbb{R}^3$ case this area does not even exist. Thus, this issue of the computation of an area bounded by a curve in $\mathbb{R}^3$ remained unresolved until about two decades ago.

One way to model the mandible as a curve in $\mathbb{R}^3$ is to view it as a "minimum surface". Soap films are one example of minimum surfaces. Consider the ring used to blow soap bubbles, if this ring (a plane) is twisted, or distorted it enters $\mathbb{R}^3$. Once dipped into a soap solution, a very thin surface will appear which will rapidly contract due to surface tension into a minimum surface before disappearing due to evaporation [19, 20; (See [21] for a discussion of minimum surfaces.)]. The development of Surface Evolver software [22] yielded an approach that allowed for the computation of the area and centroid of many curves in $\mathbb{R}^3$. The basis of the Surface Evolver is an interactive program form the study of surfaces shaped by surface tension[8].

The Surface Evolver program, into which the area and centroid values were included as a courtesy of Ken Brakke, provided the necessary area and centroid in $\mathbb{R}^3$. Initially, the Surface Evolver program was applied to the 3D mandible as a whole unit; however the values generated were not usable because of the complexity of the mandible. At the suggestion of Ken Brakke, an alternative approach was attempted, which was to divide the MD into right and left halves to minimize the complexity of the mandibular morphology. The area and centroid value was then calculated separately for each side. This approach was successful. Subsequently, the areas of each half were added and the centroid values averaged to obtain values for the entire MD.

In summary, the modified EFF digitized file was now broken up into the left and right halves. Record number 16 marked the halfway point, which placed 52 points into each half. Only the three header files needed to be repeated and adjusted for the right side data and the number of points reduced to 52 from 104 for each half. For the left side, the second set of 16 points, the record numbers (initially from 17-29) were re-numbered from 1 to 16. In addition, the number of points in the header had to be reduced from 104 to 52. Both the right and the left halves of the file are shown in Appendix IIIA and IIIB.

Surface Evolver Computations

Once the two halves of the EFF digitized file were generated, the next step was to run the Surface Evolver program to the get the $\mathbb{R}^3$ area and centroid coordinates for each half. To do this, the digitized data has to be converted first into a format that Surface Evolver recognizes. One of the EFFA DOS routines, EFFXPORT.EXE, was developed for this purpose, allowing for the conversion of the EFF digitized file into Surface Evolver format. Using the AE_R (right half of the AE file) digitized file as an example, Surface Evolver was run. The file generated was named AE_R.EV0. The 3-letter extension is critical for the newest Surface Evolver version to work. The actual content of the AE_R.EV0 file is shown in Appendix IV. After entering the

---

[8] The software is available from Kenneth A. Brakke, brakke@susqu.edu.

AE_R.EV0 file into the Surface Evolver software, it returns the AE_R.EV file as output, which contains the area and centroid values at the bottom of the file. The AE_L.EV0 file is also needed and must be run through the Surface Evolver for its area and centroid values (see Table 1). These 3D area and centroid values were then later used with the EFFA software.

Computation of EFFs

The next step used the digitized points to compute a curve-fit that closely described the boundary outline of the MD in 3D. EFFs were developed by Kuhl and Giardina in 1982 [14]; however, they limited the discussion to 2D. The current study extends their formulation to 3D. The Kuhl and Giardina parametric functions are defined in $x(t)$ as:

$$x(t) = A_0 + \sum_{n=1}^{N} a_n \cos nt + \sum_{n=1}^{N} b_n \sin nt, \tag{1}$$

and in y(t) as

$$y(t) = C_0 + \sum_{n=1}^{N} c_n \cos nt + \sum_{n=1}^{N} d_n \sin nt, \tag{2}$$

where n equals the harmonic number, N equals the maximum harmonic number and the interval is over the closed interval $[0, 2\pi]$. In 3-space, EFFs require an additional equation, which is:

$$z(t) = E_0 + \sum_{n=1}^{N} e_n \cos nt + \sum_{n=1}^{N} f_n \sin nt. \tag{3}$$

Table 1. The 3D area (added) and centroid values (averaged) generated from the Surface Evolver software. Values for the individual AE case are shown here.

Case	Area	x-coordinate	y-coordinate	z-coordinate
AE_R.EV	4408.54	54.0451	86.7971	78.4693
AE_L.EV	4295.47	110.367	88.1376	78.7797
Mean or sum	8704.01	82.2061	87.4674	78.6245

Running the EFFA Software

Once the area and centroid values in $\mathbb{R}^3$ are available from the Surface Evolver, the usual EFF runs were carried out. This involved the insertion and use of the area and centroid values in the subsequent .P and .PA files. At this step, four runs were involved:

RUN #1: Starting with the EFFA DOS routine EFFHOMOA.EXE, the first run was begun by taking the digitized file generated by Dolphin and converted to EFFA format, to compute size and shape .P and .H files. The translation to the centroid was not done during this step. This run was used only to create the initial (untranslated) .P and .H files and allow computation of the .RES file (Appendix V).

RUN #2: The .P file generated in run #1 was then used to create the .A file. The .A file mimics a digitized file, which has been area-standardized to a constant area of 100 square units.

RUN #3: Once the .A file was created, then EFFHOMOA.EXE is again executed to create the .HA and .PA files. The translation to centroid is now carried out. This is crucial for later superimposition of the control cases with the HM cases (Appendix VI).

RUN #4: For the last run, the original size and shape files (.P and .H) were run again, but this time, the translation to the centroid was done. These files were named as AE_T.P and AE_T.H. The "T" here stands for translated data. The purpose for these translated .P files, was to be able to superimpose size and shape data, as well as the shape-standardized data (RUN #3), and eventually the Procrustes superimposition data.

Control and HM Superimpositions: Various Approaches

There are three superimposition approaches available. These are "size and shape" (original data), "area-standardized" (shape only) and Procrustes superimposition. Each of these was applied to the mandibular data. The size and shape approach is self-explanatory, a superimposition of the original mean control data onto the mean HM data. The second approach, the area-standardized method requires that all cases be first standardized by increasing or decreasing the bounded area to a constant value. This value was arbitrarily determined as the square root of 10,000. The purpose of this approach is to minimize the effect of size by generating shape-only data. The third approach is Procrustes superimposition[9]. The Procrustes approach is based on a least squares algorithm in which a target outline is superimposed on a base outline. Both base and target are then manipulated to account for scaling, translation and rotation to yield a minimum configuration.

Plotting the Superimpositions

Once the .P and .PA (both translated to centroid files) were computed and available, they needed to be exported. Eventually, once the means of the controls (n=10) and the HM cases (n=2) were averaged to produce means, these files would be exported and used. EFFUTLA.EXE was executed and both files (.P and .PA) were exported in column format with no headers or line numbers. This step created an export file named as AE.EP3 (Appendix VII) for the .PA file and an AE_T.EP1 file for the translated to centroid .P file with an identical format to the .AE.EP3 file. The next step was to use PsiPlot[10] for plotting purposes, both for plots in $\mathbb{R}^2$ and $\mathbb{R}^3$. The first step in plotting is to make sure that the axis ranges are the same. The default in PsiPlot is [7.5, 5.0] for $\mathbb{R}^2$, this was altered to [6.0, 6.0] this insures that both axes are the same size. For the area-standardized plots, the axes ranges were set to [-400, 400] for each of the XY, XZ and YZ plots. For the XYZ plots, all the three axes ranges were checked to be the same. The "base" file was loaded first, which was the mean of the controls (n=10). Then the "target" file was loaded, which was

---

[9] Originally written by M. J. Calvacanti (available from maurobio@acd.ufrj.br). This routine was slightly modified by CAW.

[10] The PsiPlot graphic package is available from www.polysoftware.com.

the mean HM cases. The two files were loaded into PsiPlot and the HM mean was placed on top of the control mean. The HM mean plot was then shifted around until all three (3D) or the two (2D) axes lined up with the "0's" superimposed. This "0" signifies the centroid position. This superimposed plot was saved and exported as a .jpeg or a .tiff file for further processing, if necessary, in Photoshop Elements. The same procedures were applied for the size and shape data (ex. AE_T.P file).

RESULTS

A number of plots were generated; composed of both 2D and 3D views. Plots were identified as follows; those containing control data (n = 10) were identified as either as controls or CNTLS, while those containing HM data (n = 2) were identified with HM. The 2D plots are identified as XY (lateral view), XZ (superior view), or YZ (frontal view). The 3D data was identified by XYZ (3D view).

Mean Area-standardized Plots for the Controls (n = 10)

Figs. 3 and 4 show the area-standardized mean plots for the controls (n = 10). Fig. 3 displays the XY (lateral view) and YZ (frontal view). Fig. 4 shows the XZ (superior view) and the XYZ (3D) view. All plots are superimposed on the centroid.

The mandibular "size and shape" data for both the controls and the HM cases are not displayed because the plots are identical with the exception of the axis values. This is made clear in the example in Fig. 5 where only the axis values differ between the "size and shape" and "area-standardized" plots. Note that while the axes values differ between the plots, the mandibles seem to appear identical in "shape". This is a consequence of setting the area-standardization to a constant value for all specimens; therefore, the only difference between the two mandibular plots is one of size, which is reflected in the axes, but not in the actual plot figures. This same approach applies to the mean of the HM cases (not shown).

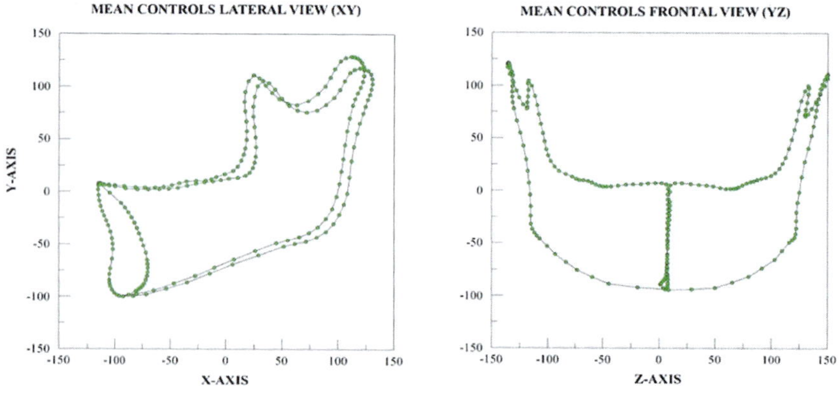

Fig. 3. Area-standardized mean plots of the controls (n = 10). The lateral and frontal views are depicted here. The plots are superimposed on the centroid.

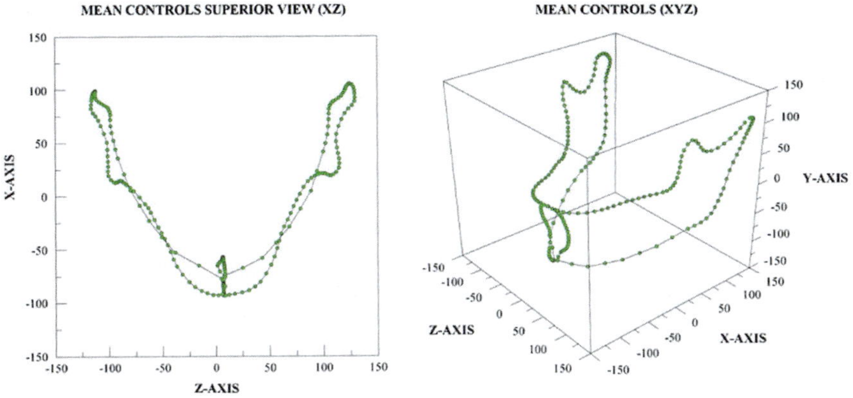

Fig. 4. Area-standardized mean plots of the controls (n = 10). The superior and 3D views are displayed here. The plots are superimposed on the centroid.

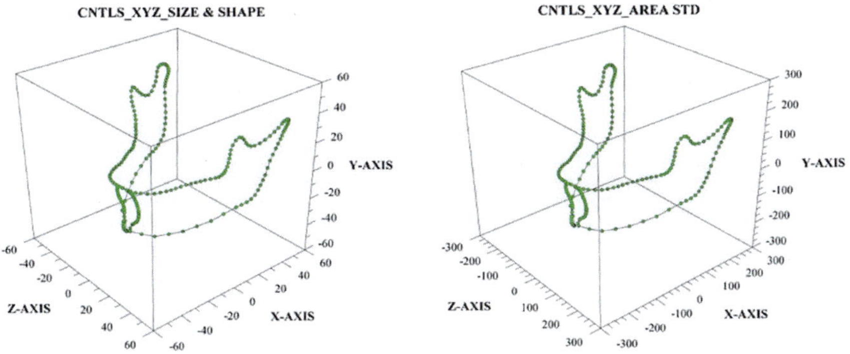

Fig. 5. An example of the effects of area-standardization is shown here. The 3D mean "size and shape" (original data) mandible is on the left and the "area-standardized" (shape only) 3D plot is on the right (refer to the text).

Mean Area-standardized Plots for the HM Cases (n = 2)

Figures 6 and 7 display the area-standardized HM cases (n = 2). Fig. 6 displays the XY (lateral view) and YZ (frontal view). Fig. 7 shows the XZ (superior view) and the XYZ (3D view). All plots are superimposed on the centroid.

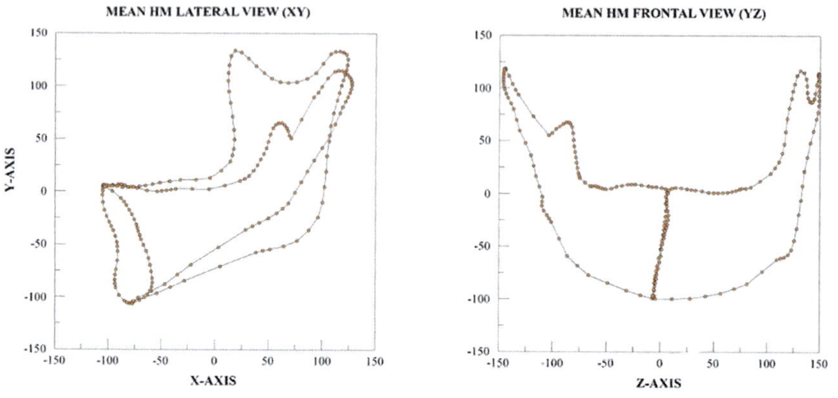

Fig. 6. Mean plot of the HM cases (n=2), area-standardized. The lateral and frontal views are depicted here. The plots are superimposed on the centroid.

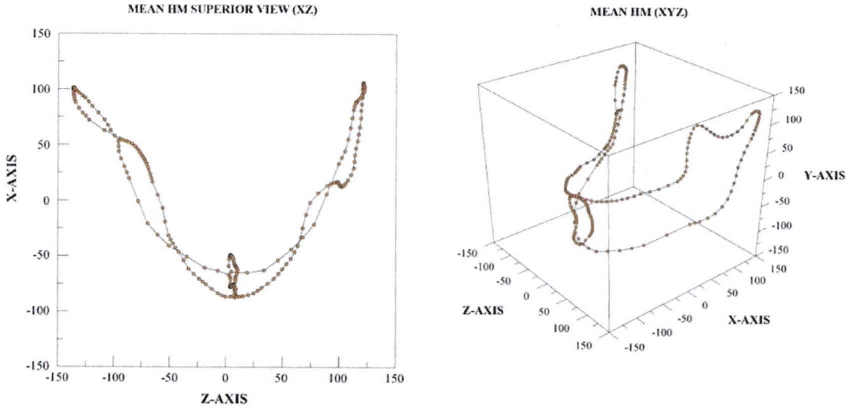

Fig. 7. Mean plot of the HM cases (n=2), area-standardized. The superior and 3D views are displayed. The plots are superimposed on the centroid.

Superimposition of Mean Controls on the Mean HM Cases (area-standardized)

Figs. 8 and 9 show the 2D and 3D plots of the area-standardized superimposition of the controls onto the area-standardized HM cases. Fig. 8 displays the plots with the XY (lateral view) and YZ (frontal view). Fig. 9 shows the XZ (superior view) and XYZ (3D) view. The superimpositions are on the centroid.

That there are negligible effects between the "size and shape" (original data) data and the "area-standardized" (shape only) data can be seen with Fig. 10. Therefore, no attempt was made to plot the size and shape component views (XY, XZ, YZ and 3D). The reason is that these mandibular cases (controls and HM) are already very similar in size so that the area-standardization adjustment was minimal. This situation would not be necessarily the case with a different data set.

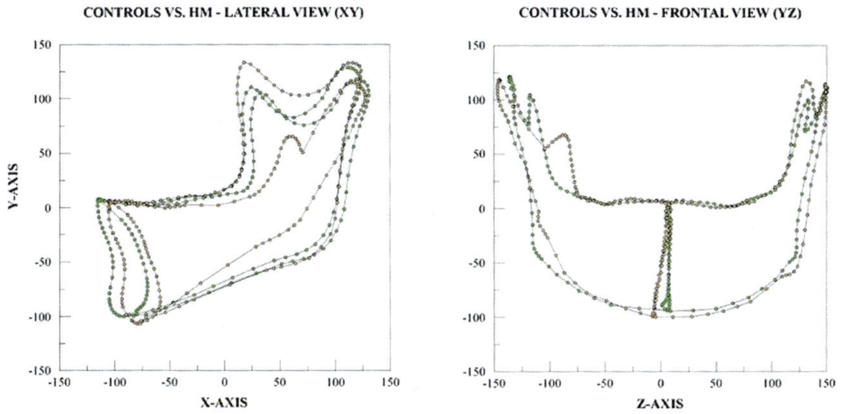

Fig. 8. Mean plot of the superimposition of the controls onto the HM cases. These plots are area-standardized and superimposed on the centroid. The XY (lateral view) and the YZ (frontal view) are shown here. The green circles refer to the controls and the orange ones to the HM cases.

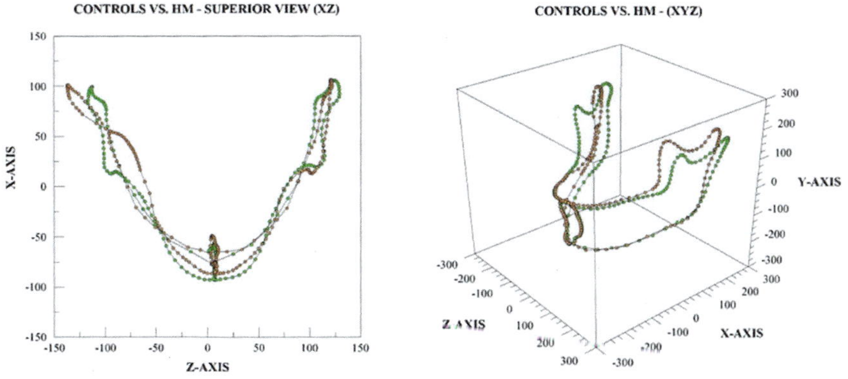

Fig. 9. Mean plot of the superimposition of the controls onto the HM cases. These plots are area-standardized and superimposed on the centroid. The XZ (superior view) and the XYZ (3D view) are displayed here. The green circles refer to the controls and the orange ones to the HM cases.

Procrustes Superimposition Plots

The initial Procrustes superimposition plots did not materially differ from either the "size and shape" plots or the "area-standardized" ones. This was a consequence of the control and HM cases being initially similar in size. However, a further objection can be raised with the use of the Procrustes superimposition with this specific data set; namely, that the Procrustes algorithm treats all points (those on the normal mandibular HM half and those on the deformed half) equally, which tends to act as a bias here. Thus, in an attempt to alleviate this bias, it was decided to treat each of the mandibular halves separately; that is, run the Procrustes superimposition on both the left and right halves independently.

This issue with Procrustes also arises in a similar way if part of a morphological structure exhibits deviation from the norm and/or displays higher variation compared to other aspects of the structure. See Daniels (this volume) for a discussion of the so-called "Pinocchio" problem.

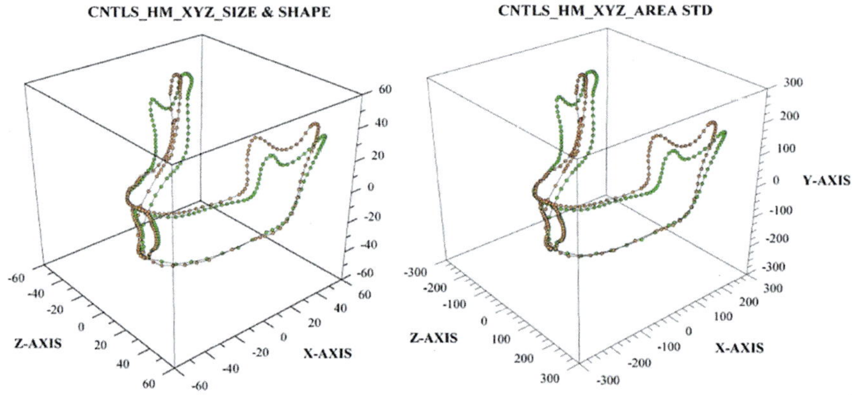

Fig. 10. This figure shows the negligible effects of area-standardization. The 3D mean "size and shape" superimposition of the controls on the HM cases is on the left and the "area-standardized" 3D plot is on the right. The green circles refer to the controls and the orange ones to the HM cases (refer to the text).

Procrustes Superimposition of the Left Half of the Mean Controls on the HM Cases

Figs. 11 and 12 display the 2D and 3D plots of the Procrustes superimposition of the left half of the controls onto the left half of the HM cases. Fig. 11 displays the plots with the XY (lateral view) and YZ (frontal view). Fig. 12 shows the XZ (superior view) and XYZ (3D) view.

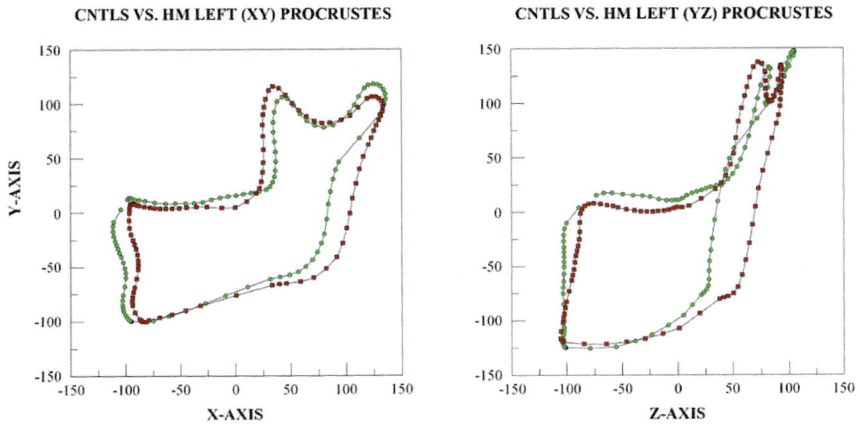

Fig. 11. Procrustes superimposition of the left half of the mean controls onto the left half of the mean HM cases. These views are NOT superimposed on the centroid. The XY (lateral view) and the YZ (frontal view) are shown here. The green circles refer to the controls and the red squares to the HM cases.

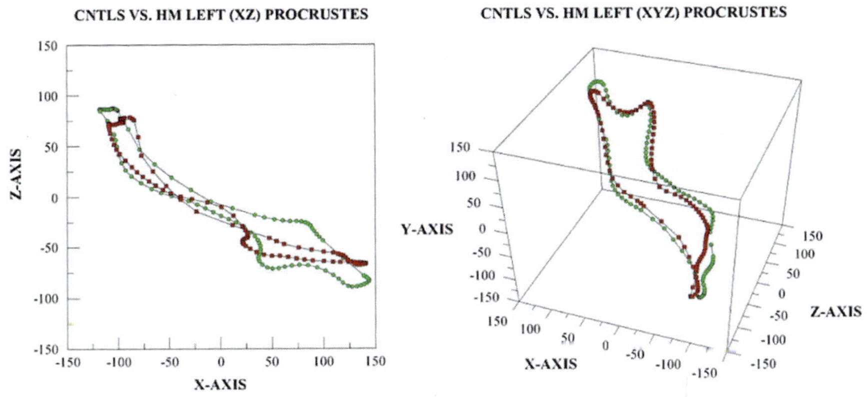

Fig. 12. Procrustes superimposition of the left half of the mean controls onto the left half of the mean HM cases. These views are NOT superimposed on the centroid. The XZ (superior view) and the XYZ (3D view) are depicted here. The green circles refer to the controls and the red squares to the HM cases.

Procrustes Superimposition of the Right Half of the Mean Controls on the HM Cases

Figs. 13 and 14 show the 2D and 3D plots of the Procrustes superimposition of the right half of the controls onto the right half of the HM cases. Fig. 13 displays the plots with the XY (lateral view) and YZ (frontal view). Fig. 14 shows the XZ (superior view) and XYZ (3D) view. Fig. 15 displays the Procrustes superimposition of both the left and right mandibular halves of the mean 3D controls and mean HM cases with the 2D components reflected into their respective planes.

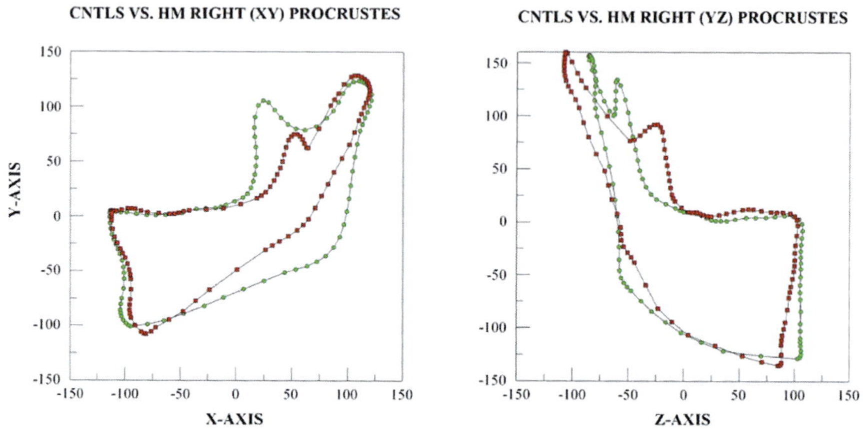

Fig. 13. Procrustes superimposition of the right half of the mean controls onto the right half of the mean HM cases. These views are NOT superimposed on the centroid. The XY (lateral view) and the YZ (frontal view) are shown here. The green circles refer to the controls and the red squares to the HM cases

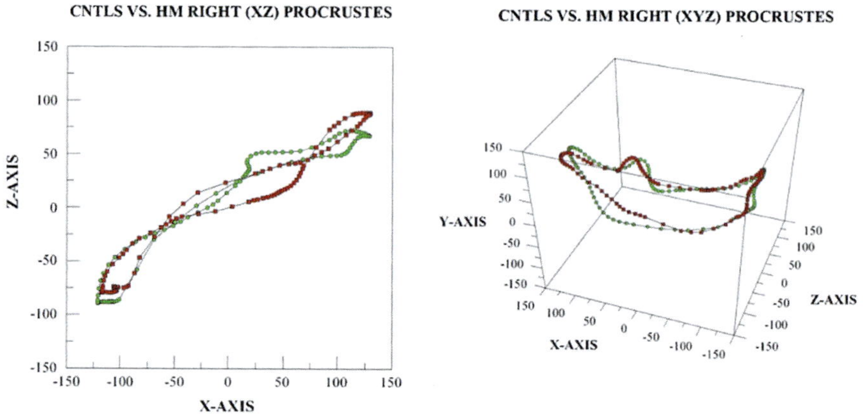

Fig. 14. Procrustes superimposition of the right half of the mean controls onto the right half of the mean HM cases. These views are NOT superimposed on the centroid. The XZ (superior view) and the XYZ (3D view) are depicted here. The green circles refer to the controls and the red squares to the HM cases.

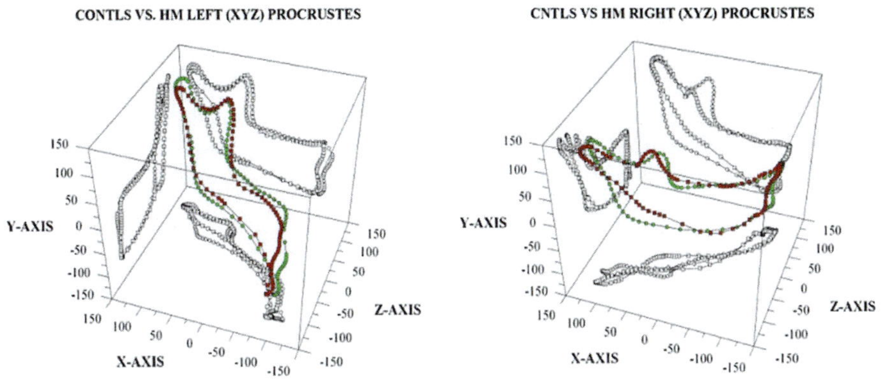

Fig. 15. Displayed here are the left and right mandibular halves of the mean 3D (XYZ) controls and HM samples with the 2D projections reflected into their respective planes. These projections are depicted in Figs. 11 and 12 for the left side and in Figs. 13 and 14 for the right side.

DISCUSSION AND CONCLUSIONS

Controls versus HM Plots

The fact that the controls and the HM cases did not vary very much in size, resulted in all three superimposition approaches yielding substantially similar results. The abnormality, HM, displayed clear shape differences on the right side as expected since the lower jaw is asymmetric for this deformity. This is clearly displayed in Fig. 8 where the area-standardized XY (lateral view) of the mean HM is substantially distorted in comparison to the mean controls. This asymmetrical deviation seen in the HM is even clearer in the Procrustes superimposition of the right half (Fig. 13).

However, it should be noted that the deviation on the right side of the mandible was not limited to that side but extended into the left side. This can be seen in Fig. 8 where the area-standardized YZ (frontal view) midline of the mean HM cases is clearly deviated compared to the controls. This is also quite apparent in Figs. 11 and 13 where the Procrustes superimposition displays the deviation regardless of whether one looks at the left or the right half of the mandible.

A quantitative assessment of these shape differences between the controls and HM cases is, unfortunately, not really feasible due to the small sample sizes with controls (n = 10) and HM (n = 2) respectively. Given larger sample sizes, the differences between the controls and any abnormality that exhibits substantial shape changes should be apparent and relatively easy to measure.

The primary purpose of this study was to establish an approach that is applicable to morphological structures that are best envisioned in $\mathbb{R}^3$. Such an approach requires the need for both: (1) superimposition methods and (2) visualization techniques. Both aspects turned out to be challenging; clearly, the step from $\mathbb{R}^2$ to $\mathbb{R}^3$ is not a trivial one.

Clinical Utility

The clinical utility of this approach is evident as seen with the various 2D and 3D plots that were generated. These plots enable the clinician to have an exact visualization of the shape changes of the normal side when compared to the affected side in HM cases. Moreover, the superimposition of both these sides provides a good tool for the diagnosis and surgical treatment planning for correction of the affected side.

Diagnosis and Treatment Planning

The traditional approach of making cephalometric measurements of HM cases is by using CMA. This method is unable to even begin to capture and describe the shape changes of the mandible on the affected side along its boundary outline. Hence, much information about the biological form is lost. The description of asymmetry in the mandibular border is not easily, if at all, quantifiable with CMA.

The application of this novel 3D approach as advocated here with the HM example shows that it is indeed an invaluable tool for diagnosis and treatment planning. It provides the clinician with reliable and accurate information about the size and shape of the asymmetry. This, in turn, will enable the clinician to precisely determine the extent of surgery needed to correct the asymmetry in 3D.

Creation of a Norm or Standard

The example here of the mean plots of the controls (n = 10) is intended to simulate an average norm. Clearly, ten cases are insufficient and this needs to be greatly increased, a higher number of cases, for example, say 100 or preferably even more, which can then be subdivided into either age or facial profile categories would be clinically useful. For example for age: 15 – 25 years, 25 – 35 years, 35 – 45 years, etc., or for the facial profile: concave, convex and straight. The mean plot for a specific category can then be designated as a norm for that specific group. Unlike the use of CMA in cephalometrics, no points or lines are used; hence, such a method should eliminate the variation in the interpretation of the results due to instability of the points or lines used in CMA.

Overcoming the Limitations of CMA

The novel 3D approach advocated here overcomes various limitations of CMA as outlined earlier. There are no mean norm values to be memorized. Instead, a graphical representation of the biological form shows the deviation from the norm. This approach also eliminates the various interpretations obtained from different analyses in conventional cephalometrics. Hence, it removes confusion while using this approach as a tool to aid in diagnosis and treatment planning, as it primarily provides the clinician with an individual graphical representation for each patient. This graphical representation provides boundary information regarding the biological form, and also enables quantification of the aspect of the biological form that deviates from the norm.

Abnormalities of Growth

As illustrated in this chapter, the approach can be used with the numerous conditions displaying abnormal growth patterns, such as abnormalities in shape and size. In sum, data from 2 patients with HM was used. Quantification and description of this data was carried out with EFFs in 2D and 3D. Precise changes in form of the affected HM side were determined, and then compared with normal cases. Size and shape, area-standardization, and Procrustes superimposition where carried out and various plots generated. The centroid was used for superimposition using the area-standardized method but not applicable to the Procrustes method. The approach advocated here enables the clinician to determine the deviation of growth patterns from the norm in size and especially in shape. This data can then be utilized further to evaluate these abnormal growth patterns.

This chapter illustrates the clinical utility of the 3D novel approach using EFFs to describe, quantify and graphically represent the mandible and its deviation from the norm in HM in 3D. In the future, this approach can be applied to disorders in the medical sciences also, for example, growth disorders or anomalies in limbs, etc. This may be considered a turning point in the representation of the biological form in 3D.

REFERENCES

[1] Lestrel PE (2000) *Morphometrics for the Life Sciences.* World Scientific Publishing Co. Pte. Ltd., Singapore.

[2] Steiner CC (1953) Cephalometrics for you and me. *Am J Orthod* **39**: 729-755.

[3] Cevidanes LH, Styner MA, and Proffit WR (2006) Image analysis and superimposition of 3-dimensional cone-beam computed tomography models. *Am J Orthod Dentofacial Orthop* **129**: 611-618.

[4] Lestrel PE (1974) Some Problems in the assessment of morphological size and shape differences. *Yearbk Phys Anthrop* **18**: 140-162.

[5] Lestrel PE (1980) A quantitative approach to skeletal morphology: Fourier analysis. *Soc Photo Inst Engrs* (SPIE) **166**: 80-93.

[6] Lestrel PE (1982) A Fourier analytic procedure to describe complex morphological shapes. **In**: *Factors and Mechanisms Influencing Bone Growth* AD Dixon and BG Sarnat (Eds.) Alan R. Liss, Inc., New York

[7] Lestrel PE (1997) Introduction and overview of Fourier Descriptors. **In**: Lestrel PE (Ed.) *Fourier Descriptors and their Applications in Biology*. Cambridge University Press, Cambridge.

[8] Kuhl FP and Giardina CR (1982) Elliptical Fourier features of a closed contour. *Comp Graph Imag Proc* **18**: 236-258.

[9] Lestrel PE, Berkowitz S and Takahashi O (1999) Shape changes in the cleft palate maxilla: A longitudinal study. *Cleft Palate – Craniofacial J* **36**: 292-303.

[10] Ferrario VF, Sforza C, Guazzi M, and Serrao G (1996) Elliptical Fourier analysis of mandibular shape. *J Craniofacial Genet Dev Biol* **16**: 208-217.

[11] Schmittbuhl M, Le Minor JM, Schaaf A, and Mangin P (2002) The human mandible in lateral view: elliptical Fourier descriptors of the outline and their morphological analysis. *Anatomischer Anzeiger* **184**: 199-207.

[12] Schmittbuhl M, Le Minor JM, Taroni F, and Mangin P (2001) Sexual dimorphism of the human mandible: demonstration by elliptical Fourier analysis. *Int J Legal Med* **115**:100-101.

[13] Lestrel PE (1989) Some approaches toward the mathematical modelling of the craniofacial complex. *J Craniofacial Genet Dev Biol* **9**: 77-91.

[14] Lowe BF, Phillips C, Lestrel PE and Fields HW Jr. (1994) Skeletal jaw relationships: A quantitative assessment using elliptical Fourier functions. *Angle Orthod* **64**: 299-308.

[15] Lestrel PE and Roche AF (1986) Cranial base shape variation with age: A longitudinal study of shape using Fourier analysis. *Hum Biol* **58**: 527-540.

[16] Khullar R, Lestrel PE, Moon W and Wolfe C (2013) Representation of the Mandible as a Curve in 3-space: A Preliminary Study using Fourier Descriptors. **In**: PE Lestrel (Ed.) *Biological Shape Analysis*: *Proc 2nd Int Symp Biol Shape Anal* (ISBSA) World Scientific Pub, Singapore.

[17] Smail LL (1953) *Analytic Geometry and Calculus*. Appleton-Century-Crofts. Inc., New York.

[18] Fulks W (1961) *Advanced Calculus, an Introduction to Analysis*. (Chapter 13) John Wiley, Sons. New York.

[19] Boys CV (1959) *Soap Bubbles*. Dover Pub., New York.

[20] Isenberg C (1992) *The Science of Soap Films and Soap Bubbles*. Dover Pub., New York.

[21] Lestrel PE, Read DW, and Wolfe C (1997) Size and shape of the rabbit orbit: 3-D Fourier descriptors. **In**; Lestrel PE (Ed.) *Fourier Descriptors and their Applications in Biology*. Cambridge University Press, Cambridge.

[22] Brakke KA (1992) The surface evolver. *Exp Math* **1**: 41-165.

[23] Read DW (1990) From multivariate to qualitative measurement: representation of shape. *Hum Evol* **5**: 417-429.

APPENDICES

Appendix I. The Digitized 3D Dolphin File AE_REDolphinfile.txt

* Units are in either Millimeters or Degrees.

---[Mandibular Analysis - Khullar] Landmark Name---		--X--	--Y--	--Z--
1	DentMidline	81.3	46.4	80.3
2	LR3Tip -2	76.6	47.6	80.0
3	LR3Tip -1	73.3	48.7	80.3
4	LR3Tip	69.1	51.1	80.4
5	MesialLR6 -2	62.7	58.3	80.2
6	MesialLR6 -1	61.0	65.2	80.5
7	MesialLR6	58.5	70.1	82.0
8	DistalLR6 -1	55.5	74.1	81.6
9	DistalLR6	51.9	80.3	82.1
10	RtAlvBorder -2	51.9	82.4	82.5
	(Columns omitted to save space)			
95	ConcavePtChin -3	82.0	58.0	47.6
96	ConcavePtChin -2	82.0	59.1	48.9
97	ConcavePtChin -1	82.0	60.2	50.3
98	ConcavePtChin	82.1	60.7	52.3
99	CEJLingual -4	82.0	58.8	58.1
100	CEJLingual -3	82.0	58.8	62.3
101	CEJLingual -2	82.0	58.2	66.6
102	CEJLingual -1	82.0	55.4	72.0
103	CEJLingual	81.9	54.0	74.7
104	DentMidline	81.3	46.4	80.3

Column numbers added for clarity (not present in the 3D Dolphin file)
Note: Point #1 is identical to Point #104

Appendix II. The Generated EFFA Digitized File AE

```
!EFFA_File_v2.txt                                                1EFFA_File_v2.txterec       1
OBS 0 0 3N 3DZ 0104   0        2:12 A02092011    2  .1 1      0       0       0       0
3  813  464  803  766  476  800  733  487  803  691  511  804 4  627  583  802
610  652  805  585  701  820  555  741  816 5  519  803  821  519  824  825  501
864  832  459  919  843 6  439  934  884  436  938  932  418  923  985  407  910
1052 7  393  937  1112  398  995  1079  396  1051  1030  386  1099  1017 8  364
1157  1053  324  1207  1104  317  1229  1157  346  1269  1168 9  339  1281  1162  321
1290  1147  346  1298  1122  317  1282  1082 10  357  1269  1049  346  1250  1012
361  1219  935  369  1205  877 11  371  1193  817  375  1179  733  386  1129  647
396  1078  622 12  418  992  598  468  903  564  512  797  523  623  641  473 13
814  561  443  815  505  452  815  498  488  814  501  506 14  814  509  547  811
508  624  813  501  661  814  494  682 15  813  480  710  817  469  731  811  461
766  813  464  803 16  813  464  803  850  465  804  898  477  810  942  504  802
17  979  547  795  1036  576  801  1065  696  816  1092  770  820 18  1125  804
830  1139  853  825  1172  918  825  1192  924  850 19  1198  935  881  1216  941
942  1216  925  1000  1227  918  1047 20  1249  940  1125  1248  987  1094  1252  1028
1047  1249  1096  1011 21  1280  1153  1020  1292  1201  1077  1317  1229  1128  1310
1282  1144 22  1350  1305  1135  1303  1314  1099  1310  1299  1058  1320  1280  1038
23  1292  1265  1002  1281  1239  947  1281  1219  886  1281  1207  800 24  1267  1179
718  1267  1130  667  1259  1085  642  1242  1016  627 25  1198  920  590  1159  802
542  1037  631  482  814  561  443 26  815  576  457  818  578  467  820  580
476  820  591  489 27  820  602  503  821  607  523  820  588  581  820  588  623
28  820  582  666  820  554  720  819  540  747  813  464  803 29
```

The data here is composed of multiple logical records of which each is one is 88 characters long. These are set up as one physical record. Note: Point #1 is identical to Point #104.

Appendix III. Right and Left Halves of the EFF AE Digitized File

IIIA. *Right half of EFF File AE*

```
!EFFA_File_v2.txt                                    1EFFA_File_v2.txterec     1
OBS 0 0 3N 3DZ 0052  0      2:12 A02092011   2  .1 1     0      0      0      0
3  813  464  803  766  476  800  733  487  803  691  511  804 4  627  583  802
610  652  805  585  701  820  555  741  816 5  519  803  821  519  824  825  501
864  832  459  919  843 6  439  934  884  436  938  932  418  923  985  407  910
1052  7  393  937  1112  398  995  1079  396  1051  1030  386  1099  1017  8  364
1157  1053  324  1207  1104  317  1229  1157  346  1269  1168  9  339  1281  1162  321
1290  1147  346  1298  1122  317  1282  1082  10  357  1269  1049  346  1250  1012
361  1219  935  369  1205  877 11  371  1193  817  375  1179  733  386  1129  647
396  1078  622 12  418  992  598  468  903  564  512  797  523  623  641  473 13
814  561  443  815  505  452  815  498  488  814  501  506 14  814  509  547  811
508  624  813  501  661  814  494  682 15  813  480  710  817  469  731  811  461
766  813  464  803 16
```

IIIB. *Left half of EFF File AE*

```
!EFFA_File_v2.txt                                    1EFFA_File_v2.txterec     1
OBS 0 0 3N 3DZ 0052  0      2:12 A02092011   2  .1 1     0      0      0      0
3  813  464  803  850  465  804  898  477  810  942  504  802 4  979  547  795
1036  576  801  1065  696  816  1092  770  820 5  1125  804  830  1139  853  825
1172  918  825  1192  924  850 6  1198  935  881  1216  941  942  1216  925  1000
1227  918  1047  7  1249  940  1125  1248  987  1094  1252  1028  1047  1249  1096
1011  8  1280  1153  1020  1292  1201  1077  1317  1229  1128  1310  1282  1144  9  1350
1305  1135  1303  1314  1099  1310  1299  1058  1320  1280  1038 10  1292  1265  1002
1281  1239  947  1281  1219  886  1281  1207  800 11  1267  1179  718  1267  1130
667  1259  1085  642  1242  1016  627 12  1198  920  590  1159  802  542  1037  631
482  814  561  443 13  815  576  457  818  578  467  820  580  476  820  591  489
14  820  602  503  821  607  523  820  588  581  820  588  623 15  820  582  666
820  554  720  819  540  747  813  464  803 16
```

Appendix IV. Surface Evolver AE_R.EV0 File

```
// EFF: AE_REFFA_File_v2.txterec       1
quantity xweight info_only method facet_scalar_integral global
scalar_integrand x
quantity yweight info_only method facet_scalar_integral global
scalar_integrand y
quantity zweight info_only method facet_scalar_integral global
scalar_integrand z


vertices
  1   81.3000031   46.4000015   80.3000031 fixed
  2   76.5999985   47.6000023   80.0000000 fixed
  3   73.3000031   48.7000008   80.3000031 fixed
  4   69.0999985   51.1000023   80.4000015 fixed
  5   62.7000008   58.2999992   80.2000046 fixed
  6   61.0000000   65.2000046   80.5000000 fixed
  7   58.5000000   70.0999985   82.0000000 fixed
  8   55.5000000   74.0999985   81.5999985 fixed
  9   51.9000015   80.3000031   82.0999985 fixed
 10   51.9000015   82.4000015   82.5000000 fixed
```

(Columns omitted to save space)

```
 45   81.4000015   50.9000015   54.7000008 fixed
 46   81.0999985   50.7999992   62.4000015 fixed
 47   81.3000031   50.1000023   66.0999985 fixed
 48   81.4000015   49.4000015   68.2000046 fixed
 49   81.3000031   48.0000000   71.0000000 fixed
```

```
 50   81.7000046    46.9000015    73.0999985 fixed
 51   81.0999985    46.1000023    76.5999985 fixed
 52   81.3000107    46.3999901    80.3000031 fixed

edges
  1    1    2
  2    2    3
  3    3    4
  4    4    5
  5    5    6
  6    6    7
  7    7    8
  8    8    9
  9    9   10
 10   10   11

 (Columns omitted to save space)

 45   45   46
 46   46   47
 47   47   48
 48   48   49
 49   49   50
 50   50   51
 51   51   52
 52   52    1

faces
 1 1  2  3  4  5  6  7  8  9  10 \
 11  12  13  14  15  16  17  18  19  20 \
 21  22  23  24  25  26  27  28  29  30 \
 31  32  33  34  35  36  37  38  39  40 \
 41  42  43  44  45  46  47  48  49  50 \
 51  52

read
print_cm := { printf "%g %g %g %g\n", total_area, xweight.value/total_area,
            yweight.value/total_area,zweight.value/total_area; }
printf "%g %g %g %g\n",total_area, xweight.value/total_area,yweight.value/total_area,
zweight.value/total_area;

print_cm >> "f:/EVOLVER/AE_R.EV"    // ">>" for append, ">>>" for overwrite

//End of File
```

This Surface Evolver file reflects only the right half of the mandible hence the 52 points. It returns the area and the centroid coordinates in the AE_R.EV file (see the values in Table 1).

Appendix V. Residual File AE.RES

```
!AE.RES                                                           1
EFFA_File_v2.txterec      1      O003NXYZ 2 104       22:47:4803-12-2013 2
                                                                  3
 -0.59200  0.58673 -0.4885  0.9661   4
  0.20464 -0.11391  0.2676  0.3556   5
 -0.08502  0.14038 -0.1291  0.2088   6
  0.14257  0.10764 -0.0123  0.1791   7
  0.21240  0.17270 -0.0449  0.2774   8
 -0.05447 -0.05966  0.2265  0.2405   9
 -0.12226 -0.02195 -0.3127  0.3365  10

   (Columns omitted to save space)

  0.68335  0.45448  0.5498  0.9878  95
  0.13155 -0.48971  0.0914  0.5152  96
 -0.19467 -0.15364 -0.1868  0.3105  97
 -0.34030  0.21247 -0.3410  0.5265  98
 -0.15212  0.17176 -0.1283  0.2629  99
  0.04259 -0.07367  0.1509  0.1732 100
  0.05466 -0.23662  0.0847  0.2572 101
 -0.01965  0.11408 -0.0157  0.1168 102
 -0.04108 -0.04985  0.0132  0.0659 103
  0.04997 -0.13211 -0.0568  0.1522 104
 -0.06699  0.03572  0.0212  0.0788 105
 -0.09013 -0.09896 -0.2351  0.2706 106
 -0.59206  0.58670 -0.4885  0.9661 107
Mean Residual is          .4090174
```

First three columns refer to the x, y, and z components, while the fourth column depicts the actual distance between the observed points and the predicted points computed from the EFF. Column numbers on the far right refer to logical records, Header records are the first 3, data records are 104. The mean residual, 0.41, is in pixel counts, which is 0.07 in mm.

Appendix VI. Size-standardized Points File AE.PA

```
!AE.PA                                              1EFFA_File_v2.txterec       1      PRE 0
1 3N 3D  0 208  0      22:46 03122013   2 3030.696 1000000 0.000000  0      0      0      0   0
0          3  -923 -19505   493  -1497 -19650   717  -2541 -19489   827  -3498 -19164   799  4 -
4420 -18773   755  -5453 -18218   760  -6277 -17607   824  -7924 -15952   847  5  -9351 -14060
742  -9903 -12550   833 -10355 -10860  1035 -10901  -9598  1244  6 -11603  -8460  1463 -12263
-7495  1584 -12934  -6497  1580 -13872  -4982  1501  7 -14587  -3437  1695 -14747  -2970
1806 -14891  -2502  1922 -15198  -1528  2126  8 -15610  -569  2237 -16626   863  2396 -17677
1983  3027 -18160  2515  3782  9 -18464  2866  4747 -18672  3012  5857 -18891  2958  7035
-19229  2709  8439  10 -19570  2382  9720 -19900  2021  11176 -20225  1890  12920 -20545
2369  14548  11 -20718  3213  15218 -20768  4340  15151 -20684  5831  14226 -20625  7226
12959  12 -20771  8606  11979 -20977  9722  11543 -21241  10871  11448 -21714  12335  12010
13 -22337  13610  13060 -23238  14985  14188 -24138  16046  15526 -24411  16637  16776  14 -
24239  17390  17853 -23919  18230  18349 -23647  19141  18365 -23599  19395  18277  15 -23571
19624  18155 -23565  19942  17892 -23602  20183  17566 -23711  20397  16997  16 -23851  20439
16318 -23963  20225  15311 -23863  19861  14420 -23562  19459  13577  17 -23165  19056  12692
-22919  18717  11896 -22826  18325  10992 -22830  17461   9079  18 -22458  16783   7225 -22152
16414   5833 -22044  16064   4390 -22020  15741   2942  19 -21936  15484   1506 -21812  15151   -
566 -21726  14633  -2537 -21465  13607  -4688  20 -21182  12172  -6620 -20982  11164  -7407 -
20699   9920  -7952 -20181   7735  -8558  21 -19557   5786  -9183 -18428   3511  -9973 -17265
1375 -10821 -16151  -1229 -11802  22 -14958  -3702 -12760 -12320  -7554 -13995  -9498 -11205 -
15154  -5064 -13368 -15939  23   -789 -15576 -16610   -312 -16660 -16488   -276 -17678 -15959   -
350 -18097 -15411  24   -423 -18298 -14654   -442 -18303 -14190   -443 -18245 -13690   -399 -
17999 -12539  25   -367 -17805 -11524   -452 -17779  -9857   -558 -17913  -7891   -525 -18001  -
6894  26   -448 -18168  -6004   -400 -18319  -5533   -359 -18507  -5092   -328 -18828  -4483  27   -
338 -19190  -3837   -378 -19467  -3287   -438 -19708  -2685   -522 -19955  -1707  28   -505 -20042
-791   -334 -20029    -88    -46 -19982    379    -46 -19982    379  29    -46 -19982    379    384 -
```

19922 725 1091 -19840 986 2472 -19659 1104 30 3829 -19291 1028 4846 -18692 911 5661 -17923 781 6631 -16967 617 31 7795 -16078 497 9094 -15203 535 10038 -14289 708 11132 -11612 1085 32 11821 -8704 1405 12443 -6919 1562 13304 -5151 1701 13913 -4211 1841 33 14511 -3336 1991 15070 -2346 2071 15538 -1137 1984 16201 663 1784 34 17032 1815 2146 17438 2170 2539 17838 2463 3099 18182 2703 3816 35 18418 2888 4601 18679 3100 6080 18962 3084 7549 19224 2832 9019 36 19305 2489 10428 19345 2291 11458 19552 2220 12553 20348 2621 14788 37 20630 3494 15784 20635 4192 15784 20683 5315 15052 20900 6543 13708 38 20914 7546 12732 20702 9027 11825 20989 10810 11127 21605 12176 11053 39 22062 13384 11578 22493 14740 12739 23094 15791 14146 23449 16502 15379 40 23629 17403 16440 23943 18505 17089 24416 19678 17273 24666 20416 17075 41 24695 20798 16790 24506 21147 16204 23939 21221 15071 23639 20980 14182 42 23598 20552 13302 23679 20194 12722 23751 19842 12191 23678 19332 11380 43 23220 18870 10456 22437 18359 9141 22214 17797 7797 22417 17167 6326 44 22403 16762 4840 22197 16487 2750 22309 16053 694 21985 15488 -1345 45 21612 14668 -3235 21686 13602 -4581 21619 12342 -5732 21385 11346 -6418 46 21151 10230 -6931 20880 8510 -7393 20388 6895 -7785 19296 4545 -8658 47 18379 2167 -9546 17291 -671 -10735 16248 -3457 -11797 13422 -7668 -13291 48 10284 -11608 -14706 5082 -13548 -15684 119 -14842 -16169 -8 -14795 -16075 49 -137 -14700 -15923 -198 -14623 -15810 -250 -14522 -15668 -284 -14411 -15517 50 -305 -14284 -15344 -306 -14021 -14977 -268 -13748 -14563 -202 -13496 -14115 51 -130 -13306 -13671 -59 -13179 -13188 -9 -13144 -12733 -4 -13444 -11404 52 -130 -13885 -9993 -173 -13985 -8969 -143 -13967 -7924 -78 -14026 -6891 53 -43 -14295 -5893 -97 -14946 -4498 -179 -15598 -3206 -192 -15922 -2630 54 -183 -16308 -2083 -234 -18265 -445 -923 -19505 493 -923 -19505 493 55

The points have been doubled to 208 and the file centroid-adjusted hence the negative values.

Appendix VII. Exported .PA File AE.EP3 for Plotting Purposes

1	-9.2299995,	-195.0500031,	4.9299998
2	-14.9699993,	-196.5000000,	7.1700001
3	-25.4099998,	-194.8899994,	8.2699995
4	-34.9799995,	-191.6399994,	7.9899998
5	-44.2000008,	-187.7299957,	7.5499997
6	-54.5299988,	-182.1799927,	7.5999999
7	-62.7700005,	-176.0699921,	8.2399998
8	-79.2399979,	-159.5199890,	8.4700003
9	-93.5099945,	-140.5999908,	7.4199996
10	-99.0299988,	-125.5000000,	8.3299999

(Columns omitted to save space)

199	-1.4299999,	-139.6699982,	-79.2399979
200	-0.7800000,	-140.2599945,	-68.9099960
201	-0.4300000,	-142.9499969,	-58.9300003
202	-0.9700000,	-149.4599915,	-44.9799995
203	-1.7900000,	-155.9799957,	-32.0599976
204	-1.9200000,	-159.2200012,	-26.2999992
205	-1.8299999,	-163.0800018,	-20.8299999
206	-2.3399999,	-182.6499939,	-4.4499998
207	-9.2299995,	-195.0500031,	4.9299998
208	-9.2299995,	-195.0500031,	4.9299998

Columns 2, 3, and 4 refer to the x, y, z coordinates respectively. Recall that the 104 points have been doubled here. Column 1 displays point numbers added for clarity (not present in the export file). Note the negative values indicating the file has been centroid-adjusted.

Utilizing a Novel 3D Surface Mapping Technology to Evaluate Craniofacial Morphology

WON MOON[1], RYAN W. MCCOMB[1], BORIS GUTMAN[2], REENA KHULLAR[1] and JAY H. SUNG[1]

Abstract

Current analyses of 3-D CBCT images used in orthodontic and orthognathic surgery diagnosis/treatment planning still rely upon 2-D linear and angular measures, along with subjective visual evaluation, to assess complex facial disharmony and other dentofacial characteristics. The overall objective of this investigation is to overcome these limitations by developing and testing protocol for mapping and averaging human skull surfaces in three dimensions. A selection of 10 female patient files ranging in age from 19 to 35 was selected as an initial sample for development of the technical protocol. Existing CBCT files in DICOM format were segmented using a triangular mesh approach. The topology of each skull was then corrected using a principal axis star map and smoothing of the extrapolated regions. Shapes were mapped onto a sphere using conformal and area preserving maps, which were then registered using a spherical patch mapping approach. Finally, an average was created using a 7-parameter Procrustes alignment. Size-standardized and non-size standardized average skull models were generated for the sample population. A single patient was then superimposed on the average and color-coded displacement maps were generated to demonstrate the clinical applicability of this protocol. The results of this investigation suggest that it is possible to average multiple shapes of highly variable topology such as the human skull. The most immediate application of this research will be rapid and detailed diagnostic imaging analysis for orthodontic and surgical treatment planning, particularly in complex cases. There is also great potential for application to areas outside orthodontics such as anthropometrics and genomics[3].

INTRODUCTION

A robust quantification of craniofacial structures is required to properly evaluate the diagnostic factors presented in orthodontics as well as with craniofacial anomalies, facial disharmony, and trauma. Accurate diagnosis and treatment in these cases necessitates tools that permit the recognition of obvious structural malformations as well as the more subtle discrepancies present in such cases.

The traditional approach of morphometric analysis of craniofacial structures in orthodontics is known as cephalometrics [1, 2], which is based on a standardized 2-dimensional radiographic image. Orthodontists and professionals in other disciplines have relied heavily upon 2-dimensional cephalometric imaging as a foundation for the analysis of the craniofacial complex in diagnosis and treatment planning. Despite its widespread use in orthodontic and orthognathic surgery, the 2-dimensional cephalometric imaging has several limitations: (1) the image is a 2-dimensional (2D) representation of a complex 3-dimensional (3D) object; (2) complex anatomical structures are superimposed obscuring details; (3) patient's head position may affect the resulting image; (4) the head position and angulation of an object relative to the

[1] UCLA School of Dentistry, Section of Orthodontics, Los Angeles, CA 90095, USA, wmoon@dentistry.ucla.edu.

[2] UCLA School of Medicine, Imaging Genetics Center, Laboratory of Neuro Imaging, Department of Neurology, Los Angeles, CA 90095, USA.

[3] Supported or funded by AAOF.

x-ray source and film result in radiographic projection error; and (5) landmark identification error by the operator can yield flawed analyses [3]. The advent of 3D Cone based Computed Tomography (CBCT) has offered an opportunity to avoid the limitations inherent in 2D cephalometric imaging, and provide new, novel methods for visualizing and analyzing the human skull in three dimensions.

The introduction of CBCT to the orthodontic specialty has increased the utilization of 3D data. Over the past decade, researchers have demonstrated the increased accuracy of 3D CBCT imaging compared to traditional 2D imaging modalities and have pursued the development of new analyses for clinical applications. The great majority of these analyses, however, has either reduced the 3D image into traditional 2D views by compressing the image, or has relied upon 2D linear and angular measurements to analyze complex 3D shapes [3 - 9]. Most of these analyses were repeating the conventional methods used in 2D radiographs where 3D CBCT images were only used for linear measurements between conventional cephalometric landmarks [10 - 15]. All of these approaches show the inherent limitations of multivariate morphometrics as with conventional 2D analyses. The absence of a true 3D quantification method lead to some literature prematurely concluding that there is no significant difference in measurements between 3D CBCT and conventional 2D cephalograms [16] and that there is no assured evidence regarding the benefits of CBCT use in orthodontics [17].

An accurate outline of anatomical structures can be described using continuous curves approximated by a larger number of landmarks along the structure. This approach was used to describe the 2D image of the human mandible successfully, showing a direct morphologic comparison between the average of a treatment group and control group [18]. A similar method was expanded into 3D where average shapes were obtained for various developmental stages of a rabbit orbit, and differences in size and shape were investigated [19]. In these studies, Fourier Descriptors (FDs) provided a set of functions that led to a smooth and continuous curve connecting the points. Recently, the same method was used to describe the human mandible with 104 points collected along the border of a 3D image and to find the 3D average of 10 subjects [20], as shown in Fig 1. Procrustes superimposition allowed the comparison of shapes under the best fit found by repositioning the structure along three axes. This particular study on the human mandible has established the possibility of representing 3D craniofacial structure by curved outlines defined by a number of point coordinates, and allowing for normalization and superimposition. Traditionally, craniofacial structures were divided into three major parts for evaluation. These were the cranium, maxilla, and mandible, since they are distinct in their features and growth patterns [21]. The final shape representation of the craniofacial structure must include all three components.

Although the 3D curve approach advocated in the chapter may be promising, the use of FDs cannot describe the surface topology of the craniofacial anatomy, which is also an important part in evaluating the facial structure.

A group from Duke University also sought to establish 3D normative models of the cranium for treating patients with craniosynostosis. The currently available objective methods tend to focus on the most prominent features associated with a given condition, and they suggested that 3D analysis should use a finite yet comprehensive dataset of defined cranial surface points.

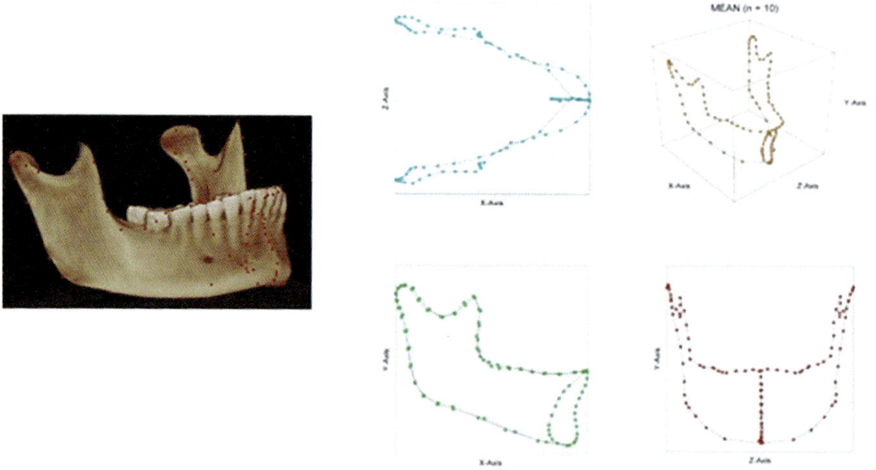

Fig. 1. Left: Outline of the mandible. Right: three isometric views and one 3D view of the normalization sample (n = 10) Superior view (blue), lateral view (green) and frontal view (red).

If a single dataset is compared to an appropriate averaged control dataset, the graphic display may demonstrate variance for each point in color-coded standard deviations from the control data [22]. They appropriately concluded that, "lack of normative data has been the obstacle to the application of three-dimensional image-based anthropometry" and "...one must understand that derivation of two-dimensional measurement indices from three-dimensional data still only provides two-dimensional measurement indices: more accurate than caliper measurements, but not any more useful [23]". Thus, the use of novel technology requires a novel measurement system [31].

The first aim of this study is to develop a way to represent the shape of the maxilla, mandible and cranial base-zygoma which will allow superimposition, comparison, and normalization of the components as well as the entire facial complex. The second aim of this study is to develop a way to represent the surface of the entire skull by the use of Procrustes superimposition assisted by the use of topology based registration that will allow for superimposition, comparison, and normalization of the entire craniofacial complex. This technology will ultimately allow for comparison of a unique patient's 3D CBCT image against a gender, age and ethnicity-matched average or standard for diagnostic and treatment planning purposes.

To overcome the shortcomings of conventional cephalometrics and provide a useful application, the new approach must:

1. Avoid excessive dependence on limited or sparsely located landmarks
2. Represent the actual shape
3. Allow for superimposition
4. Provide a for normal (average) form
5. Allow statistical results for comparison

This true 3D morphometric method for analyzing craniofacial structure will serve as a basis for more reliable diagnostic and analytic tools.

MATERIALS AND METHODS

Boundary Quantification

Collection of the coordinate information

Three dimensional CBCT images taken at the UCLA School of Dentistry by the Newtom3G CBCT scanner (Image Works) were collected, and imported into Dolphin Imaging® software. The subjects with significant craniofacial defects were excluded, resulting in a total of 10 subjects without morphologic abnormality. The image was formatted for optimum hard tissue threshold of opacity. The 3D image was sectioned into five different specific regions of interest for the purpose of defining the outline of the mandible, maxilla and cranial base (Fig. 2). Using an interface developed for a previous study [20], point coordinates were collected for each structure. The collection of coordinates was divided into three separate components, resulting in a total of 104 points, 92 points, and 117 points for mandible, maxilla and cranial base/zygoma, respectively.

The coordinate data for each component was saved as a text file (Appendices I, II and III), ready to be used for the following geometric realignment and analysis.

Shape registration and alignment of each component

Correspondence of points in each component was assumed based on manual labeling as shown in the example of the text file for the mandible (Appendix III). Given a pair of homologous weighted point sets $\{\boldsymbol{x}_i, \boldsymbol{y}_i, w_i\}, \boldsymbol{x}_i, \boldsymbol{y}_i \in \mathbb{R}^3, w_i \in \mathbb{R}$, the transformation that minimizes the sum of squared distances between each pair of points is known as a Procrustes alignment [24], and can be encoded by a 4x4 homogenous matrix A. This matrix minimizes the following functional:

$$C(A) = \sum_i w_i (A\tilde{\boldsymbol{x}}_i - \tilde{\boldsymbol{y}}_i)^2, \tag{1}$$

where $\tilde{\boldsymbol{x}}_i, \tilde{\boldsymbol{y}}_i$ are the 4-dimensional versions of the original point sets, augmented by 1 in the 4th dimension to account for translation.

Computing Average Affine Transformations

Assuming existing correspondence, we seek an average position, size, and orientation – the mean affine transformation – of the average shape, given the sample of normal subjects. A true "average" shape necessitates the idea of a shape metric. In this formalism, shapes can be represented simply as point-sets, continuous one-dimensional curves [25], or orientable 2-manifolds [26]. Each of these cases leads to a different metric formulation, $d: \mathcal{M} \times \mathcal{M} \rightarrow \mathbb{R}^{+}$ forming its own Riemannian manifold of shapes $\mathcal{M}$. In this case, the average shape is approximated simply as the Euclidean average of the individual shapes based on their embedding in the original 3-space, $\mathbb{R}^{3}$, and after removing effects of affine alignment.

A Riemannian metric naturally induces a manifold, and the task of finding a mean of a set of points on this manifold becomes a minimization problem:

$$\bar{A} = \arg\min_{A} \sum d^{2}(A, A_{i}). \tag{2}$$

This is known as the Karcher mean [27]. An equivalent, though more general definition of the Karcher mean is the point whose unit vectors along geodesics to each A_i sum to zero in the tangent space of $\bar{A}$.

Although the formulation of [25] was tried, it was found that the Euclidean average approximation with assumed correspondence is the most robust. Further, because the essential measurements of interest in clinical practice are based on Euclidean displacement from the norm, such a construction of the normative shape is most appropriate.

The essential idea behind the computation of a mean affine transformation is similar in spirit to the idea of a metric space of shapes. Developed by Roger Woods [28], the approach formulates a semi-Riemannian metric of linear transformations based on matrix exponentiation. As noted in the study [28], the latter definition of the mean can be used to drive a local optimization of (2). This is based on two facts:

1. Given an exponential map on a manifold at point p, or the geodesic path parameterized by t in the direction v, $\gamma_{v}(t) = \mathrm{Exp}_{p}(t * v)$, the condition for the mean, $\mathbf{0} = \sum_{j=1}^{N} \mathrm{Exp}_{\bar{A}}^{-1}(A_{j})$, is also the local gradient for the current approximation to the mean.

2. As affine transformation matrices form a Lie group, the exponential map on the manifold is equivalent to matrix exponentiation. This leads to an iterative approach for finding the mean affine transformation.

Reconstruction of a comprehensive 3D model

Armed with the computational methods of the previous section, we can now reposition the 3D skull components of each normal subject's average position. To do this, one can initially choose a single subject to serve as the target for all alignments. In this way, our mean affine transform will be the mean of all inverse transforms to the target. Then, each subject's average repositioning matrix is computed as $A_{j} * \overline{A^{-1}}$. However, this approach is still potentially confounded by the arbitrary choice of the target subject, which can reduce the stability of the computed mean. An alternative which avoids this confounding is to compute the full set of Nx(N-1) affine alignments, and find the corresponding mean transformation for each subject.

This is a true group-wise approach, as no target subject needs to be selected. This latter approach was used, as implemented by Roger Woods in the "AIR Reconcile" program of the AIR package, available at http://bishopw.loni.ucla.edu/air5/.

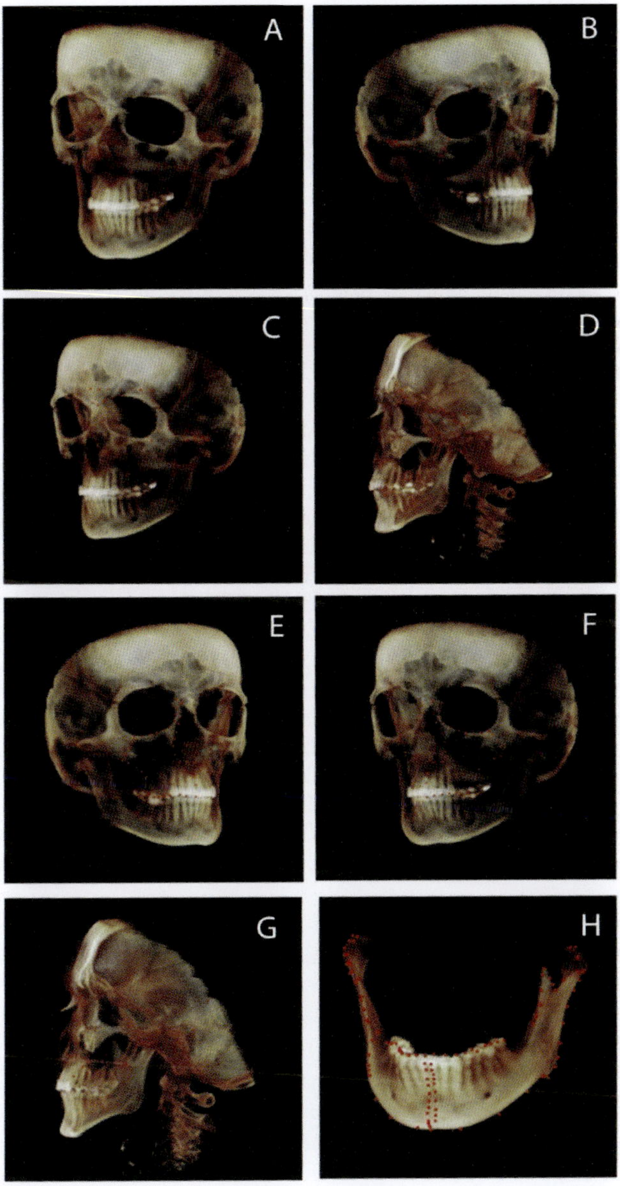

Fig. 2. A screen capture of the collection of points. (A) Subject 1 (B) Right Zygoma (C) Left Zygoma (D) Mid Cranial Base (E) Right Maxilla (F) Left Maxilla (G) Palate (H) Mandible.

Measurements and analysis by superimposition

Two subjects were selected for comparison with the normalized craniofacial complex: one internally selected from the ten sample subjects and another patient with Crouzon syndrome to represent two patients with relatively minor deviation and great variation respectively from the population used for the normalization. The coordinates of the points that define the boundary were collected, and Procrustes analysis was used to superimpose the cranial base, maxilla, mandible, and the integrated craniofacial complex of each subject with those of the normalized set. Vector form and linear distance for each landmark were used to analyze the deviation which was compared with the size-standardized or non-size-standardized data. The 117, 92, 104, and 313 points for cranial base, maxilla, mandible, and the total craniofacial complex were used, and these boundary point data were included in the computation for both the visual and numeric analyses. Every set of points as morphologic entities and analytic results were plotted in 3-dimensions using the LONI *Shapeviewer* software® developed by the UCLA Laboratory of Neuro-Imaging. However, only a limited number of points were focused on to simplify the evaluation (Table 1).

Table 1. Selected Points of Interest and Their Description

Components	Type of Landmarks	Points	Description
Cranial Base (CB)	Sagittal	Basion	108th point of Cranial base
		Nasion	45th point of Cranial base
	Asymmetric	Right Fronto-zygomatic	37th point of Cranial base
Maxilla (MX)	Sagittal	ANS	30th point of Maxilla
		A Point	29th point of Maxilla
	Asymmetric	Right Orbitale	9th point of Maxilla
		Right Lateral Nasal Cavity	34th point of Maxilla
Mandible (MD)	Sagittal	B point	29th point of Mandible
		Pogonion	33th point of Mandible
		Menton	37th point of Mandible
	Asymmetric	Right Gonial Angle	93rd point of Mandible
		Right Condylar Head	1st point of Mandible

Surface Quantification

Data Collection and surface segmentation of sample skull CBCT images

NewTom® 3G scanned images for patients at the UCLA Section of Orthodontics were evaluated, and patients demonstrating the presence of a craniofacial anomaly or any other gross facial asymmetry were excluded. The 10 images from a relatively homogenous female patient population were included in the study, and these were imported into Dolphin Imaging® software. Following import, the images was formatted for optimum hard tissue resolution, and the surface of the skull was extracted using a triangular mesh approach that identified a variable number of surface nodes to closely describe the skull shape. The number of vertices was then standardized to 16,384 for a working resolution for vertex-to-vertex correspondence between skulls during rigid alignment (Fig. 3). Files were exported in .PLY format (Polygon File Format), a format for storing 3D data of graphical objects described by a series of polygons, and these files were then uploaded into MeshLab® Software for initial viewing.

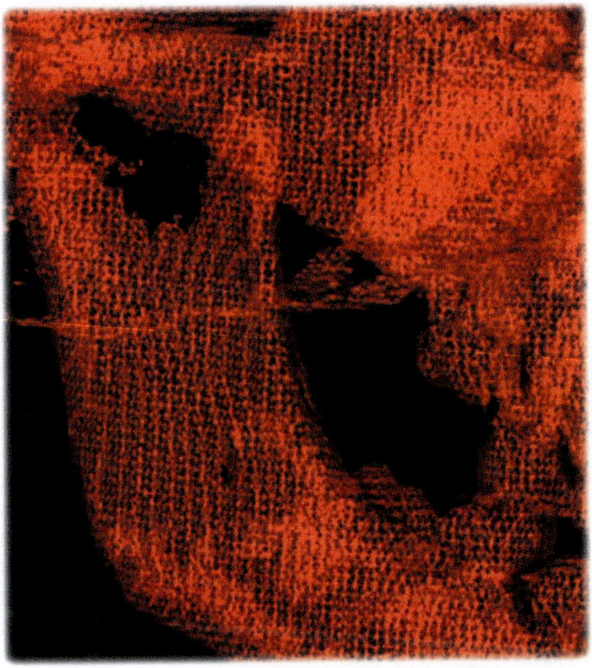

Fig. 3. This image illustrates the intricate triangular mesh surface framework generated with the Dolphin Imaging® software.

Shape registration

Numerous shape registration techniques exist for shapes of spherical topology, including those based on spherical parameterization. The method devised in this chapter is specifically for skull mapping and is independent of the existence of an orientation, sufficiently flexible, and allows for continuous regularization. In Spherical Patch Mapping the multiple corresponding landmarks are automatically identified on each skull using mean curvature maps on the sphere around each landmark. This method requires a few manually marked landmarks on a sample template skull; but once these same landmarks are automatically identified on new skulls, the skull models are registered by matching corresponding points with spherical splines (Fig. 4).

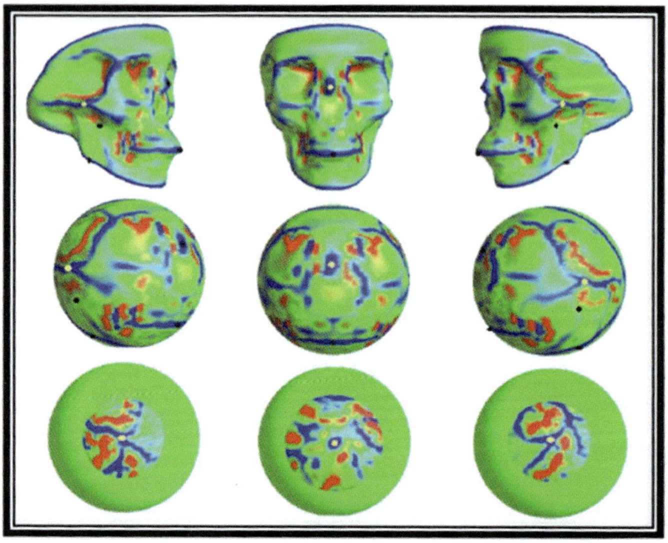

Fig. 4. Top Row: Spherical curve-based skull models with patches marked in black and yellow. Middle Row: Spherical curvature maps with corresponding landmarks shown. Bottom Row: 'Patches' around selected landmarks are displayed (see text).

Average and Distance Map Creation

Procrustes alignment

The shape models must first be aligned in their original (non-parametric) space based on the computed dense correspondence in order to compute the average skull, t. The 7-parameter Procrustes method was used [29]. The selected shape models are not an absolutely true representations of the original, and only the points that are sufficiently close to a non-extrapolated region are considered in this study. Once the final dense mapping has been computed, each source mesh is re-sampled to reflect the connectivity of the target mesh.

Distance and statistical maps

After computing the average shape of skull, the distance from the average to each individual shape at each point can be found. Distance-from-individual-to-average maps are displayed as colorized surface maps. This will eventually allow for the precise comparison of a patient's 3D CBCT skull data to a specific normative model (standard) of the skull based on age, gender, and ethnicity. Given a discrete, or continuous biological variable, it is possible to create statistical parametric maps based on distance to the average. These typically involve parametric or non-parametric statistical tests done at each point, which then localize the effect of the biological variable on the skull surface [30].

RESULTS

Boundary Quantification

Representation of individual samples

The coordinates of the boundary points for each the original ten subjects showed a well-defined boundary that represents the morphology of each component of the patient. Fig. 5 shows an example (case 1) of each component.

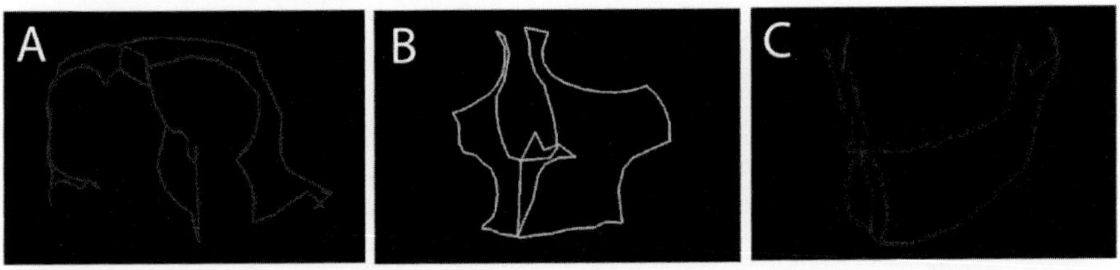

Fig. 5. Shown here are the three components of case 1. (A) cranial base/zygoma, (B) maxilla, and (C) mandible. Compare the shape irregularities in the outlines with Fig. 6.

Normalized result of each component

The 10 cases of the cranial base, maxilla, and mandible respectively, are superimposed and averaged to create a normalized form of each of the three components. Each normalized structure is plotted as shown in Fig. 6. The variation within the group of samples was quantified by the deviation of each landmark from one another. Tables 2 and 3 show the variation of the selected points in terms of their standard deviations for each component. The non-size-standardization of case 1 as well as the size-standardization one are presented here.

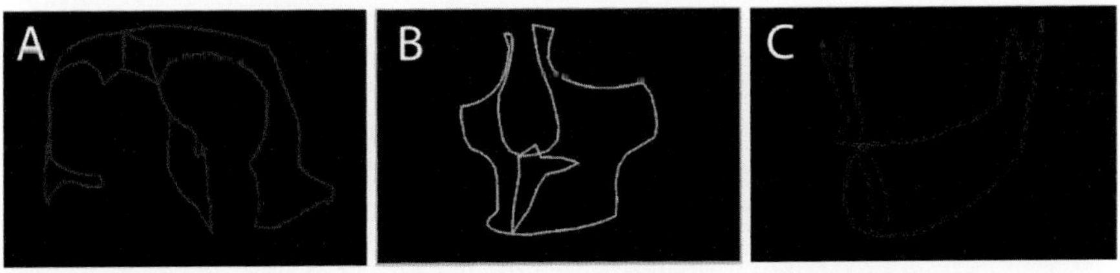

Fig. 6. Illustrated here is the averaged or normalized form ($n = 10$) for each component. (A) Normalized cranial base/zygoma, (B) normalized maxilla, and (C) normalized mandible. Compare the now smoother outlines of the normalized components with Fig. 5.

Table 2. Non-size-standardized Result of the Deviation

Component	Type of Landmark	Landmark	Non-size-standardized			
			STD DEV of Coord			STD DEV of displacement
			x	y	z	
CB	Sagittal	Basion	1.21	2.10	4.32	2.23
		Nasion	0.40	1.39	1.76	1.11
	Asymmetric	Right Fronto-zygomatic	1.97	1.86	1.03	1.51
MX	Sagittal	ANS	1.10	2.84	1.24	1.52
		A Point	0.93	1.34	1.54	1.05
	Asymmetric	Right Orbitale	2.73	1.21	1.71	1.47
		Right Lat. Nasal Cavity	0.65	1.39	2.31	1.84
MD	Sagittal	B point	0.33	1.80	1.54	1.20
		Pogonion	0.56	1.93	2.50	1.96
		Menton	0.63	1.76	1.96	1.60
	Asymmetric	Right Gonial Angle	2.30	3.93	1.97	2.74
		Right Condylar Head	3.03	1.02	2.34	2.24

Table 3. Size-standardized Result of the Deviation

Component	Type of Landmark	Landmark	Size-standardized			
			STD DEV of Coord			STD DEV of displacement
			x	y	z	
CB	Sagittal	Basion	1.18	2.90	4.32	2.44
		Nasion	0.38	1.66	1.70	1.16
	Asymmetric	Right Fronto-zygomatic	0.93	1.38	1.17	0.82
MX	Sagittal	ANS	1.08	2.99	1.24	1.64
		A Point	0.92	1.50	1.48	1.14
	Asymmetric	Right Orbitale	2.12	1.23	1.55	1.39
		Right Lat. Nasal Cavity	0.64	1.37	2.27	1.83
MD	Sagittal	B point	0.34	1.95	1.09	1.11
		Pogonion	0.59	2.04	1.46	1.38
		Menton	0.65	2.00	1.07	1.08
	Asymmetric	Right Gonial Angle	2.14	3.42	1.98	2.56
		Right Condylar Head	2.45	1.90	2.59	2.25

Average of the complex

Reconstruction of the normalized craniofacial structure was accomplished by applying the invert transformation and repositioning each component with respect to one another. The result is a set of 313 points, representing the outline of the normalized 3D craniofacial structure. The normalized form and numerical values of the coordinates are as shown in Fig. 7 and Table 4.

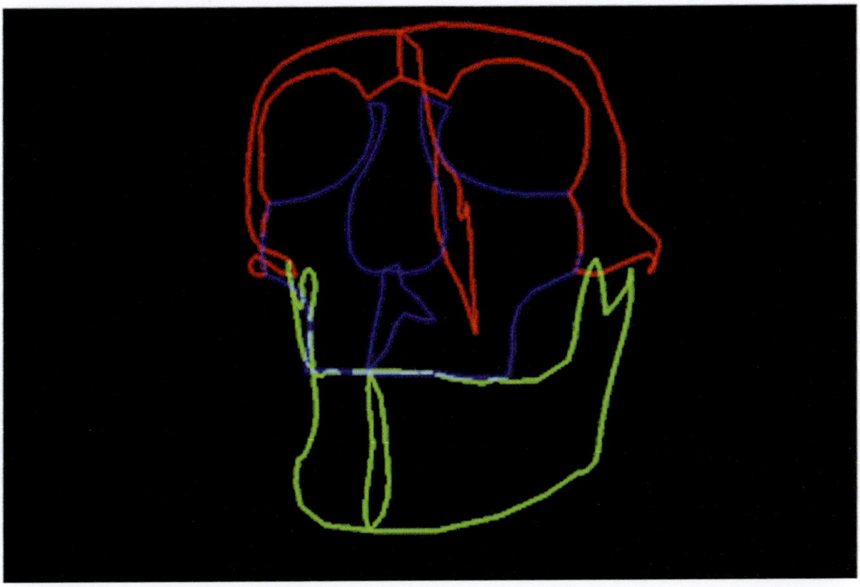

Fig. 7. The complete normalized facial complex composed of the three components.

Table 4. The Normalized Coordinate Points

Component	Type of Landmark	Landmark	Average Coordinates		
			x	y	z
CB	Sagittal	Basion	169.44	126.85	31.14
		Nasion	170.63	51.31	58.43
	Asymmetric	Right Fronto-zygomatic	123.95	65.08	52.84
MX	Sagittal	ANS	170.61	46.98	5.53
		A Point	170.38	47.07	-1.15
	Asymmetric	Right Orbitale	136.54	58.96	28.22
		Right Lateral Nasal Cavity	157.92	53.39	17.58
MD	Sagittal	B point	169.45	49.30	-38.35
		Pogonion	169.31	47.56	-52.70
		Menton	169.21	52.01	-58.16
	Asymmetric	Right Gonial Angle	123.75	112.58	-36.83
		Right Condylar Head	115.57	121.96	23.49

Superimposition of an internal subject with the norm

The first example was selected as the internal subject (case 1) to be superimposed and compared with the norm. The non-size-standardized and size-standardized superimpositions of the cranial base, maxilla, and mandible revealed that subtle deviations from the norm in the coordinates of the boundary points were detected, and the visualized comparison are shown in Figs. 8 and 9. Numerical values of the selected points of the non-size-standardized and size-standardized superimpositions are given in Tables 5 and 6.

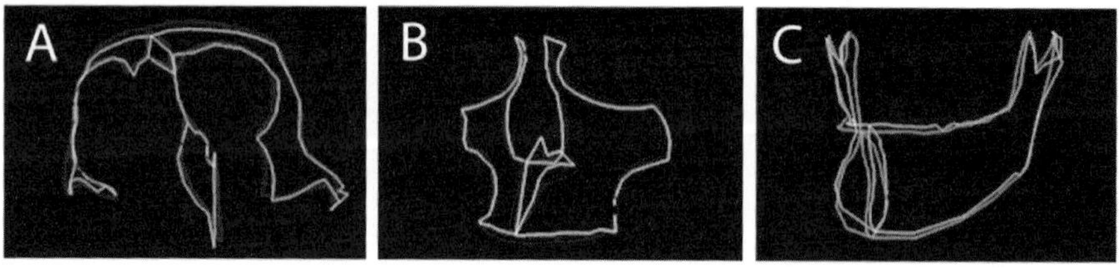

Fig. 8. Non-size-standardized superimposition of the three components of Case 1. (A) cranial base/zygoma, (B) maxilla, and (C) mandible (White: Case 1, Colored: Norm). The overall non-size-standardized shape differences of the components between Case 1 and the norm while moderate are clearly evident.

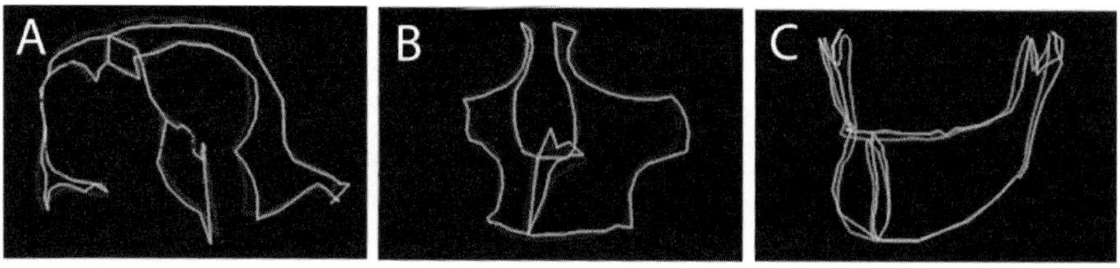

Fig. 9. Size-Standardized Superimposition of the three components of Case 1. (A) cranial base/zygoma, (B) maxilla, and (C) mandible (White: Case 1, Colored: Norm). The overall size-standardized shape differences of the components between Case 1 and the norm are slightly less than for the non-size-standardized view (Fig. 8).

Table 5. Non-size-standardized Superimposition of Selected Points for Each Component

Pt.	Subject Coordinates			Deviation of Coordinates			Deviation of Coordinates In STD DEV			Dev*	Dev in STD DEV**
	x	y	z	x	y	z	x	y	z		
1	168.51	123.42	37.56	-0.93	-3.43	6.42	-0.76	-1.63	1.49	7.33	3.29
2	170.89	52.45	59.04	0.26	1.14	0.60	0.64	0.82	0.34	1.31	1.18
3	122.24	67.43	52.96	-1.71	2.35	0.12	-1.71	2.35	0.12	2.91	1.93
4	171.76	47.88	5.72	1.14	0.90	0.19	1.04	0.32	0.15	1.47	0.96
5	171.07	46.78	-3.76	0.69	-0.28	-2.61	0.74	-0.21	-1.69	2.71	2.59
6	136.53	57.67	25.66	-0.01	-1.28	-2.56	0.00	-1.06	-1.50	2.86	1.94
7	158.34	54.13	17.65	0.42	0.75	0.07	0.64	0.54	0.03	0.86	0.47
8	169.61	52.58	-37.77	0.16	3.28	0.57	0.50	1.82	0.37	3.33	3.33
9	169.39	50.19	-50.70	0.07	2.63	2.00	0.13	1.36	0.80	3.31	3.31
10	169.54	53.13	-56.19	0.33	1.12	1.97	0.53	0.64	1.00	2.29	2.29
11	123.00	108.51	-37.58	-0.75	-4.07	-0.75	-0.33	-1.03	-0.38	4.20	1.20
12	114.54	120.04	23.96	-1.02	-1.93	0.47	-0.34	-1.88	0.20	2.23	2.23

The numbers (1-12) in the first column on the left refer to the landmarks (Pt.): 1 = Basion, 2 = Nasion, 3 = Right Fronto-zygoma, 4 = ANS, 5 = A-Point, 6 = Right Orbitale, 7 = Right Lateral Nasal Cavity, 8 = B-Point, 9 = Pogonion, 10 = Menton, 11 = Right Gonial Angle, 12 = Right Condylar head.

*Dev. = distance deviation in 3D

**Dev. in STD DEV = distance deviation in terms of STD DEV units

Table 6. Size-standardized Superimposition of Selected Points for Each Component

Pt.	Subject Coordinates			Deviation of Coordinates			Deviation of Coordinates In STD DEV			Dev*	Dev in STD DEV**
	x	y	z	x	y	z	x	y	z		
1	168.50	123.95	37.53	-0.94	-2.90	6.38	0.80	-1.00	1.48	7.07	2.90
2	170.91	52.06	59.28	0.28	0.75	0.85	0.72	0.45	0.50	1.16	1.00
3	121.62	67.23	53.13	-2.33	2.15	0.29	-2.50	1.56	0.25	3.18	3.90
4	171.78	47.67	5.58	1.17	0.68	0.05	1.08	0.23	0.04	1.36	0.82
5	171.08	46.55	-4.10	0.70	-0.52	-2.95	0.76	-0.35	-2.00	3.07	2.71
6	135.80	57.67	25.94	-0.73	-1.29	-2.28	-0.35	-1.05	-1.47	2.72	1.96
7	158.08	54.05	17.77	0.16	0.67	0.18	0.25	0.49	0.08	0.71	0.39
8	169.67	51.61	-38.42	0.22	2.30	-0.07	0.65	1.18	-0.06	2.31	2.08
9	169.43	49.14	-51.74	0.12	1.59	0.97	0.21	0.78	0.66	1.86	1.34
10	169.59	52.18	-57.40	0.39	0.17	0.77	0.60	0.08	0.72	0.88	0.81
11	121.63	109.25	-38.22	-2.12	-3.33	-1.39	-0.99	-0.97	-0.70	4.19	1.63
12	112.92	121.12	25.20	-2.65	-0.85	1.71	-1.08	-0.45	0.66	3.26	1.45

The numbers (1-12) in the first column on the left refer to the landmarks (Pt.): 1 = Basion, 2 = Nasion, 3 = Right Fronto-zygoma, 4 = ANS, 5 = A-Point, 6 = Right Orbitale, 7 = Right Lateral Nasal Cavity, 8 = B-Point, 9 = Pogonion, 10 = Menton, 11 = Right Gonial Angle, 12 = Right Condylar head.

*Dev. = distance deviation in 3D

**Dev. in STD DEV = distance deviation in terms of STD DEV units

The non-size-standardized and size-standardized superimpositions can be applied to the entire craniofacial structures as shown in Figs. 10 and 11, and the numerical data are given in Tables 7 and 8.

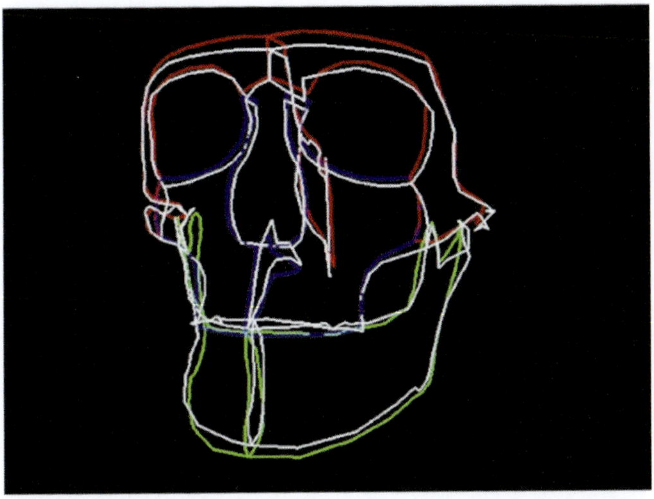

Fig. 10. Non-size-standardized superimposition of the total craniofacial complex of the example subject (Case 1) onto the normalized form (White: Case 1, Colored: Norm).

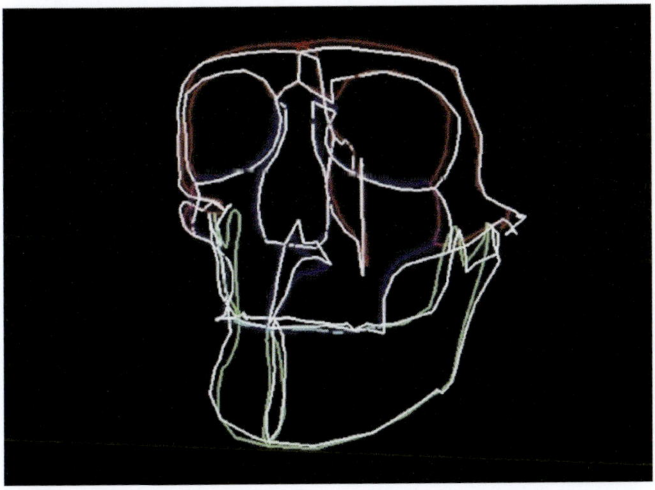

Fig. 11. Size-standardized superimposition of the total craniofacial complex of the example subject (Case 1) onto the normalized form (White: Case 1, Colored: Norm). Note that the overall differences in shape of the size-standardized components of Case 1 compared to the norm are smaller here than in the non-size-standardized comparison (Fig. 10).

Table 7. Non-size-standardized Superimposition of the Components of Case 1

Pt.	Subject Coordinates			Deviation of Coordinates			Deviation of Coordinates In STD DEV			Dev*	Dev in STD DEV**
	x	y	z	x	y	z	x	y	z		
1	169.33	117.98	32.72	-0.11	-8.87	1.58	-0.09	-3.06	0.37	9.01	3.70
2	169.40	46.38	58.14	-1.23	-4.93	-0.29	-3.21	-2.96	-0.17	5.09	4.37
3	122.19	67.20	57.24	-1.76	2.12	4.40	-1.89	1.53	3.77	5.19	6.35
4	171.67	44.20	10.25	1.06	-2.78	4.72	0.98	-0.93	3.81	5.58	3.39
5	170.71	51.19	2.34	0.33	4.13	3.49	0.36	2.75	2.36	5.41	4.77
6	137.78	57.86	23.60	1.24	-1.10	-4.61	0.58	-0.90	-2.98	4.90	3.53
7	153.89	57.83	17.29	-4.03	4.45	-0.29	-6.31	3.24	-0.13	6.00	3.28
8	168.58	47.03	-41.94	-0.87	-2.28	-3.59	-2.59	-1.16	-3.29	4.34	3.90
9	168.52	44.81	-57.46	-0.79	-2.75	-4.76	-1.34	-1.35	-3.27	5.55	4.01
10	168.35	50.75	-62.94	-0.85	-1.26	-4.78	-1.32	-0.63	-4.46	5.01	4.63
11	125.67	101.90	-35.82	1.92	-10.68	1.01	0.90	-3.12	0.51	10.90	4.25
12	126.43	123.73	22.60	10.86	1.76	-0.88	4.44	0.93	-0.34	11.04	4.92

The numbers (1-12) in the first column on the left refer to the landmarks (Pt.): 1 = Basion, 2 = Nasion, 3 = Right Fronto-zygoma, 4 = ANS, 5 = A-Point, 6 = Right Orbitale, 7 = Right Lateral Nasal Cavity, 8 = B-Point, 9 = Pogonion, 10 = Menton, 11 = Right Gonial Angle, 12 = Right Condylar head.

*Dev. = distance deviation in 3D

**Dev. in STD DEV = distance deviation in terms of STD DEV units

Table 8. Size-standardized Superimposition of the Components of Case 1

Pt.	Subject Coordinates			Deviation of Coordinates			Deviation of Coordinates In STD DEV			Dev*	Dev in STD DEV**
	x	y	z	x	y	z	x	y	z		
1	168.72	123.81	38.06	-0.72	-3.04	6.92	-0.61	-1.05	1.60	7.59	3.12
2	170.96	50.45	58.67	0.33	-0.86	0.24	0.85	-0.52	0.14	0.95	0.82
3	121.02	65.99	52.32	-2.93	0.90	-0.52	-3.15	0.65	-0.45	3.11	3.81
4	171.11	48.86	5.87	0.50	1.88	0.35	0.46	0.63	0.28	1.97	1.20
5	170.24	47.63	-3.84	-0.14	0.56	-2.69	-0.15	0.38	-1.82	2.75	2.43
6	135.59	60.65	26.63	-0.95	1.70	-1.59	-0.45	1.39	-1.03	2.51	1.81
7	157.74	56.00	18.20	-0.18	2.62	0.61	-0.28	1.91	0.27	2.69	1.47
8	169.96	50.97	-37.62	0.51	1.67	0.72	1.51	0.86	0.66	1.89	1.70
9	169.71	48.27	-50.86	0.39	0.71	1.85	0.66	0.35	1.27	2.02	1.46
10	169.80	51.19	-56.56	0.59	-0.82	1.61	0.91	-0.41	1.50	1.90	1.76
11	121.27	107.78	-38.30	-2.48	-4.80	-1.47	-1.16	-1.40	-0.74	5.60	2.18
12	112.67	120.69	24.75	-2.89	-1.28	1.26	-1.18	-0.67	0.49	3.40	1.52

The numbers (1-12) in the first column on the left refer to the landmarks (Pt.):
1 = Basion, 2 = Nasion, 3 = Right Fronto-zygoma, 4 = ANS, 5 = A-Point, 6 = Right Orbitale, 7 = Right Lateral Nasal Cavity, 8 = B-Point, 9 = Pogonion, 10 = Menton, 11 = Right Gonial Angle, 12 = Right Condylar head.

*Dev. = distance deviation in 3D

**Dev. in STD DEV = distance deviation in terms of STD DEV units

Superimposition of an external subject with the norm

The boundary information from a 3D CBCT image of a Crouzon syndrome patient with large discrepancy from the norm (Fig. 12) was used for a similar comparison. The set of points was superimposed and compared with the normalized form as was carried out for the internal sample comparison (Case 1). Cranial base, maxillary and Mandibular components were again individually superimposed with the normalized components, illustrating the regional comparison between the Crouzon subject and the norm. The visualized comparison is shown in Figs. 13 and 14, and the numerical values for the selected points are charted in Tables 9 and 10.

The cranial base, maxillary and mandibular components were individually superimposed onto the normalized components, to illustrate a regional comparison between this subject and the norm. The visualized comparison is shown in Figs. 13 and 14, and the numerical values for the selected points are charted in Tables 9 and 10.

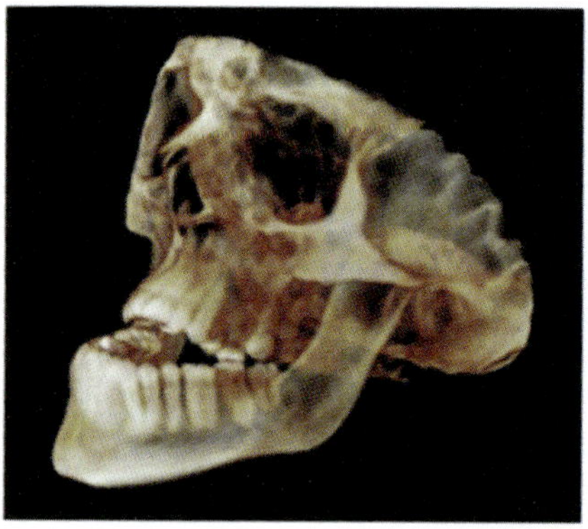

Fig. 12. CBCT image of a Crouzon syndrome patient used as an external subject is illustrated here.

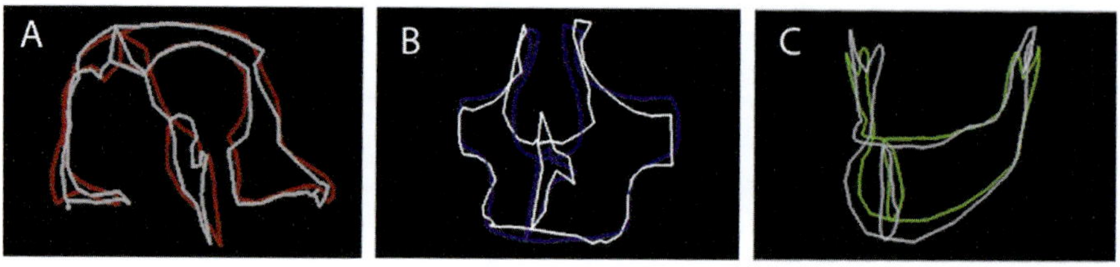

Fig. 13. Non-Size-Standardized superimposition of the components of the external subject (A) cranial base/zygoma, (B) maxilla, (C) mandible (White: external subject, Colored: Norm).

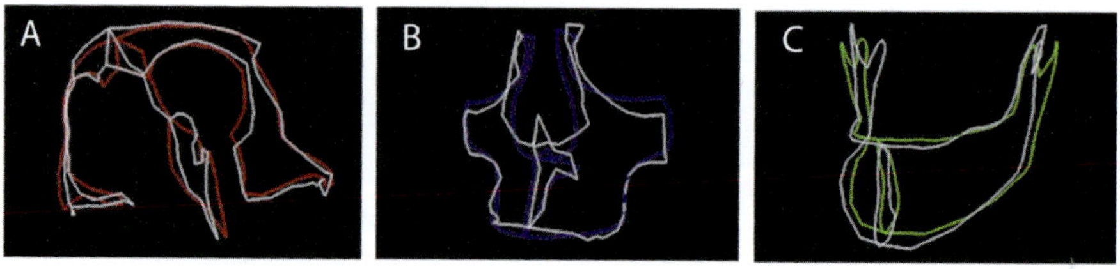

Fig. 14. Size-Standardized Superimposition of the components of the external subject (A) cranial base/zygoma, (B) maxilla, (C) mandible (White: external subject, Colored: Norm).

Table 9. Non-size-standardized Superimposition of the Components, External Subject

Pt.	Subject Coordinates			Deviation of Coordinates			Deviation of Coordinates In STD DEV			Dev*	Dev in STD DEV**
	x	y	z	x	y	z	x	y	z		
1	169.34	117.04	32.92	-0.10	-9.81	1.78	-0.08	-4.67	0.41	9.97	4.47
2	169.41	47.35	57.66	-1.22	-3.96	-0.77	-3.04	-2.86	-0.44	4.22	3.79
3	123.46	67.61	56.79	-0.49	2.53	3.94	-0.25	1.36	3.82	4.71	3.12
4	171.66	44.33	10.27	1.05	-2.66	4.74	0.95	-0.94	3.82	5.53	3.63
5	170.71	51.26	2.43	0.33	4.19	3.58	0.36	3.13	2.32	5.52	5.28
6	138.08	57.86	23.50	1.54	-1.10	-4.72	0.56	-0.91	-2.76	5.08	3.45
7	154.05	57.83	17.25	-3.87	4.45	-0.34	-5.92	3.20	-0.15	5.91	3.21
8	168.63	44.57	-43.59	-0.82	-4.73	-5.24	-2.52	-2.62	-3.40	7.11	5.93
9	168.56	42.21	-60.13	-0.75	-5.34	-7.43	-1.32	-2.77	-2.97	9.18	4.68
10	168.39	48.54	-65.97	-0.82	-3.47	-7.80	-1.30	-1.97	-3.97	8.58	5.37
11	122.91	103.04	-37.07	-0.84	-9.54	-0.24	-0.36	-2.43	-0.12	9.50	3.49
12	123.72	126.29	25.17	8.16	4.33	1.69	2.69	4.23	0.72	9.39	4.19

The numbers (1-12) in the first column on the left refer to the landmarks (Pt.):
1 = Basion, 2 = Nasion, 3 = Right Fronto-zygoma, 4 = ANS, 5 = A-Point, 6 = Right Orbitale, 7 = Right Lateral Nasal Cavity, 8 = B-Point, 9 = Pogonion, 10 = Menton, 11 = Right Gonial Angle, 12 = Right Condylar head.

*Dev. = distance deviation in 3D

**Dev. in STD DEV = distance deviation in terms of STD DEV units

Table 10. Size-standardized Superimposition of the Components, External Subject

Pt.	Subject Coordinates			Deviation of Coordinates			Deviation of Coordinates In STD DEV			Dev*	Dev in STD DEV**
	x	y	z	x	y	z	x	y	z		
1	169.33	117.98	32.72	-0.11	-8.87	1.58	-0.09	-3.06	0.37	9.01	3.70
2	169.40	46.38	58.14	-1.23	-4.93	-0.29	-3.21	-2.96	-0.17	5.09	4.37
3	122.19	67.20	57.24	-1.76	2.12	4.40	-1.89	1.53	3.77	5.19	6.35
4	171.67	44.20	10.25	1.06	-2.78	4.72	0.98	-0.93	3.81	5.58	3.39
5	170.71	51.19	2.34	0.33	4.13	3.49	0.36	2.75	2.36	5.41	4.77
6	137.78	57.86	23.60	1.24	-1.10	-4.61	0.58	-0.90	-2.98	4.90	3.53
7	153.89	57.83	17.29	-4.03	4.45	-0.29	-6.31	3.24	-0.13	6.00	3.28
8	168.58	47.03	-41.94	-0.87	-2.28	-3.59	-2.59	-1.16	-3.29	4.34	3.90
9	168.52	44.81	-57.46	-0.79	-2.75	-4.76	-1.34	-1.35	-3.27	5.55	4.01
10	168.35	50.75	-62.94	-0.85	-1.26	-4.78	-1.32	-0.63	-4.46	5.01	4.63
11	125.67	101.90	-35.82	1.92	-10.68	1.01	0.90	-3.12	0.51	10.90	4.25
12	126.43	123.73	22.60	10.86	1.76	-0.88	4.44	0.93	-0.34	11.04	4.92

The numbers (1-12) in the first column on the left refer to the landmarks (Pt.):
1 = Basion, 2 = Nasion, 3 = Right Fronto-zygoma, 4 = ANS, 5 = A-Point, 6 = Right Orbitale, 7 = Right Lateral Nasal Cavity, 8 = B-Point, 9 = Pogonion, 10 = Menton, 11 = Right Gonial Angle, 12 = Right Condylar head.
*Dev. = distance deviation in 3D
**Dev. in STD DEV = distance deviation in terms of STD DEV units

The superimpositions were applied to the entire craniofacial structures as shown in Figs. 15 and 16, and the numerical data were given in Tables 11 and 12.

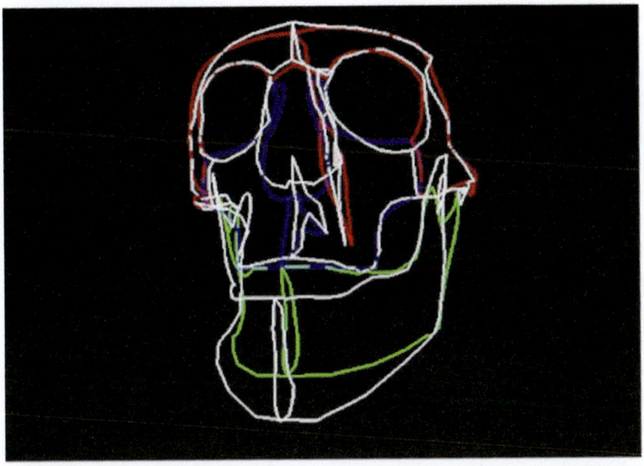

Fig. 15. Non-size-standardized superimposition of the craniofacial complex of external subject (Crouzon) with the normalized form (White: External Subject, Colored: Norm). Considerable differences in shape can be observed here, especially in the mandible.

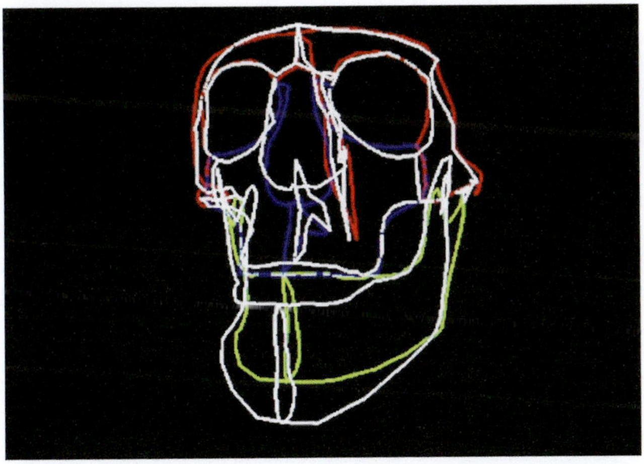

Fig. 16. Size-standardized superimposition of the craniofacial complex of external subject (Crouzon) with the normalized form (White: External Subject, Colored: Norm). For the size standardized superimposition the same shape differences can be observed.

Table 11. Non-size-standardized Superimposition of the Complex, External Subject

Pt.	Subject Coordinates			Deviation of Coordinates			Deviation of Coordinates In STD DEV			Dev*	Dev in STD DEV**
	x	y	z	x	y	z	x	y	z		
1	168.92	116.97	32.28	-0.52	-9.88	1.13	-0.43	-4.70	0.26	9.96	4.47
2	170.78	48.07	59.09	0.14	-3.24	0.65	0.35	-2.34	0.37	3.31	2.97
3	124.33	67.16	57.75	0.38	2.07	4.90	0.19	1.11	4.76	5.34	3.54
4	170.62	51.04	11.98	0.00	4.06	6.46	0.00	1.43	5.20	7.63	5.00
5	169.75	58.70	4.85	-0.63	11.64	6.01	-0.68	8.69	3.90	13.11	12.53
6	137.09	63.47	26.37	0.55	4.52	-1.85	0.20	3.73	-1.08	4.91	3.33
7	153.07	63.93	20.19	-4.84	10.55	2.61	-7.40	7.60	1.13	11.90	6.47
8	168.18	42.77	-51.98	-1.27	-6.53	-13.64	-3.91	-3.62	-8.84	15.17	12.65
9	168.32	42.74	-68.69	-0.99	-4.82	-15.99	-1.76	-2.50	-6.39	16.73	8.54
10	168.44	49.82	-73.59	-0.77	-2.19	-15.43	-1.22	-1.25	-7.86	15.60	9.78
11	124.19	101.25	-37.95	0.44	-11.33	-1.12	0.19	-2.88	-0.57	11.40	4.15
12	124.68	115.60	26.93	9.11	-6.37	3.44	3.01	-6.22	1.47	11.64	5.19

The numbers (1-12) in the first column on the left refer to the landmarks (Pt.): 1 = Basion, 2 = Nasion, 3 = Right Fronto-zygoma, 4 = ANS, 5 = A-Point, 6 = Right Orbitale, 7 = Right Lateral Nasal Cavity, 8 = B-Point, 9 = Pogonion, 10 = Menton, 11 = Right Gonial Angle, 12 = Right Condylar head.

*Dev. = distance deviation in 3D

**Dev. in STD DEV = distance deviation in terms of STD DEV units

Table 12. Size-standardized Superimposition of the Complex of the External Subject

Pt.	Subject Coordinates			Deviation of Coordinates			Deviation of Coordinates In STD DEV			Dev*	Dev in STD DEV**
	x	y	z	x	y	z	x	y	z		
1	168.92	116.56	32.08	-0.51	-10.29	0.94	-0.44	-3.56	0.22	10.35	4.25
2	170.76	48.35	58.62	0.13	-2.96	0.19	0.33	-1.78	0.11	2.96	2.55
3	124.79	67.25	57.30	0.84	2.16	4.45	0.90	1.57	3.81	5.02	6.15
4	170.61	51.29	11.99	-0.01	4.31	6.47	-0.01	1.44	5.22	7.77	4.73
5	169.74	58.88	4.94	-0.64	11.81	6.09	-0.69	7.87	4.12	13.30	11.72
6	137.41	63.60	26.23	0.87	4.64	-1.99	0.41	3.79	-1.28	5.12	3.69
7	153.24	64.06	20.12	-4.68	10.67	2.53	-7.34	7.79	1.12	11.92	6.52
8	168.19	43.11	-51.33	-1.26	-6.19	-12.98	-3.75	-3.17	-11.90	14.44	12.96
9	168.33	43.07	-67.87	-0.98	-4.48	-15.16	-1.66	-2.20	-10.42	15.84	11.44
10	168.45	50.08	-72.72	-0.76	-1.93	-14.55	-1.17	-0.96	-13.58	14.70	13.57
11	124.65	100.99	-37.44	0.90	-11.59	-0.61	0.42	-3.39	-0.31	11.64	4.54
12	125.13	115.20	26.79	9.56	-6.77	3.30	3.91	-3.56	1.28	12.17	5.42

The numbers (1-12) in the first column on the left refer to the landmarks (Pt.):
1 = Basion, 2 = Nasion, 3 = Right Fronto-zygoma, 4 = ANS, 5 = A-Point, 6 = Right Orbitale, 7 = Right Lateral Nasal Cavity, 8 = B-Point, 9 = Pogonion, 10 = Menton, 11 = Right Gonial Angle, 12 = Right Condylar head.
*Dev. = distance deviation in 3D
**Dev. in STD DEV = distance deviation in terms of STD DEV units

Surface Quantification

3D skull average

The sample population consisted of 10 female patients ranging in age from 19 to 35. The 10 CBCT scans were entered into the developed software and subjected to the methods proposed earlier, including surface segmentation, topology correction, and shape registration. The resulting average is displayed below in Fig. 17.

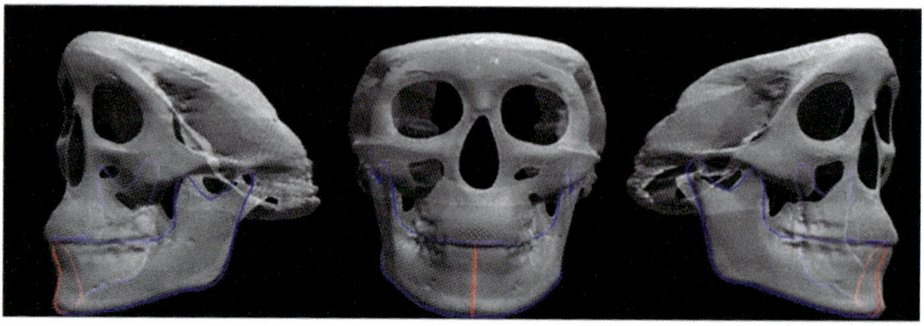

Fig. 17. The skull average (n = 10) in left lateral, frontal, and right lateral views. Note that the resulting average shows a relatively smooth and harmonious skull model. All areas here, except the actual surface, have been extracted through the protocol described in this chapter.

3D superimposition

Following the creation of an average skull model from the sample of 10 patients, the same registration technique was used to validate the ability to superimpose a patient skull on the average model. This technique is possible because even with the average skull, a mesh framework is maintained which retains point-to-point correspondence of vertices between shapes. Thus, correspondence can be established between each vertex of the average skull and each vertex of the sample patient.

Displacement maps (Fig. 18) allow clinicians to compare anatomical structures of a given patient to a demographically-matched average for diagnostic and treatment planning purposes. The users can zoom in and out of relevant areas of the skull to more closely evaluate variation at a given point or region. Opacity of the average skull can be adjusted to permit visualization of the inner regions, or areas where the sample skull lays within the average. The images below in Fig. 18 illustrate the relative ease in understanding the maxillomandibular relationship, the growth pattern, and other key features of a patient's skeletal morphology.

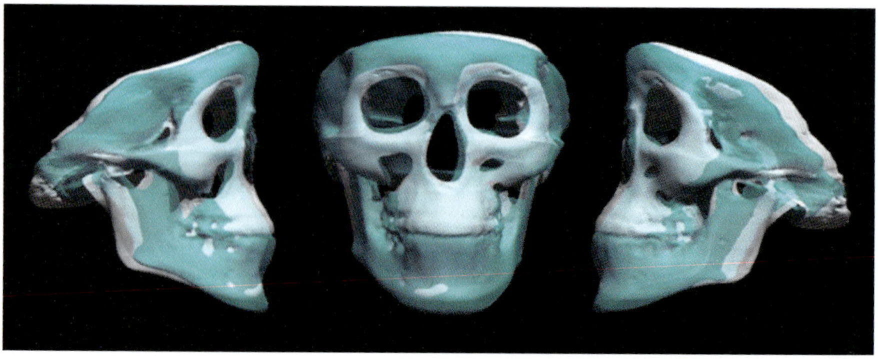

Fig. 18. The skull average (n = 10) is represented in white while a sample patient skull is represented in green. Anatomical variation from the average is now easily detectable.

Size standardization

The default of the established protocol in generating an average model was to use size-standardization. Thus, the average allowed for evaluation of shape changes, irrespective of the size of each sample skull. This resulted because the 7-parameter Procrustes alignment was used as the final stage of registration, in which two objects were optimally superimposed by rotation, translation, and uniform scaling in order to maximize correspondence. This scaling of the objects resulted in a size-standardized average. However, the creation of non-size-standardized averages was also possible. This was accomplished by taking the average determinant of each Procrustes matrix for all shapes and multiplying the size-standardized shape by this average. The result represented an accurate representation of the size and shape of the average skull. The differences in size and shape, versus shape only are displayed in Fig. 19. One may opt for either one method or the other depending upon the circumstances.

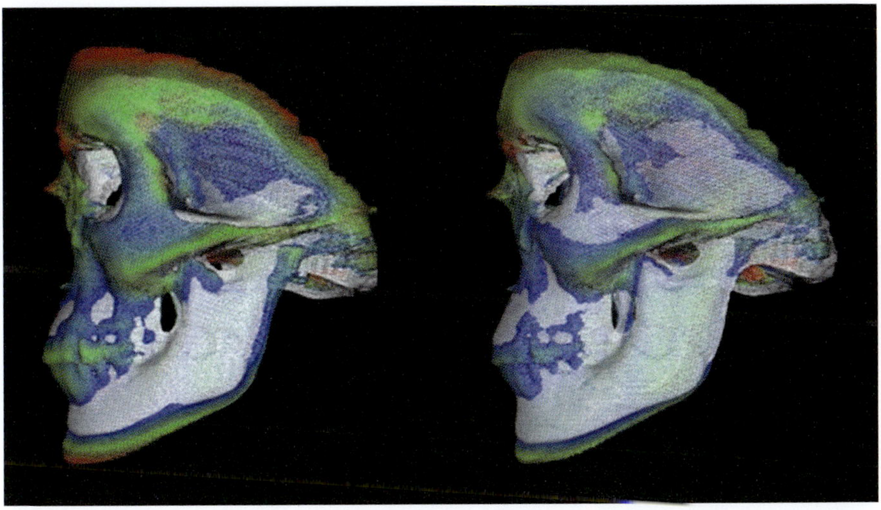

Fig. 19. Relative comparison of the non-size-standardized (left) and size-standardized (right) average skull is displayed. When size is taken into account, the average skull is multiplied by the average determinant of each Procrustes matrix used in the averaging process, resulting in a model that accurately represents both size and shape.

DISCUSSION AND CONCLUSIONS

In orthodontics and orthognathic surgery, one of the primary goals is to arrive at a proper diagnosis and an appropriate treatment plan. For decades, clinicians have relied upon 2Dcephalometric films and the normative data, based on linear and angular measurements, for a better understanding of the patient's skeletal and dental relationships. In recent years, there has been a steady increase in the use of 3D CBCT images, particularly in complex cases involving severe skeletal and dental anomalies. Despite this advancement in 3D technology, little progress has been made in developing new analyses that take full advantage of the 3D information provided by CBCT images. Most studies have applied the 2D linear and angular measures to 3D images in an attempt to better define structures in 3D space but without much success in capturing the true anatomical 3D form.

Normalization of a group of subjects is an essential step in the analysis and comparison of the irregularities or deviations present in an individual subject. Traditionally in cephalometrics, the normalized data of a population was given as numerical values derived from multivariate analysis based on edge-superimposition. Due to the unique characteristics of different anatomic structures during individual growth and development, this technique poses inherent problems. This chapter attempted to reconstruct the entire craniofacial complex by combining boundary and surface quantifications.

The boundary quantification reconstructed the entire craniofacial complex by combining the normalized components of three different segmented anatomical structures. The process of superimposition and normalization of the components inevitably accompanies an incoherent affine transformation of the components that result in their separation and misalignment in space. This irregular arrangement of the normalized components is shown in Fig. 20. The components were highly coordinated after the inverse transformation method was applied, which provided for a reconstructed craniofacial complex (Fig. 21). Using this method, the entirety of craniofacial structures is averaged providing for a better fit. Most importantly, each component corresponding to cranial base, maxilla, and mandible is a normalized form in itself. This 3D averaging process of the human craniofacial morphology has never been previously accomplished. This model will serve as the first 3D norm for comparing either individual components or the whole craniofacial complex.

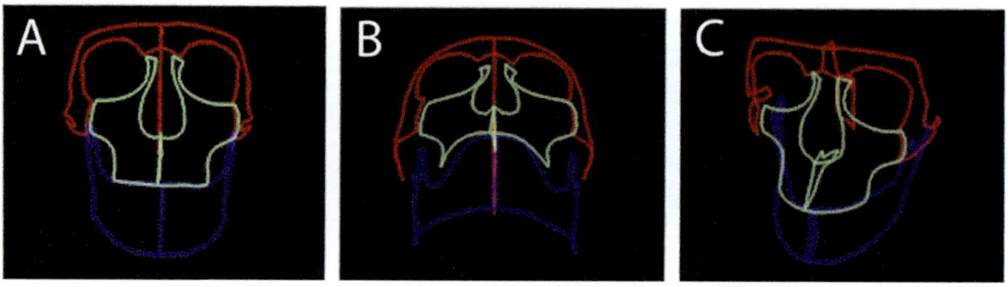

Fig. 20. Averaged components before realignment.

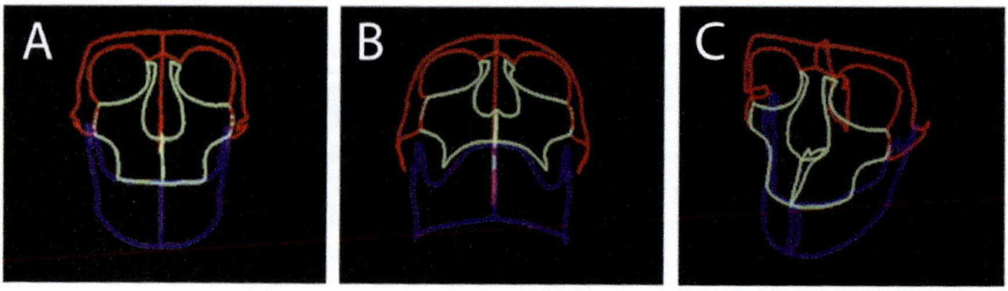

Fig. 21. Averaged complex reconstructed with the inverse transformation method.

The superimposition of each component showed the anatomical variation of the cranial base, maxilla, and mandible, while the superimposition of the whole craniofacial complex shows the morphologic variations when the relative positioning of the components is also taken into consideration. The comparison was available not only for defining the small variation of an internal subject (Case 1) but also for the great deviation of an external subject (Crouzon) from the norm.

Non-size-standardized and size-standardized superimpositions were generated in this study. With non-size-standardized superimposition, the procedure allowed rigid transformations of the structures without size correction, and it is useful when the original size of the structures need to be preserved, as in growth studies. On the other hand, size-standardized superimposition eliminates the size factor, and the variation of shape alone is clearly revealed. This new approach in boundary quantification has the following advantages:

1. The new method has eliminated the excessive dependence on limited or sparse landmarks; the analysis considers each boundary point with equal importance.

2. The geometric figure used in the analysis represents the actual shape. The craniofacial structure remains intact as a 3D entity, and the shape is preserved throughout the process.

3. The analysis provides for superimposition of a normative form with individual subjects.

4. The analysis allows for statistical results to determine the distinct characteristics of an individual subject or of a sample of cases.

In this analysis, every point in a 3D structure is used in the registration process for the superimposition. This is useful for finding an overall deviation of the structure, when there is no reliable information of stable points. However, for a subject that has a localized defect, the irregularity is evened out over the entire structure and therefore does not show the desired regional contrast. This problem can be managed only if selective boundary points can be used for the Procrustes superimposition. Another limitation of the boundary quantification by itself is that the surface topology of the craniofacial anatomy, which is also an important part in evaluating the facial structure, is ignored.

Surface based 3-dimensional averages of the human skull have demonstrated the potential for creating average skulls specific to a patient's age, gender, ethnicity, and other unique characteristics. As with the boundary quantification, the ability to superimpose a patient's surface map on the normative model and to produce colorized displacement maps will allow for 3D evaluation of the morphological variation of a given patient in greater detail. This novel approach may prove to be particularly useful in cases with notable asymmetries, craniofacial anomalies, or extreme growth patterns in which visual inspection and 2D analyses may be inadequate. This technology permits point-to-point correspondence mapping to evaluate variability at specific vertices, and it also allows for volumetric evaluation of regional variation, evaluation of mean difference between two samples, and tensor-based morphometric evaluation [28]. The following conclusions can be drawn from the surface quantification:

1. This study developed the protocol necessary to segment the surface of the human skull, correct the topology, and register skulls of highly variable topology utilizing a spherical patch mapping approach.

2. Ten skulls from patients with varying demographic and skeletal characteristics were then successfully averaged using the 7-parameter Procrustes alignment.

3. A sample patient skull was superimposed on this average, creating a colorized displacement map and illustrating the potential clinical utility of this protocol in diagnosis and treatment planning of patients with complex skeletal conditions.

These analyses are parts of the development of a comprehensive morphologic analysis. Ultimately, the boundary and surface quantification will be combined to establish the final 3D analysis of the human skull. A schematic picture of this is rendered in Fig. 22.

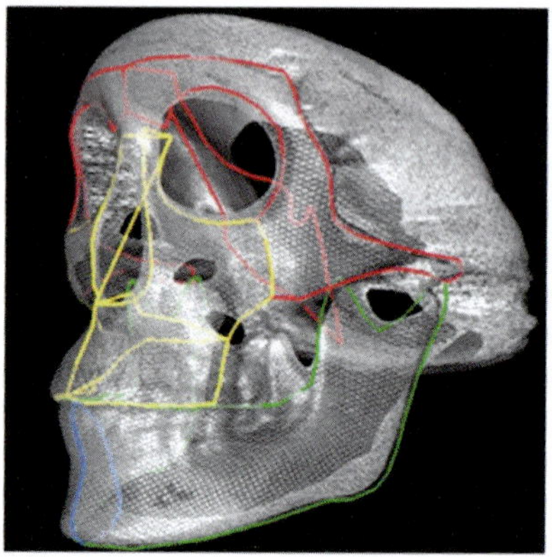

Fig. 22. An example of a normalized 3D craniofacial structure with both the surface and boundary attributes displayed.

With these novel approaches, averages of the craniofacial components within the craniofacial complex can be obtained, utilizing both the boundary and surface information. The normative craniofacial structure can then be superimposed with those of individual subjects for comparison. These approaches provide a basis for a new morphologic analysis that fulfills the requirement of a true 3D analysis and eliminates the limitations of conventional cephalometry.

ACKNOWLEDGEMENTS

The senior author wishes to thank Charles Wolfe for his programming efforts and Pete Lestrel for his editing of this chapter.

REFERENCES

[1] Miyashita K and Dixon AD (Eds.) (1996) *Contemporary Cephalometric Radiography*. Quintessence Publishing Co., Tokyo.

[2] Athanasiou AE (Ed.) (1996) *Orthodontic Cephalometry*. Mosby, Chicago.

[3] Harrell WE (2003) Limitations of Two Dimensional Cephalometric Analysis in Orthodontic Diagnosis and Treatment Planning: The need for three dimensional diagnosis. *Am Assoc Dent Maxillofac Radiograph Tech Newsletter*.

[4] Bholsithi W, Tharanon W, Chintakanon K, Komolpis R, and Sinthanayothin C (2009) 3D vs. 2D cephalometric analysis comparisons with repeated measurements from 20 Thai males and 20 Thai females. *Biomed Imaging Intervention J* **5** (Published online).

[5] Gribel BF, Gribel MN, Frazão DC, McNamara JA, and Manzi FR (2011) Accuracy and reliability of craniometric measurements on lateral cephalometry and 3D measurements on CBCT scans. *Angle Orthod* **81**: 26-35.

[6] Matteson SR, Bechtold W, Phillips C, and Staab EV (1989) A method for three-dimensional image reformation for quantitative cephalometric analysis. *J oral Maxillofac Surg* **47**: 1053-1061.

[7] Park SH, Yu HS, Kim KD, Lee KJ, and Baik HS (2006) A proposal for a new analysis of craniofacial morphology by 3-dimensional computed tomography. *Am. J Orthod Dentofacial Orthop* **129**: 601-604.

[8] Wong RWK, Chau ACM, and Hägg U (2011) 3D CBCT McNamara's cephalometric analysis in an adult southern Chinese population. *Int J Oral Maxillofac Surg* **40**: 920-925.

[9] Yáñez-Vico RM, Iglesias-Linares A, Torres-Lagares D, Gutiérrez-Pérez JL, and Solano-Reina E (2010) Three-dimensional evaluation of craniofacial asymmetry: an analysis using computed tomography. *Clinical Oral Invest* **15**: 729-736.

[10] Wong RWK, CACM, and Hagg U, 3D CBCT (2011) McNamara's cephalometric analysis in an adult southern Chinese population. *Int. J. Oral Maxillofac Surg* **40**: 920-925.

[11] de Moraes MEL, HLG, Chen CSK, Moraes LC, and Balducci I (2011) Evaluating craniofacial asymmetry with digital cephalometric images and cone-beam computed tomogrphy. *Am J Orthod Dentofacial Orthop* **139**: 523-531.

[12] Gribel BF, GMN, Manzi FR, Brooks SL, and McNamara Jr JA (2011) From 2D to 3D: an algorithm to derive normal values for 3-dimensional computerized assessment. *Angle Orthodont* **81**: 3-10.

[13] Brown AA, SWC, Scheetz JP, Silveira AM, and Farman AG, (2009) Linear accuracy of cone beam CT derived 3D images. *Angle Orthodont* **79**: 150-157.

[14] Bayome M, PJH, and Kook YH (2013) New three-dimensional cephalometric analyses among adults with a skeletal class I pattern and normal occlusion. *Korean J Orthod* **43**: 62-73.

[15] Zumpano MP and RJ (2003) Growth-related shape changes in the fetal craniofacial complex of humans (Homo sapiens) and pigtailed macaques (Macaca nemestrina): a 3D-CT comparative analysis. *Am J Phys Anthropol* **120**: 339-351.

[16] Nur MKS, Bayram M, Celikoglu M, Kilkis D, and Sezgin O (2012) Conventional frontal radiographs compared with frontal radiographs obtained from cone beam computed tomography. *Angle Orthodont* **82**: 579-584.

[17] van Vlijmen, KM, Berge S, Schols J, Maal T, Breuning H, and Kuijpers-Jagtman, (2012) Evidence Supporting the Use of Cone-Beam Computed Tomogrphy in Orthodontics. *J Am Dent Assoc* **143**: 241-252.

[18] Moon W (1997) A numerical and visual approach for measuring the effects of functional appliance therapy: fourier descriptors. **In**: *Fourier Descriptors and their Applications in Biology.* PE Lestrel (Ed.) New York: Cambridge University Press.

[19] Lestrel PE, Read DW, and Wolfe CA (1997) Size and shape of the rabbit orbit: 3-D Fourier descriptors. **In**: *Fourier Descriptors and their Applications in Biology.* PE Lestrel (Ed.) New York: Cambridge University Press.

[20] Kullar R, Lestrel PE, Moon W, and Wolfe CA (2013) Representation of the mandible as a curve in 3-space: A preliminary study using Fourier descriptors. **In**: *Biological Shape Analysis: Proceedings of the 2nd international symposium of biological shape analysis*, PE Lestrel (Ed.), Singapore: World scientific.

[21] Enlow DH and Hans MG (1996) *Essentials of Facial Growth.*, W.B. Saunders Company: Philadelphia.

[22] Marcus JR, Domeshek LF, Das R, Marshall S, Nightingale R, Stokes TH, and Mukund S Jr (2008) Objective three-dimensional analysis of cranial morphology. *Eplasty* **8**. Accessed from the Internet August 20, 2012 at http://www.ncbi.nlm.nih.gov/pmc/articles/PMC2323201/.

[23] Marcus JR, Domeshek LF, Loyd AM, Schoenleber JM, Das RR, Nightingale RW and (2009) Use of a Three-Dimensional, Normative Database of Pediatric Craniofacial Morphology for Modern Anthropometric Analysis. *Plastic Reconstruct Surg* **124**: 2076-2084.

[24] Gower JC and Dijksterhuis GB (2005) *Procrustes problems. Psychometrika* **70**: 799-801.

[25] Joshi SH, Klassen E, Srivastava A, and Jermyn I (2007) An efficient representation for computing geodesics between n-dimensional elastic shapes *IEEE Proc Comput Vision Patt Recog*.

[26] Kurtek S, Klassen E, Ding Z, Avison M, and Srivastava, A (2011) Parameterization-invariant shape statistics and probabilistic classification of anatomical surfaces. *Proc Info Processing Med Imaging* (IPMI). **22**: 147-158.

[27] Karcher H (1977) Riemannian center of mass and mollifier smoothing. *Commun Pure Appl Math* **30**: 509-541.

[28] Woods RP (2003) Characterizing volume and surface deformations in an atlas framework: theory, applications, and implementation. *Neuroimage* **18**: 769-788.

[29] Everson R (1997) Orthogonal, but not Orthonormal, Procrustes Problems. *Adv Comput Math* (available online).

[30] Thompson PM, Hayashi KM, De Zubicaray GI, Janke AL, Rose SE, Semple J, Hong MS, Herman DH, Gravano D, Doddrell DM, and Toga AW (2004) Mapping hippocampal and ventricular change in Alzheimer disease. *Neuroimage* **22**: 1754-1766.

[31] Kau CH, Richmond S, Zhurov A, Ovsenik M, Tawfik W, Borbely P, and English JD (2010) Use of 3-dimensional surface acquisition to study facial morphology in 5 populations. *Am J Orthod Dent Orthop* **137**: Suppl S56.

APPENDICES

Appendix I. Point Coordinate Text File of Subject 1 (Cranial Base/Zygoma)

Landmark Name---		--X--	--Y--	--Z--
1	Frontal	165.1	44.0	50.5
2	RFrontal	156.2	43.6	50.6
3	RFrontal -1	146.4	45.8	50.4
4	RFrontal -2	136.4	48.0	49.6
5	RFrontal -3	126.9	52.3	48.7
6	RFrontal -4	118.9	59.1	47.0
7	RTemporal	115.9	59.4	42.3
8	RTemporal -1	113.4	62.3	34.7
9	ROutZygoFront	112.3	66.3	28.2
10	ROutZygoFront -	110.7	69.5	25.0
11	ROutZygoFront -	111.3	69.5	20.5
12	ROutZygoFront -3	112.0	70.5	15.8
13	RUpZygoma	111.2	73.1	12.3
14	RUpZygoma -1	107.6	79.6	9.2
15	RUpZygoma -2	105.1	85.4	8.8
16	RUpZygoma -3	102.3	95.5	7.3
17	RUpZygoma -4	102.3	100.3	7.1
18	RUpZygoma -5	105.2	114.4	7.4
19	RLowZygoma	107.9	116.4	4.2
20	RLowZygoma -1	106.8	107.7	6.9
21	RLowZygoma -2	104.1	104.7	2.5
22	RLowZygoma -3	104.8	96.8	3.6
23	RLowZygoma -4	105.2	90.6	4.2
24	RLowZygoma -5	106.6	85.6	3.5
25	RLowZygoma -6	108.9	79.9	0.9
26	RLowZygoma -7	111.7	76.3	-1.0
27	RLowZygoma -8	115.7	73.1	-6.0
28	RLowZygoma -9	119.1	70.9	-7.8
29	RKeyridge	123.5	67.4	-9.2
30	RZygoMax -3	124.6	63.7	-4.9

(Columns omitted to save space)

103	Sella -2	164.0	107.6	23.4
104	Sella -3	164.6	109.6	22.4
105	Sella -4	165.7	112.7	21.0
106	Sella -5	163.9	114.8	29.4
107	PostCranial	164.0	116.3	27.4
108	PostCranial -1	164.2	116.6	22.1
109	PostCranial -2	165.1	117.7	14.3
110	PostCranial -3	166.0	122.0	-0.1
111	Basion	166.6	126.3	-15.6
112	LowEthmoid	167.4	112.7	-2.7
113	LowEthmoid -1	165.9	102.6	-0.3
114	LowEthmoid -2	164.9	89.6	5.4
115	LowEthmoid -3	164.5	81.4	13.5
116	LowEthmoid -4	163.6	83.9	22.6
117	UpEthmoid	163.8	84.6	31.9

Appendix II. Point Coordinate Text File of Subject 1 (Maxilla)

Landmark Name---		--X--	--Y--	--Z--
1	RInfOrbitalrim	155.7	50.8	36.7
2	RInfOrbitalrim -1	157.3	53.4	31.0
3	RInfOrbitalrim -2	157.3	53.1	26.1
4	RInfOrbitalrim -3	156.3	52.6	20.9
5	RInfOrbitalrim -4	153.0	52.9	16.9
6	RInfOrbitalrim -5	149.1	53.4	14.1
7	RInfOrbitalrim -6	143.9	53.7	12.1
8	RInfOrbitalrim -7	138.8	54.0	9.9
9	ROrbitale	132.8	55.2	8.5
10	RZygoMax	127.8	56.9	7.9
11	RZygoMax -1	128.2	59.6	3.9
12	RZygoMax -2	126.6	61.6	-0.1
13	RZygoMax -3	124.6	63.7	-4.9
14	RKeyridge	123.5	67.4	-9.2
15	RKeyridge -1	128.8	68.8	-10.5
16	RKeyridge -2	134.1	66.8	-13.5
17	RKeyridge -3	138.3	66.9	-16.2
18	RKeyridge -4	141.2	65.9 -	20.5
19	RKeyridge -5	142.5	64.9	-26.4
20	RU6	141.9	63.6	-32.9
21	RU6-1	144.2	58.6	-34.8
22	RU6-2	144.7	51.5	-36.2
23	RU6-3	147.6	47.9	-35.9
24	RU6-4	150.4	44.0	-35.0
25	RU6-5	156.7	41.1	-34.1

(Columns omitted to save space)

71	ANS	168.5	44.0	-9.8
72	ANS -1	176.5	50.1	-9.5
73	ANS -2	178.6	51.2	-6.3
74	ANS -3	179.0	50.4	-1.4
75	LPiriform	178.7	49.3	5.0
76	LPiriform -1	177.8	47.5	11.4
77	LPiriform -2	175.8	49.3	15.3
78	LPiriform -3	173.0	48.2	19.9
79	LPiriform -4	170.6	49.3	25.9
80	LPiriform -5	170.0	50.4	29.8
81	LPiriform -6	169.7	48.2	35.9
82	LInfOrbitalrim	175.3	54.2	34.2
83	ANS	168.5	44.0	-9.8
84	Palate	165.6	58.8	-0.7
85	IncisCan	167.1	62.0	-7.9
86	IncisCan -1	166.4	74.9	-6.6
87	PNS	166.9	89.2 -	13.3
88	PNS -1	168.2	72.4 -	11.7
89	PalatalRidge	167.0	62.6 -	13.2
90	PalatalRidge -1	166.8	57.4 -	16.8
91	PalatalCEJ	168.0	51.5	-24.6
92	Uppermid	168.6	39.6 -	35.1

Appendix III. Point Coordinate Text File of Subject 1 (Mandible)

Landmark Name---		--X--	--Y--	--Z--
1	RtTopCondyle	110.5	113.6	6.0
2	RtTopCondyle -1	113.9	110.5	6.6
3	RtTopCondyle -2	112.7	108.9	4.5
4	RtTopCondyle -3	110.2	107.7	-2.3
5	RtCorNotch	118.8	97.1	-11.1
6	RtCorNotch -1	118.9	87.9	-5.0
7	RtCorNotch -2	118.2	86.6	-1.3
8	RtCorTip	118.5	84.9	0.2
9	RtCorTip -1	119.9	83.2	-1.0
10	RtCorTip -2	121.8	84.2	-10.2
11	RtCorTip -3	124.5	85.0	-16.7
12	RtCorTip -4	125.7	85.0 -	23.1
13	RtAlvBorder	127.9	84.6 -	29.1
14	RtAlvBorder -1	132.1	86.1	-34.4
15	RtAlvBorder -2	140.4	74.1	-35.7
16	DistalLR6	144.1	69.6 -	34.9
17	DistalLR6 -1	147.8	64.9 -	33.4
18	MesialLR6	150.3	62.7 -	35.0
19	MesialLR6 -1	152.1	58.9 -	35.2
20	MesialLR6 -2	153.4	52.9 -	33.7
21	LR3Tip	156.4	47.4 -	33.4
22	LR3Tip -1	160.7	45.0 -	33.5
23	LR3Tip -2	164.4	44.3 -	32.9
24	DentMidline	168.9	43.2 -	32.6
25	DentMidline -1	169.0	42.5 -	36.2
26	CEJ	169.3	43.2 -	39.5
27	CEJ -1	169.6	44.3 -	44.3
28	CEJ -2	169.6	46.0 -	47.7
29	B-point	169.8	46.5 -	52.1
30	B-point -1	169.8	46.1 -	55.8
71	MesialLL6 -1	186.3	59.3 -	31.8
72	MesialLL6 -2	184.6	54.1 -	32.5
73	LL3Tip	182.5	48.1 -	32.1
74	LL3Tip -1	177.9	46.1 -	32.2
75	LL3Tip -2	172.5	45.1 -	31.9
76	DentMidline	168.9	43.2 -	32.6

(Columns omitted to save space)

94	RtGonAngle -1	121.3	104.8	-50.4
95	RtGonAngle -2	121.7	108.7	-44.2
96	RtGonAngle -3	119.2	108.2	-34.1
97	RtGonAngle -4	115.9	109.3	-24.9
98	RtConNeck	112.7	110.9	-16.0
99	RtConNeck -1	109.8	111.9	-12.5
100	RtConNeck -2	107.7	113.4	-8.6
101	RtConNeck -3	106.5	114.9	-4.4
102	RtConNeck -4	106.2	114.8	-0.9
103	RtConNeck -5	108.4	115.3	2.8
104	RtTopCondyle	110.5	113.6	6.0

A Quantitative Analysis of Whole Body Outlines: Fourier Descriptors

PETE E. LESTREL[1], NORIKO MIYAKE[2], MASAMI ISHIHARA[3], CHARLES A. WOLFE[4] and ALBERT BODT[5]

Abstract

Studies of the human body form based on standardized photos, were initiated by W. H. Sheldon in the 1940's. These led to the familiar anthropometric archetypes of ectomorph, endomorph, and mesomorph to describe body physique. Recent studies of body form are either largely descriptive or based on indirect measures like the BMI. Quantitative assessments of total body images have been rarely attempted because of the lack of available methods. In contrast, Fourier Descriptors (FDs) provide for: (1) a precise quantitative as well as visual depiction of body outlines and (2) an analysis of shape thereby controlling for the effect of size. Elliptical Fourier functions (EFFs), a particular FD, was used to fit the body shape of Japanese female college students in the frontal and lateral views (n = 144). The mean age was 19.63±1.15 years. The goodness-of-fit or residual of the EFF with the actual observed frontal outline was 0.96±0.04 pixel units (roughly 0.17± 0.01 mm). Similar results were found with the lateral view. Using the BMI as a criterion, comparison of the body shape of the 10th percentile with the 90th percentile produced statistically significant results utilizing a MANOVA and displayed clear body outline differences in both the frontal and side views. Superimposition approaches were implemented to provide a visual assessment. These include size-standardized data and Procrustes superimposition. Assessment of these superimposed images was judged with a shape difference index (SDI) specifically designed for this purpose. It is proposed that whole body images analyzed with EFFs can provide useful information for: (1) the effects of nutrition and disease on body shape, (2) body perception studies, (3) body shape in health compromised situations such as *Anorexia Nervosa*, etc., (4) the quantification of developmental obesity, and (5) provide for the building of reference standards for studies of body shape.

INTRODUCTION

Body composition can be viewed as a continuum consisting of a large percentage of normal individuals and two extremes at the tails of the distribution. While obesity represents the extreme right of the continuum of body weight distribution, it is not the only problem group. At the extreme left of the body weight distribution are those who display extreme weight loss, a condition that can be life threatening. For both distribution extremes, the individuals involved tend to view the solution as one that, curiously enough, entails weight loss. That is, for those who are morbidly thin, the solution is still more weight loss. Clearly, this emphasis on weight loss implies a negative self-perception.

Body perception is generally defined in the way an individual perceives their *body image*. It reflects one's mental view of one's external appearance. An unduly

[1] Formerly, UCLA - School of Dentistry, Section of Oral Biology, Los Angeles, CA 90024-1668, USA, plestrel@earthlink.net or peter.lestrel@gmail.com.

[2] Department of Sport Psychology, Tokyo International University 2509 Matoba, Kawagoe-shi, Saitama 350-1198 Japan.

[3] Tokyo Metropolitan University, Graduate School of Science, Department of Health Promotion Sciences Minamiosawa 1-1, Hachioji-shi, Tokyo 192-0397, Japan.

[4] C. Wolfe Software Engineering, 13376 Dronfield Avenue, Sylmar, CA 91342, USA.

[5] Department of Medicine, Division of Nephrology, Kaiser Permanente, Lancaster, CA 93534, USA.

negative body image can lead to psychosomatic illnesses such as various eating disorders e.g., *Anorexia Nervosa*, *Bulimia Nervosa*, etc. In fact, with respect to *Anorexia Nervosa*, a serious mental disorder for which treatment is only moderately successful, the death rate is 10% [1, 2].

It has become increasingly apparent that obesity has reached epidemic proportions in the U.S. and other countries. Reasons for this were initially posited on poor health habits, lack of self-discipline, lack of exercise and frequent visits to fast foods restaurants. While partially correct, this view is nevertheless an oversimplification that has now been questioned with the realization that obesity is not only multifactorial but also quite difficult to treat [3].

Further, from a health perspective, it needs to be emphasized that malnutrition and its counterpart, morbid obesity, both carry serious health risks. In the case of obesity, this can lead to reduced life expectancy and increased health problems. Extreme weight loss, on the other hand, can lead to a higher risk of hip fracture in women, susceptibility to infection, organ dysfunction, and a higher mortality rate [4]. Excess fat can lead to Type II diabetes, hypertension, cardiovascular diseases, gall bladder disease, and cancer [5]. Thus, body perception can have significant psychological effects on how one perceives one's own body as well as influence how others view us [6, 7]. Body perception is also mediated by gender [8]. Finally, the conception of body image is also culturally influenced [5, 9, 10, 11, and 12].

Despite the emphasis of marketing on dieting, given that obesity is of epidemic proportions with 65% of Americans, 20 years or older, mildly to grossly obese [19]; there is some doubt as to the efficacy of such advertisements. Whether over the counter dietary products such as pills, have more than a slight effect remains questionable. There is, however, little argument that dieting under the care of a physician in conjunction with an exercise program can lower body weight. Moreover, in some cases surgery may be indicated; although, in many cases genetics may play a substantial role and mitigate some of these efforts.

Numerical assessments of the human body have been generally based on a set of disconnected measurements, or a set of indirect measures such as height, weight, BMI, etc. Quantitative assessments of the shape of the human body as an entire unit have rarely been attempted because of the paucity of available methods. One of the early workers attempting to quantify the total body composition or physique was W. H. Sheldon who, despite a number of serious deficiencies, developed the technique of somatology.

Development of Somatology

To alleviate this problem, W. H. Sheldon, in the 1940s, developed *somatology* as a classifying procedure of body physique by using standardized photographs in the frontal aspect.

From his work, the three-way typology composed of *ectomorphs* (thinness), *mesomorphs* (muscular), and *endomorphs* (fatness) was developed. Sheldon's attempt to quantify these three typological categories led to somatotyping. This involved a numerical triplet where the first number referred to thinness, the second one to muscularity and the third one to fatness. According to Sheldon, each somatotype contained aspects of all of the three body types and that such limits could be defined with following numbers: ectomorphs (117), mesomorphs (171), and endomorphs (771). For Sheldon these limits were intended to completely describe the variability found in human body physique (Fig. 1) [13, 14].

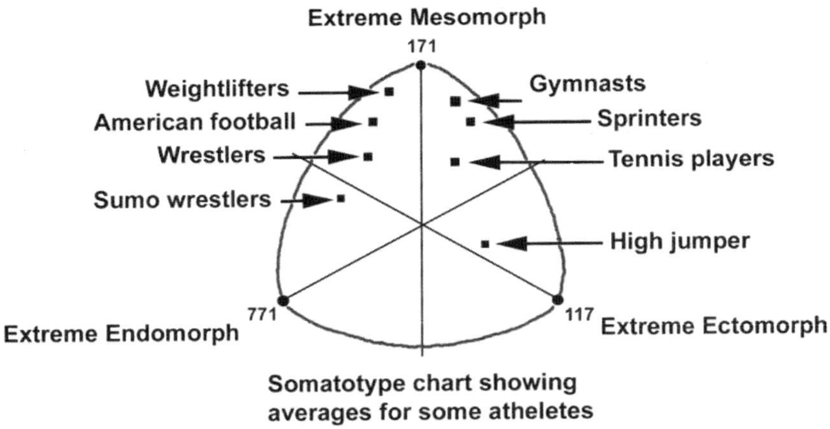

Fig. 1. Some averages for athletes computed using Sheldon's somatotyping approach (adapted from [15]).

Nevertheless, there are serious shortcomings with Sheldon's approach that need elucidation. These include the following objections: (1) the criteria that Sheldon proposed for the extreme somatotypes (117, 171, and 771) were largely conjectural and rarely existed with real data samples, (2) The system was closed and a somatotype was considered as effectively constant and did not materially change with age and other environmental factors, (3) The closed seven-point scale was too limited and does not mirror the actual variability encountered in body physique, (4) changes in the somatype did not reflect changes in weight, and (5) difficulty in predicting an older somatotype from one at a younger age. All these issues required considerable modifications in the original approach as advocated by Sheldon (Figs. 2 and 3) and these were subsequently developed by Carter and Heath [15].

Body physique then can be considered as a central component of body perception (body image) studies. However, body perception studies rarely use body images that reflect actual body shapes or use the subjects' own body physique. Moreover, one can argue that body shape, or more precisely body outline, plays an important role in compromised health conditions such as *Anorexia Nervosa*, and in developmental obesity. Thus, a drawback with both Sheldon's somatological approach and body perception studies in general, is that neither approach is actually based on a depiction or visualization of the actual total observed body outline.

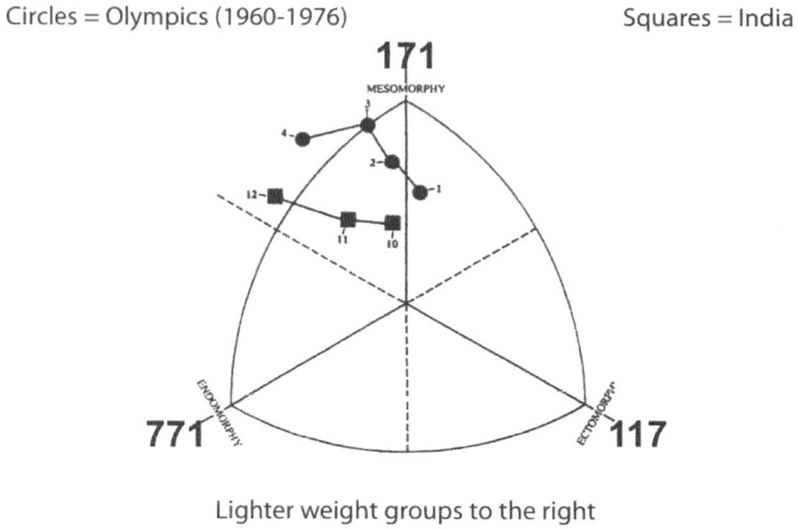

Fig. 2. An example of wrestling somatotypes is shown here. Note that two of the heavier groups fall outside of Sheldon's proscribed limits (modified from [15]).

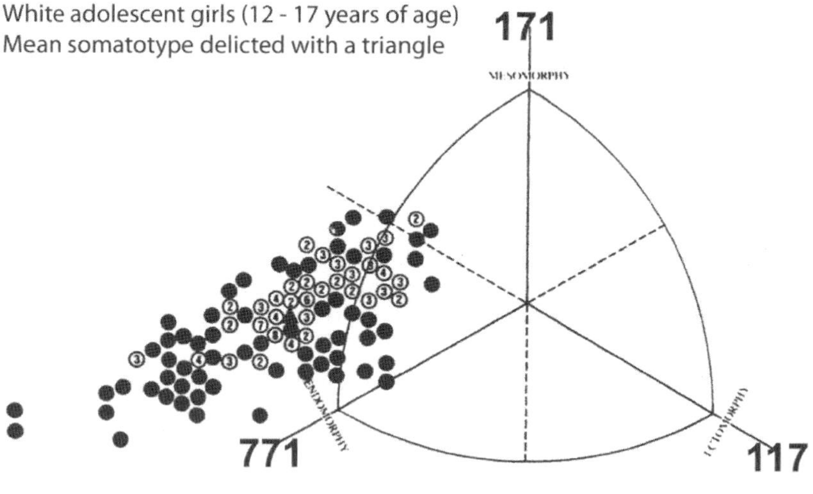

Fig. 3. Somatotypes of a sample (n = 180) of female obesity. The mean of the group, a black triangle, can be barely discerned here. Most of the group falls outside of Sheldon's limits again showing one of the serious limitations of the method (modified from [15]).

Many of the studies dealing with body perception and body satisfaction use illustrations (7-9 views) that represent a continuum of images in the frontal view starting with extreme thinness on the far left and moving to obese on the far right (Fig. 4). These range from an assumed idea of excessive thinness to obese [16, 17, 18, and 19]. In this approach, the subjects are asked to choose one of the figures that best characterize their mental view of what they would think or desire to look like. While this approach has produced useful data and continues to be widely used; it nevertheless, does not allow for the subjects to view their *own* body and judge specific aspects of it. Another limitation is that the scales being used may be too coarse resulting in: (1) limiting response options and (2) placing restrictions on the range of body outlines [20].

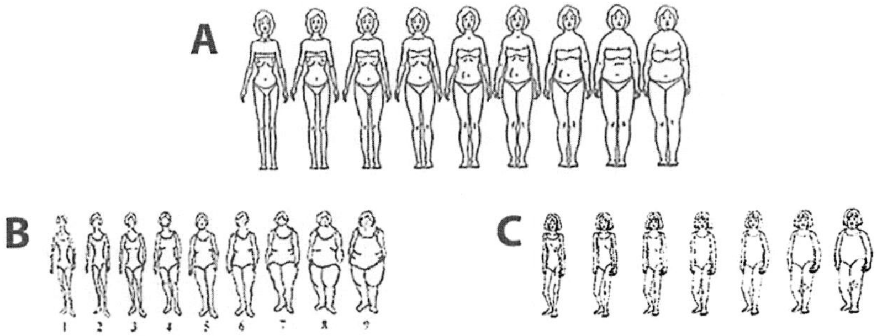

Fig. 4. Three diagrams commonly used to assess body satisfaction. The subject is asked to choose one of the figures that best characterize how they look or how they would like to look appear. (A) Adapted from [17], (B) adapted from [18], and (C) adapted from [19].

The Body Mass Index

The most common measure of body image remains the body mass index or BMI. It was developed by the Belgian Adolphe Quetelet in 1832 [21]. The BMI is defined as the ratio of weight divided by the height squared. However, it was not widely used in medical practice until the 1970's. Studies have suggested that the BMI tends to misrepresent individuals below 20 and above 27 [22]. Like many indices that reduce complex issues to a simple single measure, it provides a rough estimate of body composition and shows correlations with body measurements; however, like all indices, caution is indicated in its usage as it may give misleading results under some circumstances. A problem with the BMI is that only height is being considered in the calculation; that is, no estimate of either body width or depth is taken into account with the BMI. Also, since it is a ratio of weight divided by height squared; that is, kg/(ht in meters)^2, it does not distinguish muscle mass from fat tissue. Thus, individuals with high muscle mass (athletes) show BMI values that do not differ substantially from those with a predominance of body fat. Therefore, how the weight is actually distributed within the body outline plays a role that is not being accounted for.

A study of body fat in Asians versus Europeans, demonstrated that Asians have a lower BMI but a higher percent of body fat than whites. This led to the conclusion that BMI measurements of body fat can lead to substantial errors with respect to Asian populations [4]. This led others [23] to advocate that the BMI be based on two rather than just one component; That is, a fat mass index component (FM/ht^2) and a fat-free mass index component (FFM/ht^2). Fig. 5 displays two examples of female body shapes, one at the thin end and the other at the obese end of the body shape distribution with their respective BMI values (Fig. 5).

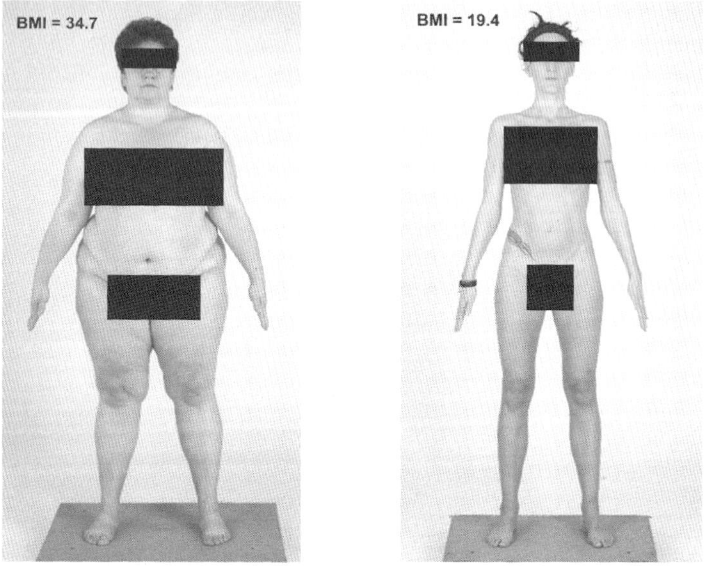

Fig. 5. Shown here are the body shapes of two individual females. These extreme body composition examples are taken from a large Swedish database[6] available on the Internet. The figure on the left has a BMI of 34.7 while the figure on the right has a BMI of 19.4.

Measurement of Body Outlines

One question that has arisen is the relationship of the mental image of body form to reality. That is, how well does one's body image perception relate to the actual body form. Body perception researchers have been dealing with this conundrum now for some time; although, most of the data used for this purpose has been qualitative and body perception studies tend to be viewed in descriptive terms.

Shifting now to issues associated with quantitative measurement, the central focus here, one can consider two basic aspects characteristic of any biological form: *size and shape*. Any distance between two selected points is considered a measure of size. Therefore, it has been presumed that size is an easily measured variable. For example, dealing with body outlines, consider a measurement such as waist width or thigh width. Such measures, when viewed separately or in isolation, cannot easily distinguish individuals; for example, a tall person compared to a shorter one may have exactly the same waist width. Thus, waist width or thigh width, or any other such distance, by itself, is of limited utility. What one really needs is a measure of the whole body outline and this requires a measurement of the total shape. One procedure that has been successfully applied to numerous diverse datasets is Fourier Descriptors (FDs). A specific FD, elliptical Fourier functions (EFFs) is utilized in the current study [24, 25, 26, and 27]. A recent study also applied EFFs to describe body outlines [28].

---

[6] Photos are available from www.3D.sk.

To illustrate the advantage of measuring whole body outlines using FDs, specifically, EFFs (to be described in detail in the Material and Methods section), Fig. 6 depicts the first harmonic ellipse superimposed on the centroid of the same two body outlines, displayed in Fig. 5. The black circles represent the EFF fit to the outline of the body outlines in Fig. 5, while the gray squares depict the first harmonic computed from the EFF. The ratio of the major to minor axis of these ellipses (shown inside each ellipse) can be considered as a rough index of the differences in body outline shape in an analogous way to the BMI.

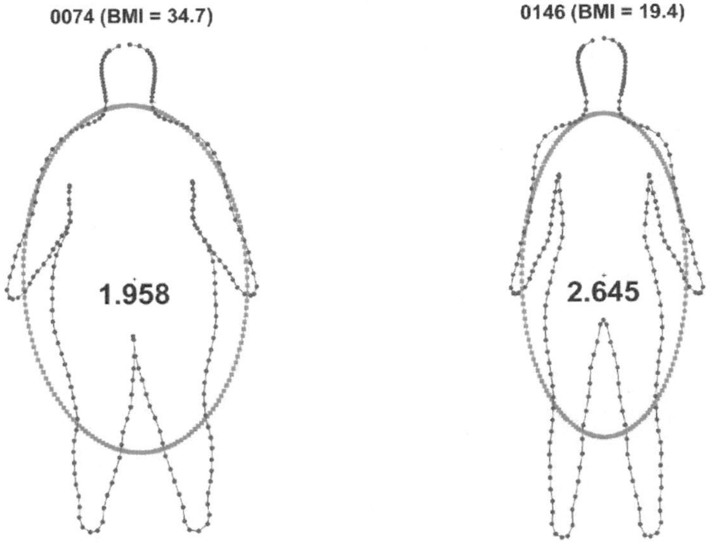

Fig. 6. Displayed here are the same two subjects as in Fig. 5 but now superimposed with the first harmonic ellipse generated from the FD (to be described subsequently). Each ellipse was superimposed on the centroid of the body outline. Also shown inside each ellipse, is the ratio of the length of the major axis to the minor axis.

MATERIALS AND METHODS

Data Collection

The data for current study was made available by Prof. Noriko Miyake now at Department of Sport Psychology, Tokyo International University. The data utilized was initially analyzed at the Department of Health Promotion Sciences (formerly, Department of Kinesiology), Tokyo Metropolitan University. A preliminary study based on this female data has been published in 2013 [29].

The purpose of that study included the: (1) development of a quantitative method to describe more completely the outline of the human body in the frontal and lateral views, (2) comparison of photos taken of subjects with their subsequent line drawing executed from memory and (3) evaluation of the relationships between photo and line drawing in an attempt to relate them to psychological and cultural influences. Of a total sample (n = 144) of female students available only a pilot sample (n = 37) was initially utilized and that 2011 study was limited to the frontal view. The present study has been expanded to include the total sample (n = 144) as well as both the frontal and lateral views.

Study Participants

The current study is based on a total sample of 144 Japanese female college students from four institutions. These four institutions where: (1) Tokyo Metropolitan University (n = 73), (2) Tokyo Women's College of Physical Education (n = 35), (3) Jissen Women's University (n = 25), and (4) Yokohama National University (n = 11). Two views in an anatomically standardized position of the whole body were available. One is the frontal view and the other is the lateral view.

Subject Photography

At the start of each photo session, each participant was asked to change into leotards designed for photography and had her photo taken in the frontal and lateral poses at a constant distance from the digital camera. Distance from camera to subject was held constant at three meters. Fig. 7 illustrates the photo of one of the participants with the frontal aspect on the left and the lateral one on the right.

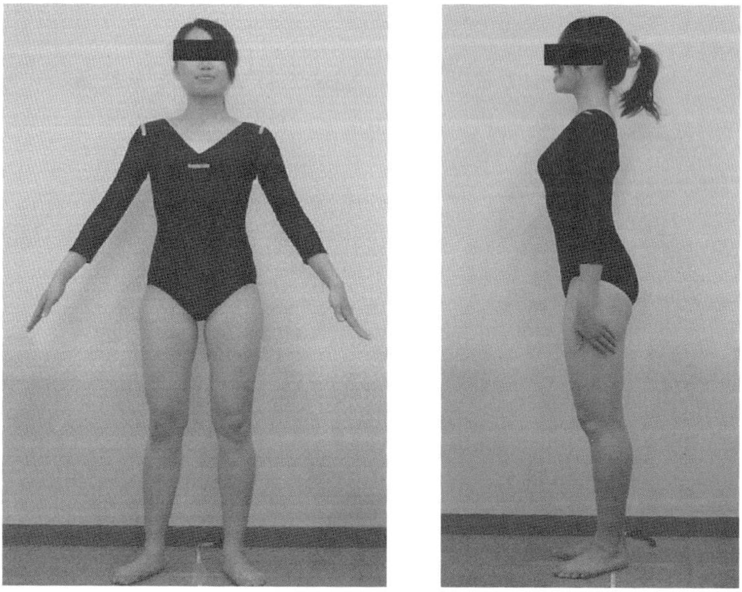

Fig. 7. Frontal and lateral photos of an individual subject.

Pre-processing of the Frontal and Lateral Digital Photographs

Before the elliptical Fourier function (EFF) of the frontal and lateral aspects of the body outline, can be computed and Fourier Descriptors (FDs) generated, a geometry has to be devised, which will allow the whole body outlines to be represented with a series of closely placed points. For the frontal view, this geometry is generated with an initial 16 points that define key aspects of the body outline as shown in Fig. 8. These 16 points are defined as follows: (1) top of head, (2, 3) left and right sides of the intersection of the face with the neck, (4, 5) left and right inflection points (maximum curvature) describing the shoulder, (6, 7) left and right inner arm junction with the body as seen on the frontal view, (8, 9) outer and inner delineation of the left wrist, (10, 11) inner and outer delineation of the right wrist, (12) junction of the left and right inner thighs, (13, 14) outer and inner delineation of the left ankle, and (15, 16) inner and outer delineation of the right ankle. These 16 points describe key aspects of the body outline that can be readily observed.

In an analogous way, the lateral view is defined with 11 points as follows: (1) top of head, (2) intersection of the lower mandible and the neck, (3) most anterior aspect of at the base of the head and adjoining neck, (4) most anterior position of the structure of the breast, (5) most posterior aspect of the upper back, (6) most posterior indentation at the hip aspect, (7) most posterior aspect of the backside, (8) most anterior aspect in the vicinity of the knee, (9) most posterior aspect of back of the lower leg, (10), intersection of the superior surface of the foot and the anterior surface of the lower leg, and (11) most anterior aspect of the posterior aspect of the ankle.

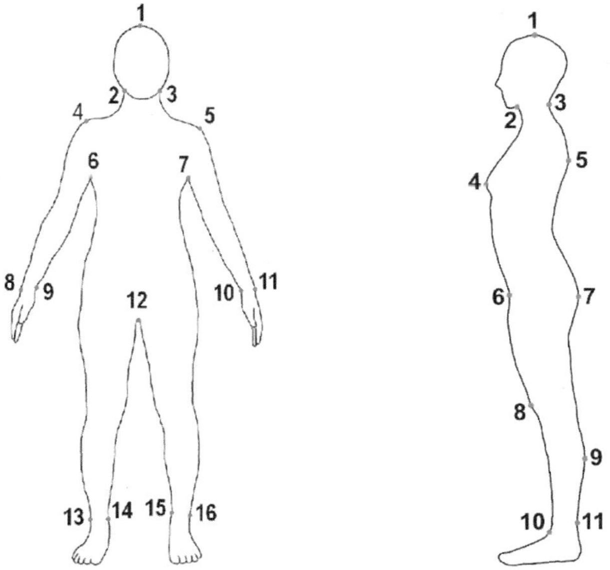

Fig. 8. The initial 16 frontal and the 11 lateral points used to construct the geometry to depict the body outlines.

Once the 16 points have been identified on the frontal view and the 11 points on the lateral view, these points are connected with a series of lines to circumscribe the two body outlines (Fig. 9). These 12 lines on the frontal view and the 10 lines on the lateral view, form the basis for a closely-spaced point system to be used to describe the whole boundary outline in the frontal and lateral aspects.

For both the frontal and lateral system of lines, line 2 is not used. It only serves to delimit the inferior margin of line 1.

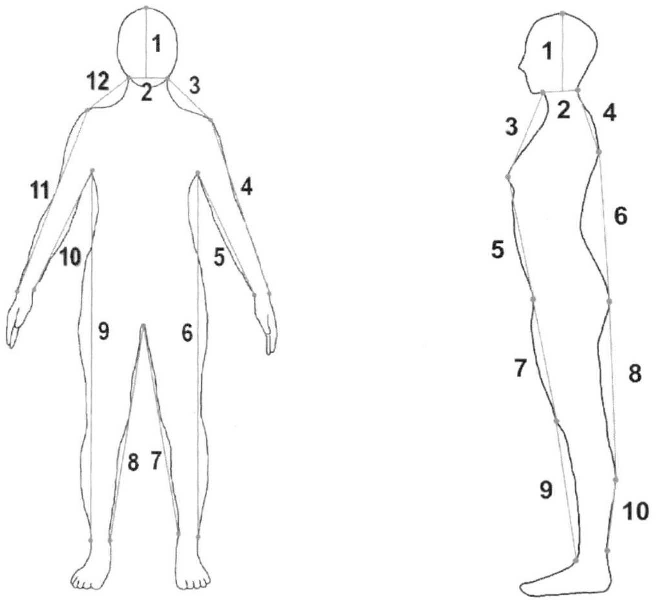

Fig. 9. The 12 lines used to delineate the frontal view as defined by the 16 points (Fig. 8) and the 10 lines used to circumscribe the lateral view as defined by the 11 points.

Fig 10 illustrates the next step, which consists of subdividing each of the 12 lines of the frontal view and each of the 10 lines in the lateral view, into a set of specific divisions depending on the length of the line. For example, for the frontal view, lines 3 and 12 contain three subdivisions, while lines 4, 7, 8, and 11 all have eight subdivisions. Lines 5 and 10 are the longest ones and contain sixteen divisions.

For the lateral view, line 4 contains five divisions; line 10 contains six divisions, lines 3, 5, 7, and 9 contain seven divisions, while lines 6 and 8 contain nine divisions.

The aim here is to generate a set of lines with approximately equal subdivisions. Finally, two issues need to be noted. The first one is that the steps shown pictorially in Figs. 8, 9, 10, 11, and 12 are carried out with specially written software developed by the authors (PEL and CAW) and called MLmetrics[7] (to be described

[7] MLmetrics software is available at a nominal cost from the senior author.

subsequently). The second one is that the images, photo and line drawing, were initially reduced to approximately 600 x 800 pixels (109 x 145 mm) prior to pre-processing and digitizing so that they would fit into the viewport of MLmetrics.

The final step is to insure that the subdivided lines are extended sufficiently to intersect the actual bounded outline as this identifies the locations of the 108 frontal view points and 84 lateral view points to be subsequently digitized.

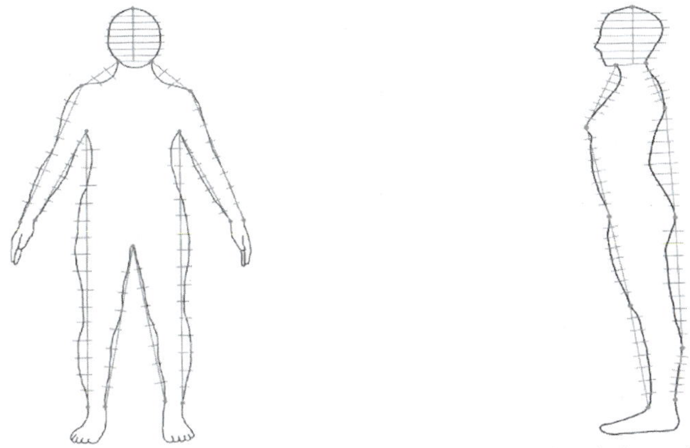

Fig. 10. Shown here are the 11 of the 12 lines of the frontal view being subdivided into equal intervals. Equivalently, 9 of the 10 lines used to delineate the lateral view are also subdivided into equal intervals.

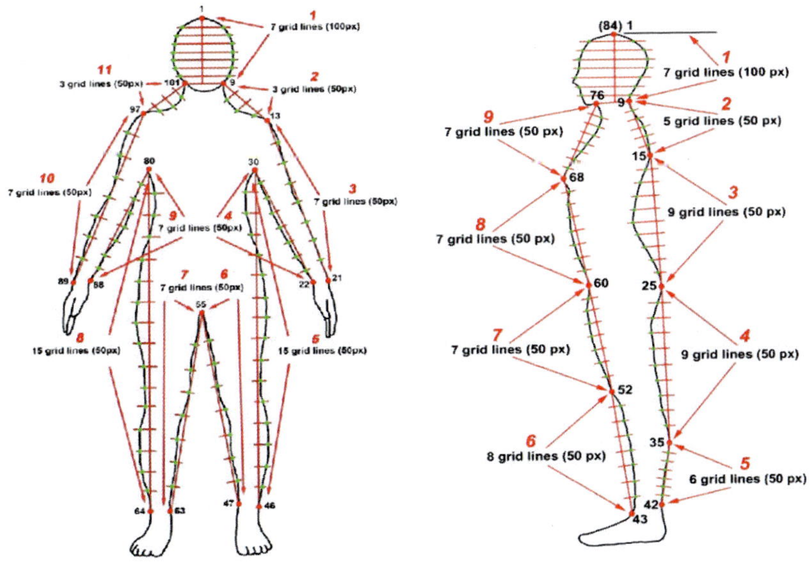

Fig. 11. The 108 points describing the frontal view and the 84 points describing the lateral view are displayed here, both views shown with line IDs prior to digitization.

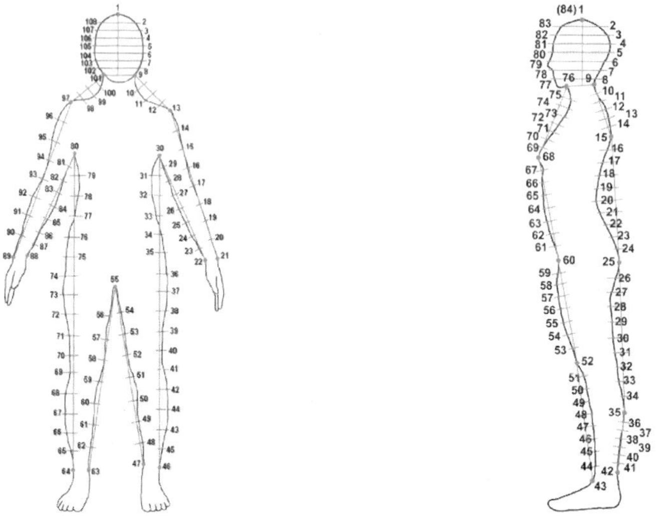

Fig. 12. Shown here are the completed 108 digitized points describing the frontal view and the 84 digitized points describing the lateral view.

Figs. 13 to 15 illustrate the online digitizing process of the frontal and lateral views using the viewport within the MLmetrics software.

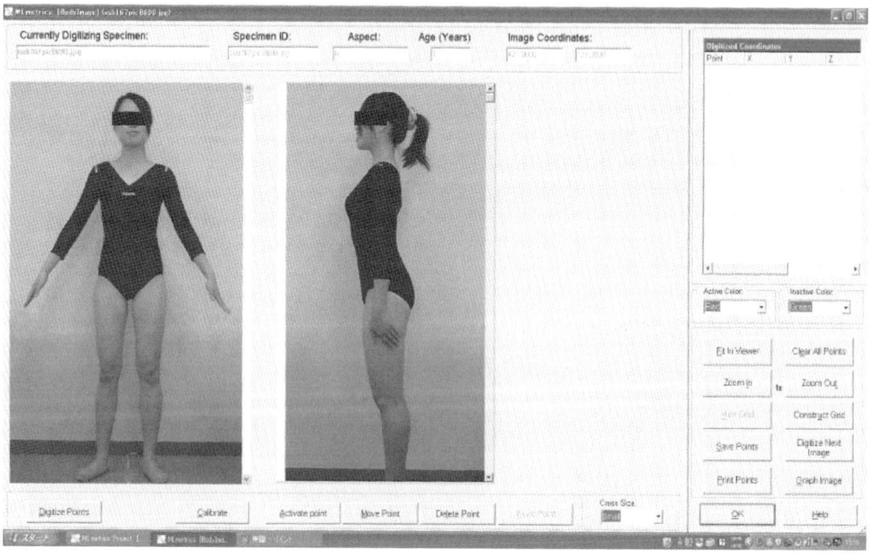

Fig. 13. Shown here is the viewport in the MLmetrics program displaying a subject's photo prior to digitizing. Digitizing was carried out separately for the frontal and lateral views.

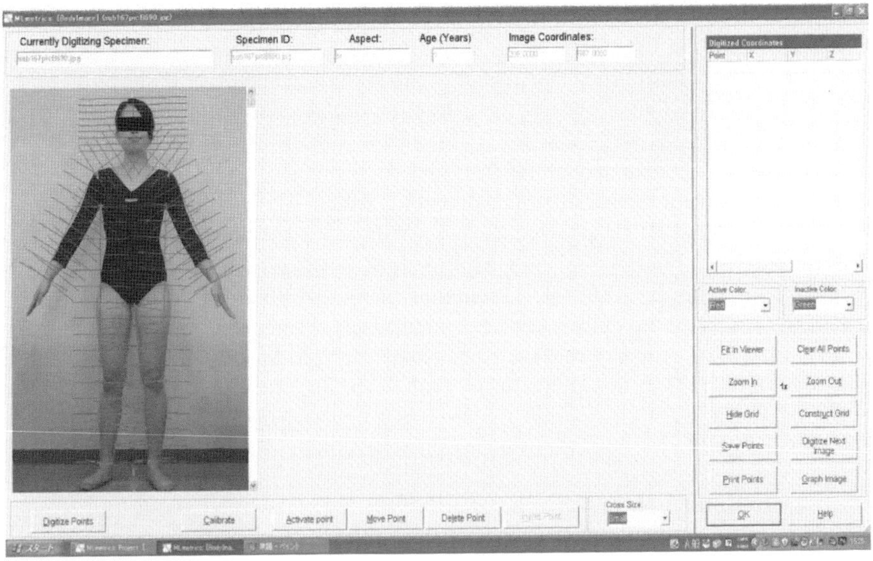

Fig. 14. Illustrated is the subject's photo within the MLmetrics viewport with the subdivided lines prior to actual digitizing. Only the frontal view is shown.

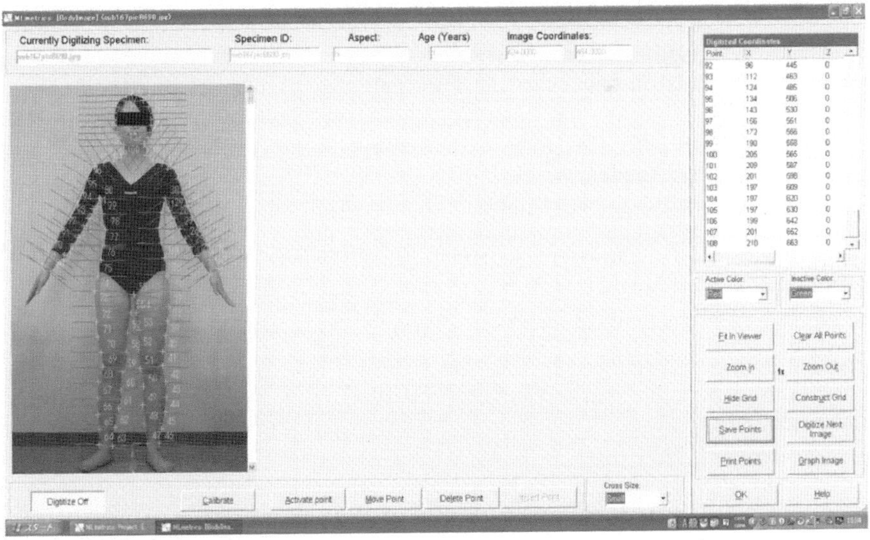

Fig. 15. Displayed here is the subject's photo within the MLmetrics viewport with the actual 108 digitized points. Again, only the frontal view is shown.

Quantitative Procedures

All forms can be viewed as containing *size* and *shape* information. With respect to the *size* and *shape* attributes, *shape* often provides more information, which is readily seen visually. Precise measurements of *shape*, however, have been difficult to obtain due to the irregularities in the biological outline. *Shape* is defined here as the form's boundary outline. Thus, what is needed is a quantitative description of the whole body outline that allows for the extraction of the *shape* information [30, 31].

Elliptical Fourier Functions (EFFs)

EFFs represent a parametric approach, which allows analysis of contour outlines that cannot be simply represented as single-valued functions (a requirement with conventional Fourier analysis). Multi-valued functions often arise with complex irregular morphologies [30, 31]. Acceptance of multi-valued functions and unequal spacing of points on the outline confer particular advantages to EFFs.

An attractive property of FDs is that they are *information-preserving contour representations* [33] in that they not only allow for the precise reconstruction of the biological outline, but also permit the re-creation, at any time, of the form from the EFF coefficients, in the absence of the original specimen [32].

Computation of EFFs

While EFFs are useful for numerically characterizing the boundary of irregular forms, and providing *global shape estimates* of the form, they have always had one serious constraint in parallel with other FDs. That is, the presence of any *localized* features along the contour cannot be identified using variables such as amplitudes, power, etc., normally generated from the EFFs. A new procedure, based on wavelets, nicely circumvents this inability of identifying localized features and in addition, provides for a totally objective identification of the localized features [19]. This problem can be also be alleviated, at least partially, by taking distances from the centroid to the points on the boundary outline or between specific points after controlling for size [20, 21]. However, this comes at the cost, since these distances have to be subjectively identified.

EFFs were developed by Kuhl and Giardina [35] but anticipated by [34]. The Kuhl and Giardina parametric functions are defined in $x(t)$ as:

$$x(t) = A_0 + \sum_{n=1}^{N} a_n \cos nt + \sum_{n=1}^{N} b_n \sin nt, \tag{1}$$

and in $y(t)$ as

$$y(t) = C_0 + \sum_{n=1}^{N} c_n \cos nt + \sum_{n=1}^{N} d_n \sin nt, \tag{2}$$

Where n equals the harmonic number, N equals the maximum harmonic number, and the interval is over 2π. EFFs have also been extended to $\mathbb{R}^3$ allowing the description of a biological form as a curve in 3-space [26].

Goodness-of-Fit

The EFF represents a series solution such that as the number of terms (harmonics) in the series is increased, it leads to a closer fit between the EFF equation and the points that define the form under consideration (a convergent series). However, this fit is subject to the Nyquist frequency requirement, which limits the number of harmonics (terms in the EFF) to ½ the number of digitized points [15]. The closeness of the fit can be assessed by computing the residual or difference between the expected points derived from the EFF and the observed data points on the boundary outline. For the body image data here, the mean residual (averaged over the 108 points) and over the pilot sample of (n=37) was for the photo in pixel units: 0.96±0.04 px. This translates to approximately 0.17±0.01 mm. This value can be considered a close and quite adequate fit of the EFF to the body outline.

In visual terms, Figs. 16 and 17 display this goodness-of-fit of the EFF to the body outlines. The EFF curves are demarcated with red squares and the actual body outline for this specific subject is displayed with black circles. For the frontal view, with one harmonic: A) the residual is 71.18 px (12.91 mm) and for the lateral view, the residual is 22.42 px (4.16 mm); B) with eight harmonics, the residual drops to 15.20 px (2.76 mm) for the frontal view and to 4.59 px (0.83 mm) for the lateral view; C) with twenty harmonics, the residual for the frontal view is 3.40 px (0.61 mm) and for the lateral view 1.14 (0.20 mm); and finally, D) with forty harmonics, the residual value for the frontal view is 1.31 px (0.23 mm) and for the lateral view is 0.52 px (0.09 mm). By this time, the actual body outline composed of the set of 108 closely spaced points is indistinguishable from the EFF curve fit (Fig. 16). Clearly, this reduction in residual value with increasing harmonic terms is an asymptotic process with diminishing returns.

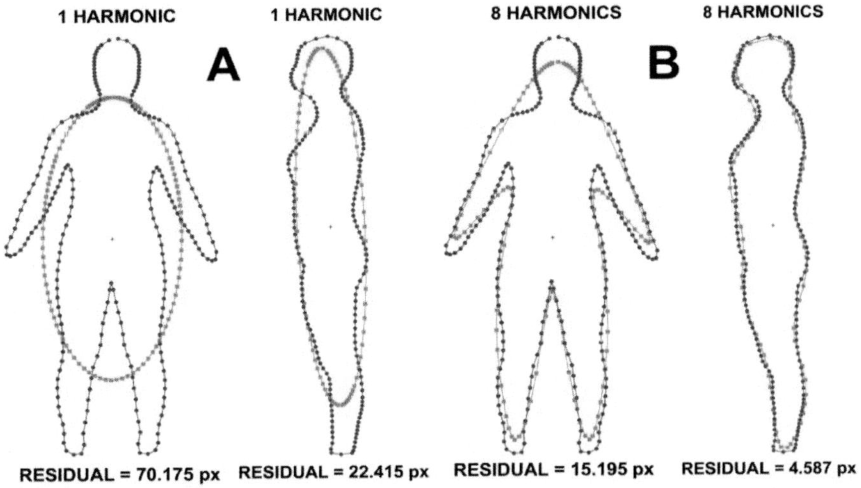

Fig. 16. Curve fit of the EFF to the body outline with 108 points showing the convergence with the EFF: (A) With one harmonic, and (B) With eight harmonics (see text).

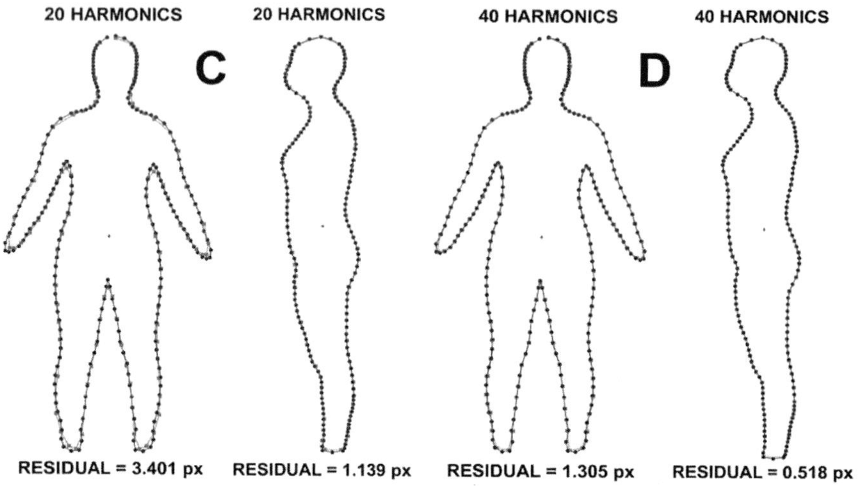

Fig. 17. Curve fit of the EFF to the body outline with 108 points showing the convergence with the EFF: (C) With 20 harmonics and (D) With 40 harmonics (see text).

Replicability Study

To assess the reliability of the digitizing process, the technician was asked to repeat the digitizing process of the photo for eight cases. The time difference between the two digitizations was six months. Each body outline photo was digitized at the two different periods, and then run through the EFFA routine to generate an EFF with 54 harmonics. This function was then used to generate distances from the centroid to each of the 108 points. These centroid distances were then averaged for each case. The difference in mean distance from the centroid for the first and second digitization was then computed. This *within-observer* error yielded means and standard deviation values of 0.46±0.40 px (0.08±0.07 mm) for the digitized body outlines. These values are in line with previous studies carried out by the senior author and can be considered reliable.

Superimposition Approaches

Prior to being able to analyze and visually access the body composition data and make comparisons, a procedure is needed that allows for the superimposition of images. This entails a methodology that controls for possible differences in size (scaling) and position (orientation) in $\mathbb{R}^2$ (2-space), in order that shape differences can be recognized and evaluated. Three superimposition procedures have been suggested. The first one sets the major semi-axis of the first ellipse to 1.0 as a scaling factor [35]. The drawback with this measure is that it is one-dimensional and, does not take into account the effect of the minor semi-axis. For example, it does not clearly distinguish between slim and fat objects of the same length.

The second one uses the area within the boundary as the scaling factor. This is accomplished by using the EFF to adjust the area within the bounded outline depicting the form so that the is equal for all comparisons.

The third one is called Procrustes superimposition. It takes into account scaling, translation, and rotation of forms during superimposition. The Procrustes approach, utilized here, is based on a least squares algorithm, in which the target mean outline, is superimposed on a base mean outline. The Procrustes algorithm[8] was modified to run in a Windows O/S environment and to read EFF data files used with this body composition data.

A Shape Difference Index

A simple index was devised to measure how different two superimposed forms were. This is the shape difference index (SDI). This index was developed to provide a simple number, which roughly corresponds with the differences seen visually. That is, if large shape differences are present, the SDI should yield a large number; and conversely, if small differences are involved, the index should be small. The index is computed as follows: (1) superimpose 2-D outlines on the centroid, (2) rotate to attain a common superimposition, (3) find areas not held in common, (4) compute areas within each section, (5) find the average area and divide by the total area, and (6) divide by a weighing factor, which is based on the number of sections. This weighting factor, based on the number of sections involved, was derived by using four sections as base and forming a ratio of the actual sections found over the base. This resulted in values such as 1.3333 (4/3), 1, (4/4), 0.6667 (4/6), etc.

This weighting factor was used to insure consistent values; that is, values that reasonably corresponded to a visual assessment of the superimposed images. SDI values can range from below 1.0 to above 10.0. Values below 1.0 indicate a close correspondence between superimposed images [36]. However, it should be cautioned that these SDI values are, in part, dependent of the complexities of the morphology under consideration.

RESULTS

The Earlier Pilot Study

To test the feasibility of: (1) the EFF procedure and (2) the two superimposition approaches, area-standardization and Procrustes, a pilot study (n = 37) was initiated. This pilot study validated the various procedures involved and produced data useful for body composition studies. This pilot study has since been published [18]. The procedures developed with the pilot study were then applied to the complete sample (n = 144) and forms the data of the current study.

Descriptive Statistics

The descriptive statistics include the mean and standard deviation of the following four variables: age, height, weight, and BMI, indicating a sample that is largely normal for young Japanese females of college age. The values for the BMI ranged from 16.8 to 30.3. Only one female had a BMI above 30, making her only marginally obese. Descriptive statistics of the pilot sample compared to the total sample are displayed in Table 1.

---

[8] The Procrustes routine was written by M. J. Calvacanti and available from maurobio@acd.ufrj.br.

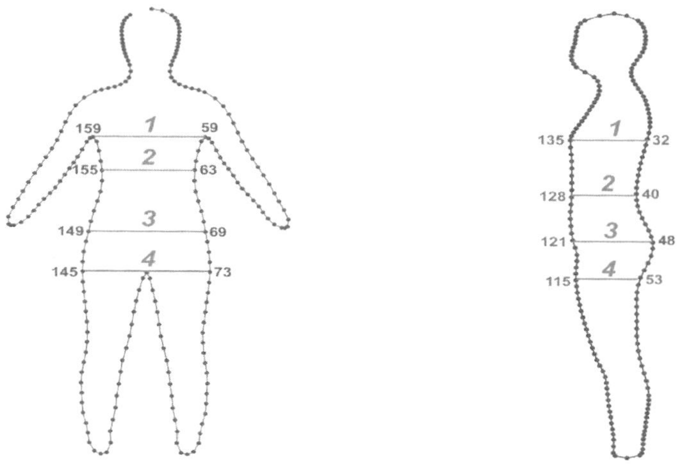

Fig. 18. The four frontal view and four lateral view distances used for statistical analysis are displayed here. Numbers shown refer to the digitized points but multiplied by two (216 points for the frontal view and 168 points for the lateral view).

Table 1. Descriptive Statistics of the Pilot Sample and the Total Sample Showing the Means and Standard Deviations for the Four Parameters

	Pilot Sample (n = 37)	*Total Sample (n = 144)*
Measure	*Mean±Std Dev*	*Mean±Std Dev*
Age in years	19.76±0.94	19.63±1.15
Height (cm)	159.32±5.26	159.37±5.06
Weight (kg)	54.41±6.32	53.58±6.23
BMI	21.50±2.50	21.10±2.30

The Set of Four Distances

A set of four distances describing aspects of the boundary outline of the frontal and lateral views to reflect possible changes that might be present was utilized. These measures were chosen to reflect width and depth aspects of the torso, regions presumed to exhibit the majority of changes with age (Fig. 18, Table 2). It should be noted that height measures were found to be largely insignificant.

Table 2. Descriptive Statistics of the Four Distances taken used to Describe the Frontal View. This comparison between the pilot study and the total sample was limited to the frontal view, as data for the lateral view was not yet available when the pilot study was completed.

	Pilot Sample (n = 37)	*Total Sample (n=144)*
Measurement	*Mean±Std Dev*	*Mean±Std Dev*
1. Armpit width	46.99±2.39	46.88±2.77
2. Waist width	40.31±2.03	39.72±1.83
3. Hip width	47.03±2.43	46.36±2.16
4. Thigh width	53.49±1.74	53.26±1.75

Extremes of the Body Composition

In an effort to begin to visually assess the variability in this body outline data, it was decided to compare the extremes of the body outline distribution (n = 144). For this purpose, the 10th and 90th percentiles were selected using the BMI as the selection criterion.

A One-Way MANOVA

To test for differences between the two groups, the 10th and 90th percentiles, a one-way MANOVA was run using the four distances generated from the frontal view and the lateral view of the body outline data (Table 3). The main effect of one-way MANOVA was found to be statistically significant. As the one-way MANOVA was significant, this allows a look at the individual or univariate F-tests associated with the four distance variables. Of the four distance variables utilized, three of the four frontal distances were found to be statistically significant, while all of the lateral distances were significant. The results of the univariate F-tests are shown in Table 4.

Table 3. Overall One-way MANOVA using the Four Distances Computed from the EFFs for the 10th and 90th Percentiles

Main Effect	*Wilk's Lambda*	*Rao's R*	*p-level*
Frontal view	0.292	15.187	p < 0.001
Lateral view	0.180	32.751	p < 0.001

Table 4. Univariate F-statistics for the Four Distances from the One-way MANOVA

Dependent variable	*F-value*	*p-level*
	Frontal view	
1. Armpit width	0.77	ns
2. Waist width	31.42	p < 0.001
3. Hip width	46.89	p < 0.001
4. Thigh width	7.67	p < 0.001
	Lateral view	
1. Armpit depth	61.14	p < 0.001
2. Waist depth	31.37	p < 0.001
3. Hip depth	58.72	p < 0.001
4. Thigh depth	111.35	p < 0.001

A Visual Assessment of the Area Standardized Superimposition

Fig. 19 displays the superimposition of the 10th percentile on the 90th percentile for the body composition data using the area-standardization procedure. The 10th percentile data is depicted with black circles while the 90th percentile is shown with red squares. The superimposition here is on the centroid.

The visual assessment demonstrates that the 10th percentile data is thinner as well as taller than the 90th percentile group. This is clearly shown for both the frontal and lateral mean views. For this particular superimposition, the SDI values were moderate at 3.59 for the frontal aspect and 4.66 for the lateral view.

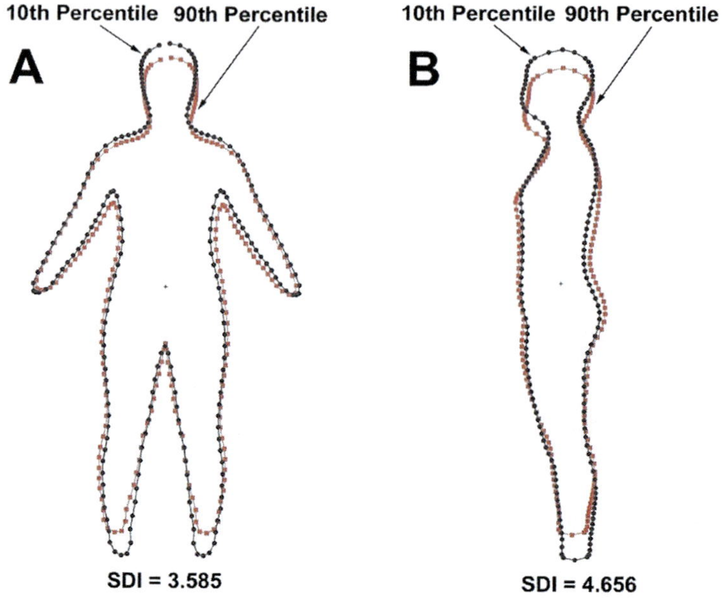

Fig. 19. Superimposition of the 10th percentile (black circles) and the 90th percentile (red squares) using the area-standardized procedure (refer to text).

A Visual Assessment of the Procrustes Superimposition

Fig. 20 displays the superimposition of the 10th percentile on the 90th percentile for the body composition data using the Procrustes superimposition. The 10th percentile data is depicted with a black circle while the 90th percentile is shown with red squares. That the 10th percentile lies within the boundary of the 90th percentile is clearly apparent. The superimposition here is *not* on the centroid. The thinness of the 10th percentile, in contrast to the 90th percentile, is nicely illustrated by the Procrustes superimposition, for both the frontal and lateral aspects reinforcing the significant statistical results. Again, the SDI values are moderate for both body views.

Which superimposition is preferable? This remains an open question as both superimpositions provide interesting data. While the Procrustes approach is perhaps more visually appealing, the area-standardization procedure clearly demonstrates that the 10th percentile sample is not only thinner but also taller than the 90th percentile sample.

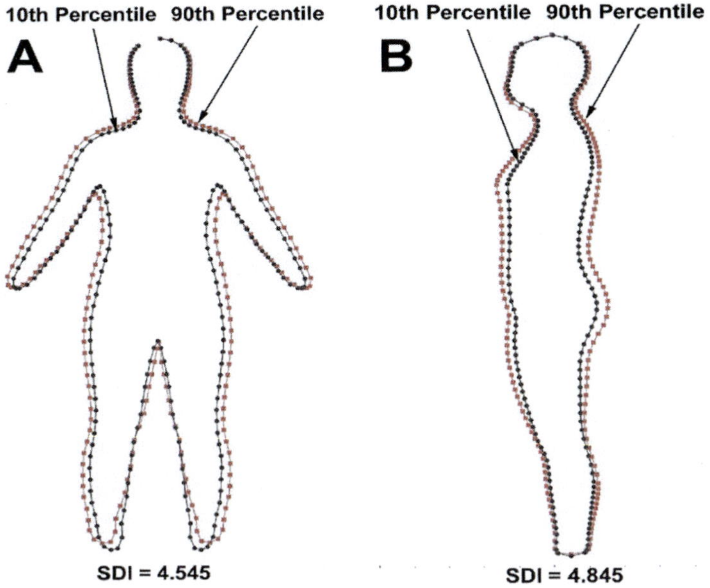

Fig. 20. Superimposition of the 10th percentile (black circles) and the 90th percentile (red squares) using the Procrustes procedure. The SDI values display the close correspondence of the Procrustes fit. (Refer to text).

DISCUSSION AND CONCLUSIONS

The 10th and 90th percentiles, the extremes of the sample (n = 144) distribution, are statistically significant for the four variables in both the frontal and lateral views as determined by the one-way MANOVA (Table 3). Of the eight comparisons that were computed, only one, arm pit width, reflected non-significance with respect to the individual univariate F-tests (Table 4). This statistical significance is re-enforced with the visual depiction of the superimpositions displayed in Fig. 19 and Fig. 20 where distinct differences between the 10th and 90 percentiles for both the frontal and lateral views can be seen. Moreover, most of these differences between the 10th and 90th percentiles are present regardless of whether the superimposition was the area-standardized one or the Procrustes one. In should be noted that these results are not unexpected, after all, one might expect that the body shape of those on the left end of the BMI distribution (thinness) would be significantly different from those on the right end of the BMI distribution (fatness). Thus, one of the advantages of EFFs is that they allow one to nicely visualize these differences in body shape.

The use of FDs as developed in this chapter are particularly useful for analyzing complex and irregular shapes of the kind encountered in the biological sciences. It is suggested that this Fourier approach may be useful for a number of purposes: [1] viewing shape changes in body composition as function of physical performance in athletes, [2] the effects that nutrition and disease have on body shape, [3] body image and perception studies, [4] The use of FDs to characterize the outline of body shape in health-compromised studies such as for example, *Anorexia Nervosa* or developmental obesity, and [5] the use of EFFs, as fitting functions of whole body outlines, are useful in the development of visual reference studies; that is, as standards or norms. Such standards may prove useful for the treatment of psychosomatic illnesses and provide a more accurate and sophisticated assessment of body composition than is possible with the current reliance on simple measures such as the BMI.

ACKNOWLEDGEMENTS

We gratefully acknowledge the assistance of the following individuals and institutions for allowing access to the data. Mieko Ae, Tokyo Women's College of Physical education; Masaru Ochiai, Yokohama National University; Kuniyasu Imanaka, Tokyo Metropolitan University; and especially to Sumie Komuro, for her precisely digitizing the photographic data used in this study.

REFERENCES

[1] Vocks S, Busch M, Schulte D, Grönermeyer D, Herpertz S, and Suchan B (2010) Effects of body image therapy on the activation of extrastriate body area in *Anorexia Nervosa*: An fMRI study. *Psychol Res Neuroimaging* **183**: 114-118.

[2] Vocks S, Legenbauer T, Rüddel H, and Troje NF (2007) Static and dynamic body image in *Bulimia Nervosa*: Mental representation of body dimensions and biological motion patterns. *Int J Eating Disorders* **1**: 59-66.

[3] Ivar Carniero JR, Gavina da Cruz G, do Valle Quaresma J, Xerez D, and Paulo de Oliveira JE (2012) Impact of an Exercise Program for the Morbidly Obese Patients on Quality of Life and on Clinical and Metabolic Profile. *J Obesity Weight Loss Therapy* **2**: 2-7.

[4] Dulloo AG, Jacquet J, Solinas G, Montani J-P, and Schutz Y (2010) Body composition phenotypes in pathways to obesity and metabolic syndrome. *Int J Obesity* **34**: S4-S17.

[5] Hossain MG, Bharati P, Aik S, Lestrel PE, Abeer A, and Kamarul T (2012) Body mass index of married Bangladeshi women: Trends and association with socio-demographic factors. *J Biosocial Sci* **44**: 1-15.

[6] Benson PJ, Emery JL, Cohen-Tovée EM, and Tovée MJ (1999) A computer-graphic technique for the study of body size perception and body types. *Behavior Res Meth Inst Comput* **31**: 446-454.

[7] Tovée MJ, Benson PJ, Emery JL, Mason SM, and Cohen-Tovée EM (2003) Measurement of body size and shape perception in eating-disoriented and control observers using body-shape software. *Brit J Psych* **94**: 501-516.

[8] Frederick DA, Forbes GB, Grigorian KE, and Jarcho JM (2007) The UCLA Body Project I: Gender and Ethnic Differences in Self-Objectification and Body Satisfaction Among 2,206 Undergraduates. *Sex Roles* **57**: 317-327.

[9] Gittelsohn J, Harris SB, Thorne-Lyman AL, Hanley AJG, Barnie A, and Zinman B (1996) Body image concepts differ by age and sex in an Ojibway-Cree community in Canada. *J Nutrition* **126**: 2990-3000.

[10] Frederick DA, Forbes GB, and Berezovskaya A (2008) Female body dissatisfaction and perceptions of the attractive female body in Ghana, the Ukraine, and the United States. *Psychol Topics* **17**: 203-219.

[11] Forbes GB and Frederick DA (2008) The UCLA Body Project II: Breast and Body Dissatisfaction among African, Asian, European, and Hispanic American College Women. *Sex Roles* **58**: 449-457.

[12] Forbes GB, Doroszewicz K, Card K, and Adams-Curtis L (2004) Association of the Thin Body Ideal, Ambivalent Sexism, and Self-Esteem with Body Acceptance and the Preferred Body Size of College Women in Poland and the United States. *Sex Roles* **50**: 331-345.

[13] Sheldon WH, Stevens SS, and Tucker WB. (1963) *The Varieties of Human Physique*. Hafner Publishing Company, New York (originally published in 1940).

[14] Sheldon WH, Dupertius CW, and McDermott E (1954) *Atlas of Men*. Gramercy Publishing Company, New York.

[15] Carter JEL and Heath BH (1990) *Somatotyping: Development and Applications*. Cambridge University Press, New York.

[16] Thompson MA and Gray JJ (1995) Development and Validation of a New Body-Image Assessment Scale. *J Pers Assessment* **64**: 258-269.

[17] Stunkard AJ, Sörensen T, and Schulsiger F (1983) Use of the Danish Adoption Register for the study of obesity and thinness. **In**: S Kety, L Roland, R Sidman and S Matthysse (Eds.) *The Genetics of Neurological and Psychiatric Disorders*. Raven Press, New York.

[18] Sörensen T, Stunkard AJ, Teasdale TW, and Higgins MW (1983) The accuracy of reports of weight: Children's recall of their parents' weights 15 years earlier. *Int J Obesity* **7**: 115-122.

[19] Adkins AN and Stivers K (2001) Behavioral and psychological factors predicting body satisfaction among adolescents. Hanover College, Dept. of Psychology 2006 Senior thesis. Retrieved from the Internet, November 30, 2012. http://psych.hanover.edu.

[20] Gardiner RM, Friedman BN, and Jackson NA (1998) Methodological concerns when using silhouettes to measure body image. *Percept Motor Skills* **86**: 387-395.

[21] Eknoyan G (2008) Adolphe Quetelet (1796–1874) — the average man and indices of obesity. *Nephrol Dialy Transpl* **23**: 47-51.

[22] William D, Ross WD, and Eiben OG (2002) BMI: A critique of its use in human biology and the health professions. **In**: OG Eiben and ÉB Bodzsár (Eds.). *Children and youth at the beginning of the 21st century. Hum biologia Budapestinensis* **27**: 47-56.

[23] Wang J, Thornton JC, Russell M, Burastero S, Heymsfield S, and Pierson Jr. RN (1994) Asians have lower body mass index (BMI) but higher percent body fat than do whites: comparisons of anthropometric measurements. *Am J Clin Nutrition* **60**: 23-28.

[24] Lestrel P E (1974) Some problems in the assessment of morphological size and shape differences. *Yearbk Phys Anthrop* **18**: 140-162.

[25] Lestrel PE (1980) A quantitative approach to skeletal morphology: Fourier analysis. *Soc Photo-opt Instr Engrs* (SPIE) **166**: 80-93.

[26] Lestrel PE (1997) *Fourier Descriptors and their Applications in Biology.* Cambridge University Press, Cambridge.

[27] Lestrel PE (2000) *Morphometrics for the Life Sciences.* World Scientific Publishing Co. Pte. Ltd., Singapore.

[28] Courtiol A, Ferdy JB, Godelle B, Raymond M, and Claude J (2010) Height and body mass influence on human body outlines: A quantitative approach using an elliptic Fourier analysis. *Am J Phys Anthrop* **142**: 22-29.

[29] Lestrel PE, Miyake N, Ishihara M and Wolfe CA (2013) Assessment of Body Image perception: A Preliminary Study using Elliptic Fourier Descriptors. **In**: PE Lestrel (Ed.) *Biological Shape Analysis*: *Proc 2nd Int Symp Biol Shape Anal* (ISBSA) World Scientific Pub, Singapore.

[30] Lestrel PE, Cesar Jr RM, Takahashi O, and Kanazawa E (2004) A Fourier-wavelet representation of 2-D shapes: sexual dimorphism in the Japanese cranial base. *Anthrop Sci* **112**: 3-28.

[31] Lestrel PE, Ohtsuki F, and Wolfe CA (2010) Cranial vault shape in fossil hominids: Fourier Descriptors in *Norma lateralis. J Comp Hum Biol* (*HOMO*) **61**: 287-313.

[32] Lestrel PE, Kanazawa E, and Wolfe CA (2011) Sexual dimorphism using elliptical Fourier analysis: Shape differences in the craniofacial complex *Anthrop Sci* **119**: 213-229.

[33] van Otterloo PJ (1991) *A Contour-oriented approach to Shape Analysis.* Prentice-Hall, New York.

[34] Granlund GH (1972) Fourier preprocessing for hand print character recognition. *IEEE Trans Comput* **C-21**: 195-201.

[35] Kuhl FP and Giardina CR (1982) Elliptic Fourier Features of a Closed Contour. *Comp Graph Imag Proc* **18**: 236-258.

[36] Lestrel PE, Wolfe CA, and Bodt A (2013) Mandibular shape analysis of Plio-Pleistocene hominins: Fourier Descriptors in Norma lateralis. **In**: PE Lestrel (Ed.) *Biological Shape Analysis*: *Proc 2nd Int Symp Biol Shape Anal* (ISBSA) World Scientific Pub, Singapore.

INDEX